170 Topics in Current Chemistry

Electrochemistry V

Editor: E. Steckhan

With contributions by
J. Bersier, P. M. Bersier, L. Carlsson,
T. Fuchigami, E. Steckhan, J. Yoshida

With 70 Figures and 47 Tables

Springer-Verlag
Berlin Heidelberg GmbH

This series presents critical reviews of the present position and future trends in modern chemical research. It is adressed to all research and industrial chemists who wish to keep abreast of advances in their subject.

As a rule, contributions are specially commissioned. The editors and publishers will, however, always be pleased to receive suggestions and supplementary information. Papers are accepted for "Topics in Current Chemistry" in English.

ISBN 978-3-662-14917-1   ISBN 978-3-540-48324-3 (eBook)
DOI 10.1007/978-3-540-48324-3

Library of Congress Catalog Card Number 74-644622

© Springer-Verlag Berlin Heidelberg 1994

Originally published by Springer-Verlag Berlin Heidelberg New York in 1994
Softcover reprint of the hardcover 1st edition 1994

Typesetting: Macmillan India Ltd., Bangalore-25
Offsetprinting: Saladruck, Berlin; Bookbinding: Lüderitz & Bauer, Berlin
SPIN: 10101086   51/3020 - 5 4 3 2 1 0 - Printed on acid-free paper

Preface to Volume V

After an intermission of 4 years, the editor of the electrochemistry series in Topics in Current Chemistry has felt the necessity to cover recent important developments in the field of organic and bioorganic electrochemistry in a fifth volume. The four contributions demonstrate typical directions of current research in organic electrochemistry which can be described as:
- Improvement of selectivity by using heteroatom functionalities for the control of the electrochemistry and reactivity during electrochemical transformations;
- improvement of regio- and enantioselectivity by using electro-enzymatic redoxreactions;
- environmental protection by use of electrochemical procedures.

The first two contributions deal with elemental organic chemistry and the demand for highly selective processes to generate complex organic compounds. In both cases, heteroatomic functions such as fluorine substituents or silyl groups together with oxygen, nitrogen, or sulfur functions are applied for the activation of the molecules and the introduction of selectivity. Thus, electrochemical methods for the conversion and functionalization of fluoro-organic and organosilicium compounds are described.

The third contribution presents the combination of electrochemistry and enzymatic synthesis for the selective formation of complex molecules. This quite young field of research is developing rapidly because the application of the reagent-free electrochemical procedure combined with the regio- and stereoselectivity of enzymes offers the possibility of establishing new environmentally friendly process even on a technical scale.

The aspect of environmental protection by using electrochemical methods is also stressed in the last contribution. Electrochemistry not only offers new methods for waste treatment (curative environmental protection) but even more important also the possibility of preventing waste formation by using electrochemical production processes (preventive environmental protection). In the latter case, even closed systems can be developed much easier than with other techniques. In this context, electrochemistry is a very promising field of research and future applications may be foreseen.

Bonn, November 1993   Eberhard Steckhan

Attention
all "Topics in Current Chemistry" readers:

A file with the complete volume indexes Vols.22 (1972) through 169 (1994) in delimited ASCII format is available for downloading at no charge from the Springer EARN mailbox. Delimited ASCII format can be imported into most databanks.

The file has been compressed using the popular shareware program "PKZIP" (Trademark of PKware Inc., PKZIP is available from most BBS and shareware distributors).

This file is distributed without any expressed or implied warranty.

To receive this file send an e-mail message to:
SVSERV@VAX.NTP. SPRINGER.DE
The message must be:"GET/CHEMISTRY/TCC_CONT.ZIP".

SVSERV is an automatic data distribution system. It responds to your message. The following commands are available:

HELP	returns a detailed instruction set for the use of SVSERV
DIR (name)	returns a list of files available in the directory "name",
INDEX (name)	same as "DIR",
CD \<name\>	changes to directory "name",
SEND \<filename\>	invokes a message with the file "filename",
GET \<filename\>	same as "SEND".

For more information send a message to:
INTERNET:STUMPE@SPINT. COMPUSERVE.COM

Table of Contents

Electrochemical Reactions of Fluoro Organic Compounds

Toshio Fuchigami

Department of Electronic Chemistry, Tokyo Institute of Technology, Nagatsuta, Midori-ku, Yokohama 227, Japan

Table of Contents

Topics in Current Chemistry, Vol. 170
© Springer-Verlag Berlin Heidelberg 1994

In spite of increasing interest in the chemistry of fluoro organic compounds, with the exception of electrochemical perfluorinations electrochemical reactions were not recognized as a notable method for fluoro organic synthesis until about ten years ago. This review deals with the recent remarkable advances in electrochemical methods for the conversion and functionalization of fluoro organic molecules. The effect on fluorine atoms on the reduction and oxidation potentials of organic compounds is first discussed. Subsequently, recent applications of the electrochemical technique to fluoro organic synthesis will be briefly explained.

1 Introduction

Among halogens, fluorine is quite characteristic and specific since it has the largest electronegativity (4.0 vs 3.5 for oxygen) and the sterically second smallest van der Waals' radius (1.35 Å vs. 1.20 Å for hydrogen). A carbon-fluorine bond is also stronger than a carbon-hydrogen bond (485 kJ/mol vs 414 kJ/mol). Therefore, fluoro organic compounds have unique chemical and physical properties. Recently, there has been increasing interest in the chemistry of fluoro organic compounds, which have wide application in various fields such as material science, medicinal, and theoretical chemistry. However, their preparation is not straightforward because ordinary synthetic methods are not always applicable to the preparation of fluorinated organic compounds.

From these view points, a number of new synthetic methods and techniques have been developed, for example, perfluoroalkylating and fluorinating reagents, ultrasound irradiation, enzymatic methods, and so on. On the other hand, electrochemical reactions have been shown to be highly efficient new tools in organic synthesis. However, only a limited number of successful examples of electrosynthesis of fluoro organic compounds had been reported, except for the well-established anodic perfluorination and anodic trifluoromethylation, up to the end of the 1970s.

This review covers the recent remarkable growth in *"Electro Organic Fluorine Chemistry"* during the last decade. From this review, you will learn how the electrochemical technique can be used in organic fluorine chemistry, particularly in the conversion and functionalization of fluoro organic molecules.

2 Cathodic Reduction of C–F Bonds

The ease of reduction of a carbon-halogen bond decreases in the order J > Br > Cl > F. The carbon-fluorine bond is the most difficult to reduce due to its having the highest electronegativity.

In general, simple alkyl fluorides are electrochemically irreducible. However, some specially substituted fluorides can be reduced, as can be seen with fluorides bearing electron-withdrawing groups such as $PhCOCH_2F$, CF_3COOEt ($E_{p1/2} = -2.36$ V vs SCE in DMF, Hg cathode) [1], and $CF_3CONHPh$ ($E_p = -2.32$ V vs -1.90 V for CCl_3CN in MeCN, Hg cathode) [2]. Trifluoromethylbenzenes are reducible, particularly trifluoromethylbenzenes bearing an electron-withdrawing ester or cyano group at the *para* position are readily reduced rather efficiently (Eq. 1) [3].

$$p\text{-}XC_6H_4CF_3 \xrightarrow[\text{Pb cathode}]{\text{6e, 6H}^+} p\text{-}XC_6H_4CH_3 \tag{1}$$

$$X = CN \quad (-1.8 \text{ V}): 58\%$$
$$= COOMe \ (-2.0 \text{ V}): 60\%$$

On the other hand, poly tetrafluoroethylene (PTFE) has been shown to be electrochemically reducible [4]. Very recently, electrochemical reduction of saturated perfluoroalkanes has been observed on an analytical scale and it was found that the reduction potentials of perfluorocycloalkanes are only slightly more negative than those of the corresponding perfluoroaromatics: perfluorodecaline ($E_p = -2.60$ V vs $Ag/0.01M$ $AgClO_4$) vs. octafluoronaphthalene (-2.58 V); perfluoromethylcyclohexane (-2.9 V) vs. perfluorotoluene (-2.75 V) [5].

Kariv-Miller and Vajtner have successfully carried out selective defluorination of 1,3-difluorobenzene to fluorobenzene by cathodic reduction at a mercury cathode in diglyme containing Bu_4NBF_4 and a small amount of dimethylpyrrolidinium (DMP^+) salt [6]. In this reaction, DMP^+ is first reduced to form an amalgam, which reduces difluorobenzene catalytically as shown in Scheme 2.1.

$$DMP^+ + e + nHg \longrightarrow DMP(Hg)_n$$

$$C_6H_4F_2 + DMP(Hg)_n \longrightarrow [C_6H_4F_2]^{-} \cdot + DMP^+ + nHg$$

$$[C_6H_4F_2]^{-} \cdot \xrightarrow[-F^-]{e + H^+} C_6H_5F \quad 85\%$$

Scheme 2.1

3 Cathodic Reduction of Polyfluoro Organic Halides

3.1 Direct Reduction

Perfluoroalkyl halides are cathodically reduced much easier than the corresponding nonfluorinated halides. Rozhkov et al. have performed a polarographic reduction study of perfluoroalkyl halides at a platinum cathode in MeCN [7]. The reduction potentials are summarized in Table 1. The reduction potentials are greatly affected by the molecular structure: a) cyclic halides are much easier to reduce than open-chain halides b) the reduction potentials decreases with the increasing length of the Rf groups. The ease of the reduction has the following order:

$$(Rf)_3CI > (Rf)_3CBr > (Rf)_2CFI > RfCF_2I \approx (CF_3)_2CFBr \approx (Rf)_3CCl$$

Reduction potentials also depend greatly on the nature of the cathode material as shown in Table 2 [8]. For example, primary perfluoroalkyl iodides are ca. 0.3 V easier to reduce at mercury than at platinum due to the strong interaction of RfI with mercury. In fact, cathodic reduction of RfI at a mercury cathode provides RfHgI [9].

Table 1. Reduction potentials of perfluoroalkyl halides

R_fX	$E_{1/2}$ V vs SCE
C_3F_7I	-1.00
$(CF_3)_2CFI$	-0.66
$(CF_3)_3CI$	$+0.14$
$CF_3CF_2CF_2C(CF_3)_2I$	$+0.32$
C_4F_9I	-2.30
$(CF_3)_2CFBr$	-1.10
$(CF_3)_3CBr$	-0.14
	$+0.71$
$(CH_3)_3CBr$	-2.51
	-1.02
$CF_3CF_2CF_2C(CF_3)_2Cl$	-0.97

In 0.1M Et_4NBF_4/MeCN, Pt cathode.

Direct electrochemical reduction of perfluoroalkyl halides generates perfluoroalkyl radicals or anions depending on the electrolytic conditions and starting halides. Calas et al. have performed the electrocatalytic addition of perfluoroalkyl iodides to 3-hydroxyalkynes in aqueous KCl emulsion using a

Table 2. Reduction potentials (peak potentials, Ep vs SCE) of CF_3Br and CF_3I at various cathodes [a]

Compd.	Glassy carbon	Stainless steel	Pt	Ni	Hg	Au	Cu
CF_3Br[b]	− 2.07	− 1.90	− 1.55	− 1.33	− 1.25	− 1.23	− 1.18
CF_3I[c]	− 1.52	—	− 0.95	—	− 0.65	− 0.70	—

[a] In 0.1M Bu_4NBF_4/DMF; 0.2 V/s.
[b] at 25 °C.
[c] at 5 °C.

carbon fiber cathode. The reaction proceeds via a radical chain reaction and the electrolytic products are readily converted into perfluoroalkyl acetylenes (Scheme 3.1) [10, 11].

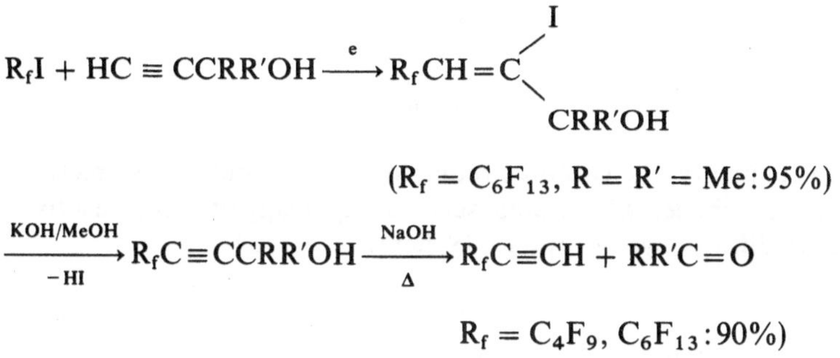

$$R_fI + HC \equiv CCRR'OH \xrightarrow{e} R_fCH = C\begin{matrix} I \\ \diagup \\ \diagdown \\ CRR'OH \end{matrix}$$

$$(R_f = C_6F_{13}, R = R' = Me : 95\%)$$

$$\xrightarrow[-HI]{KOH/MeOH} R_fC \equiv CCRR'OH \xrightarrow[\Delta]{NaOH} R_fC \equiv CH + RR'C = O$$

$$R_f = C_4F_9, C_6F_{13} : 90\%)$$

Scheme 3.1

Recently, Shono and Kise et al. have successfully carried out the electro-reductive coupling of halofluoro compounds with aldehydes in the presence of chlorotrimethylsilane (CTMS) as shown in Eqs 2–4 [12]. The yields are fairly good.

$$CCl_3CF_3 + RCHO \xrightarrow[DMF/CTMS]{2e} \underset{\underset{OH \quad 1}{|}}{RCHCCl_2CF_3} \tag{2}$$

$$(R = Alkyl, Ph, Allyl)$$

$$CClF_2COOMe + RCHO \xrightarrow[DMF/CTMS]{2e} \underset{\underset{OH}{|}}{RCHCF_2COOMe} \tag{3}$$

$$(R = Alkyl, Ph, Allyl)$$

$$R_fX + RCHO \xrightarrow[DMF/CTMS]{2e} \underset{\underset{OH}{|}}{RCHR_f} \tag{4}$$

$$(CF_3Br, C_4F_9I)$$

$$(R = Alkyl, Ph)$$

5

The coupling product, of type **1** is also cathodically converted to various fluoro compounds as shown in Scheme 3.2.

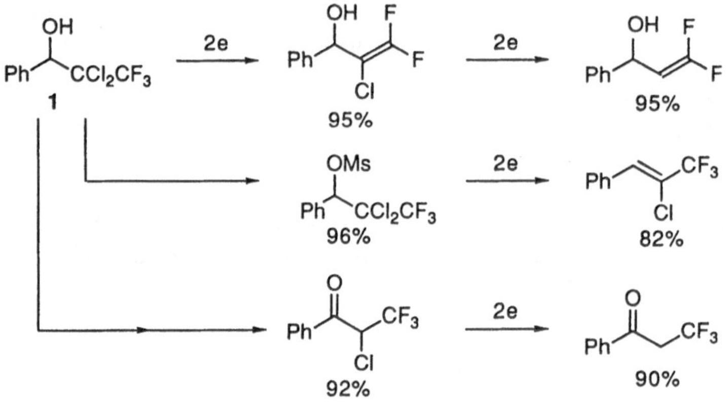

Scheme 3.2

On the contrary, the cathodic reduction of dibromodifluoromethane generates difluorocarbene, which was successfully trapped with reactive olefines yielding 1,1-difluorocyclopropanes (Scheme 3.3.) [13].

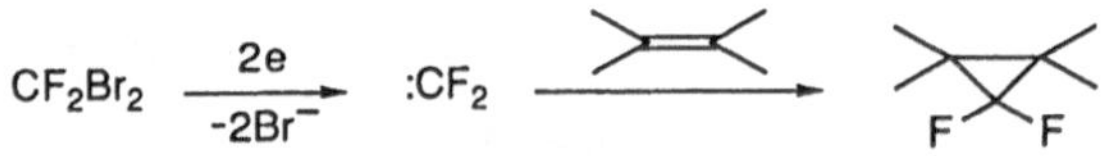

Scheme 3.3

3.2 Indirect Reduction

In contrast to the direct reduction as described above, the indirect electrochemical reduction of perfluoroalkyl halides is a versatile and novel method for generating perfluoroalkyl radicals selectively. Saveant et al. have demonstrated many successful examples. Using terephthalonitrile as a mediator, the indirect reduction of CF_3Br in the presence of styrene leads to the dimer of the radical adduct obtained by the attack of $CF_3\cdot$ on styrene. On the other hand, in the presence of butyl vinyl ether, the mediator reacts with the radical adduct obtained by the attack of $CF_3.$ on the olefin (Scheme 3.4) [14].

Nitrobenzene redox catalyzed electrolysis of $C_6F_{13}I$ in benzonitrile provides 4-perfluorohexylbenzonitrile as a main product (Scheme 3.5) [8].

Although the reduction of CF_3Br by cathodically generated aromatic anion radicals gives rise to purely catalytic currents, cathodically generated $SO_2^-\cdot$ does not give rise to catalytic currents upon reaction with CF_3Br but produces trifluoromethyl sulfinate according to an overall two electron per molecule

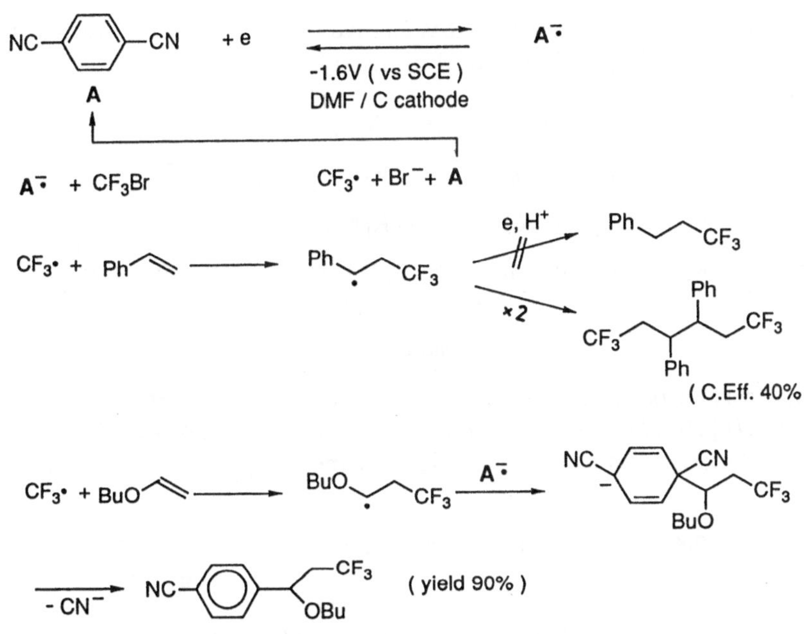

$$NC-\langle \rangle-CN \quad +e \quad \xrightleftharpoons[\substack{-1.6V\ (\text{vs SCE}) \\ \text{DMF}\ /\ \text{C cathode}}]{} \quad A^{\overline{\cdot}}$$

A

$A^{\overline{\cdot}} + CF_3Br \qquad\qquad CF_3{}^{\cdot} + Br^{-} + A$

$CF_3{}^{\cdot} + Ph\diagup\!\!\!\equiv \longrightarrow Ph\diagdown\!\!\diagup_{\!\!\cdot}CF_3 \xrightarrow{\;e,\ H^+\;} Ph\diagdown\!\!\diagup CF_3$

$\times 2 \longrightarrow$

(with product showing Ph, CF$_3$, Ph substituents)

(C.Eff. 40%)

$CF_3{}^{\cdot} + BuO\diagdown\!\!\diagup \longrightarrow BuO\diagdown\!\!\diagup_{\!\!\cdot}CF_3 \xrightarrow{A^{\overline{\cdot}}} $ (cyclohexadiene product NC, CN, BuO, CF$_3$)

$\xrightarrow[-\ CN^{-}]{} NC-\langle \rangle-\overset{CF_3}{\underset{OBu}{\diagup\!\!\diagdown}}$ (yield 90%)

Scheme 3.4

$$PhNO_2 \quad \xrightleftharpoons[\substack{-1.25V\ (\text{vs SCE}) \\ \text{PhCN}}]{} \quad PhNO_2^{\overline{\cdot}}$$

$PhNO_2^{\overline{\cdot}} + C_6F_{13}I \longrightarrow C_6F_{13}{}^{\cdot} + I^{-} + PhNO_2$

$C_6F_{13}{}^{\cdot} + PhCN \longrightarrow p\text{-}C_6F_{13}C_6H_4CN + \overset{C_6F_{13}}{\underset{}{\langle\;\;\rangle}}\!\!\!\overset{H}{\underset{CN}{}} + C_6F_{13}H + C_{12}F_{26}$

$(43\%) \qquad\qquad (7\%) \qquad (7\%) \quad (9\%)$

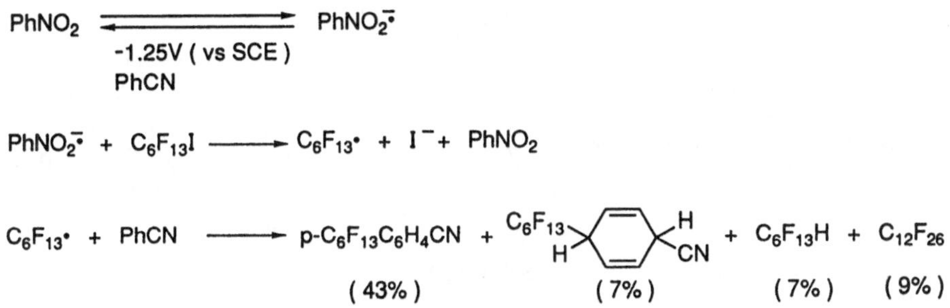

Scheme 3.5

stoichiometry [15]. In the latter case, $SO_2^{-}\cdot$ abstracts a bromine atom from CF_3Br to give the CF_3 radical which further reacts with $SO_2^{-}\cdot$ to give $CF_3SO_2^{-}$ (Eq. 5)

$$SO_2 \xrightarrow{\;e\;} SO_2^{-}\cdot \xrightarrow{\;CF_3Br\;} CF_3{}^{\cdot} + BrSO_2^{-}$$

$$CF_3{}^{\cdot} + SO_2^{-}\cdot \longrightarrow CF_3SO_2^{-} \quad (\text{C. Eff. } 60\%) \tag{5}$$

Furthermore, Saveant et al. have shown elegant examples of electrochemically induced nucleophilic substitution of perfluoroalkyl halides. The reaction mechanism is a slightly modified version of the classical $S_{RN}1$ mechanism in

7

which the reaction is triggered by dissociative electron transfer, not involving the intermediacy of the anion radical of the substrate as shown in Scheme 3.6 [16, 17].

$$M_{ox} + e \rightleftharpoons M_{red}$$

$$R_f X + M_{red} \longrightarrow R_f{}^{\cdot} + X^- + M_{ox}$$

$$R_f{}^{\cdot} + Nu^- \longrightarrow R_f Nu^{-\cdot}$$

$$\frac{R_f Nu^{-\cdot} + M_{ox} \rightarrow R_f Nu + M_{red} \ or \ R_f Nu^{-\cdot} + R_f X \rightarrow R_f Nu + R_f{}^{\cdot} + X^-}{R_f X + Nu^- \rightarrow R_f Nu + X^-}$$

Scheme 3.6

Thus, perfluoroalkylated imidazoles are obtained in excellent to good yields (Scheme 3.7) [16, 17]. Similarly, a hindered phenolate anion provides a fluoroalkylated dimeric product in high yield (Scheme 3.8) [17].

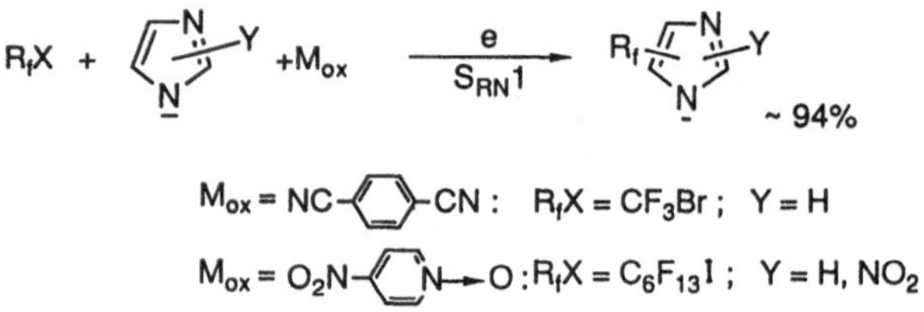

$$M_{ox} = NC-\!\!\!\langle\bigcirc\rangle\!\!\!-CN : \quad R_f X = CF_3 Br \ ; \quad Y = H$$

$$M_{ox} = O_2 N-\!\!\!\langle\bigcirc\rangle\!\!\!N\!\!\rightarrow\!\!O : R_f X = C_6 F_{13} I \ ; \quad Y = H, \ NO_2$$

Scheme 3.7

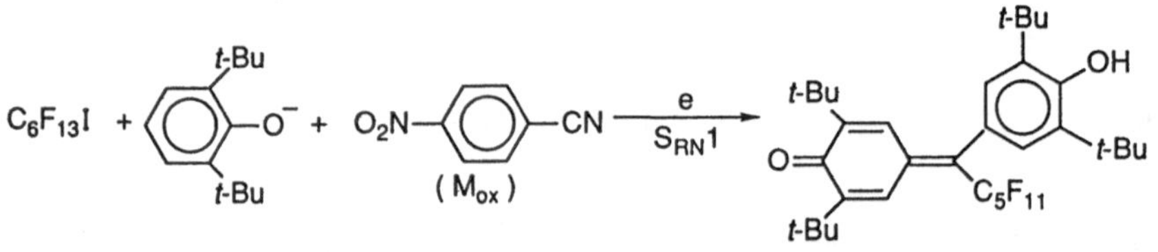

Scheme 3.8

3.3 Utilization of Sacrificial Anodes

Recently, the electrochemical functionalization of organic halides using sacrificial anodes has been remarkably developed [18]. The general mechanism can be schematically shown for the generation of a divalent cation as shown in Scheme 3.9.

At the anode : $M \longrightarrow M^{2+} + 2e$

At the cathode : $RX + 2e \longrightarrow R^- + X^-$

In solution : $R^- + E^+ \longrightarrow RE$

Overall reaction : $RX + M + E^+ \xrightarrow{\text{electricity}} RE + MX^+$

Scheme 3.9

The reactions appear to be similar to organometallic synthesis, where the reduction is performed by the metal instead of electricity. However, these reactions have been shown to be essentially different from the corresponding organometallic reactions. This method has valuable advantages. As the anode reaction is controlled, an undivided cell can be used, the reaction occurs in one-step, the conditions are quite simple, and so on. Sibille and Perichon et al. have found that the sacrificial zinc anode is quite effective for trifluoromethylation of aldehydes to form trifluoromethylated alcohols in almost quantitative yields (Eq. 6) [19]. The reaction proceeds via the reduction of Zinc(II) salts, followed by a chemical reaction between the reduced metal, CF_3Br, and aldehyde.

$$CF_3Br + RCHO \xrightarrow[\text{DMF/Zn anode}]{e} R\overset{\overset{\displaystyle OH}{|}}{\underset{}{\diagup}}CF_3 \qquad 80\text{--}90\% \qquad (6)$$

(R = Alkyl, Aryl)

With ketones, unreactive organozinc species CF_3ZnBr and CF_3ZnCF_3 are mainly formed. In this case, the use of DMF/TMEDA (7:3) as a solvent supresses the formation of the organozinc species and promotes the carbonyl attack, providing tertiary alcohols in moderate yields (Eq. 7) [20].

$$CF_3Br + \underset{R^1}{\overset{\overset{\displaystyle O}{\|}}{}}\overset{}{\underset{}{C}}R^2 \xrightarrow[\text{Zn anode}]{\underset{\text{DMF/TMEDA}}{e}} \underset{R^1}{\overset{HO\quad CF_3}{\diagdown\diagup}}\overset{}{\underset{R^2}{}} \qquad (7)$$

$$R^1 = R^2 = Ph: 57\% \text{ (solvent = DMF only)}$$
$$R^1 = Ph, R^2 = Me: 37\% \text{ (mixed solvent)}$$

Formylation can be also achieved when DMF is used as an electrophile. Thus, the cathodic reduction of CF_3Br in DMF using an aluminum anode provides trifluoroacetaldehyde in good yield (Eq. 8) [21].

$$CF_3Br \xrightarrow[\text{DMF/Al anode}]{e} CF_3CHO \quad 75\% \qquad (8)$$

Even trifluoromethylbenzene can be used as a starting material, which gives various *gem*-difluoro compounds in one-step (Scheme 3.10) [22].

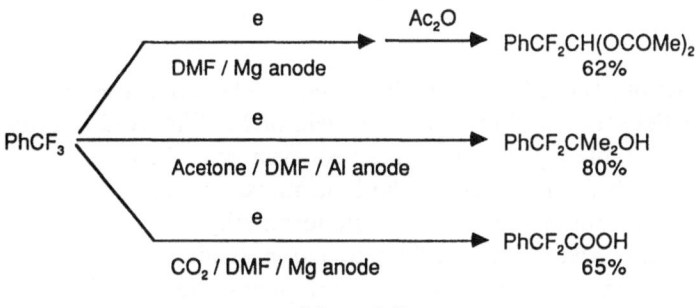

Scheme 3.10

9

One-pot electrosynthesis of trifluoromethanesulfinic acid is also achieved at sacrificial anodes in good current efficiency (Eq. 9) [23].

$$CF_3Br + SO_2 \xrightarrow[\text{DMF/Zn or Mg anode}]{e} CF_3SO_2^- \quad (60\text{--}70\%) \qquad (9)$$

Nedelec et al. have achieved the electrochemical cross-coupling of CF_3CCl_3 with $PhCH_2Br$ by using a sacrificial aluminum anode (Eq. 10) [24].

$$CF_3CCl_3 + PhCH_2Br \xrightarrow[\text{THF}]{e} CF_3CCl_2CH_2Ph \qquad (10)$$
$$60\%$$

The cross-coupling of CF_3Br with aromatic and heteroaromatic halides has also been achieved using a sacrificial copper anode (Eq. 11) [25].

$$CF_3Br + ArX \xrightarrow[\text{DMF/Ph}_3\text{P}]{e} ArCF_3 \qquad (11)$$

ArX = 4-Iodoanisole   : 90%   3-Bromoquinoline : 98%
4-Iodonaphthalene : 98%   2-Bromothiophene: 88%
3-Bromopyridine   : 98%

Furthermore, 2,2-difluoro-3-hydroxyesters are readily obtained from $ClCF_2COOMe$ and carbonyl compounds by electrolysis in a one-compartment cell using a sacrificial zinc anode and a nickel-complex catalyst [26]. The catalytic cycle is shown in Scheme 3.11 and nickel zinc exchange is a key step.

$$Ni(II) + 2e \longrightarrow Ni(0)$$
$$Ni(0) + ClCF_2COOMe \longrightarrow ClNiCF_2COOMe$$
$$ClNiCF_2COOMe + Zn(II) \longrightarrow ClZnCF_2COOMe + Ni(II)$$
$$ClZnCF_2COOMe + \underset{/}{\overset{\backslash}{C}}=O \longrightarrow ClZnO\overset{|}{\underset{|}{C}}CF_2COOMe$$

Scheme 3.11

In this reaction, the CH_2Cl_2/DMF solvent (9:1) suppresses the undesirable Claisen condensation and increases the yield of 2,2-difluoro-3-hydroxyesters. It is notable that high yields are obtained even with ketones and enolizable aldehydes, which do not undergo the Reformatsky reaction.

In addition, homocoupling of o-trifluoromethyl chlorobenzene [27] and p-fluorobromobenzene [28] leading to o,o'-bis(trifluoromethyl)biphenyl and p, p'-difluorobiphenyl, respectively, and the preparation of various trifluorom-ethylarenes [29, 30] have been performed using sacrificial anodes.

4 Application of Electrogenerated Bases to Fluoro Organic Synthesis

Electrogenerated bases have been shown to be useful for fluoro organic synthesis. For example, Fuchigami and Nakagawa have found that an electrogenerated 2-pyrrolidone anion deprotonates (trifluoromethyl)malonic ester to form the corresponding stable enolate, which undergoes efficient alkylation with alkyl halides (Scheme 4.1) [31]. This is notable since α-CF$_3$ enolates are generally unstable and facile defluorination takes place prior to trapping with electrophiles (Scheme 4.2). In this reaction, the quarternary ammonium counter cation of the electrogenerated pyrrolidone anion is essential while ordinary metal counter cations such as Na$^+$ are not effective.

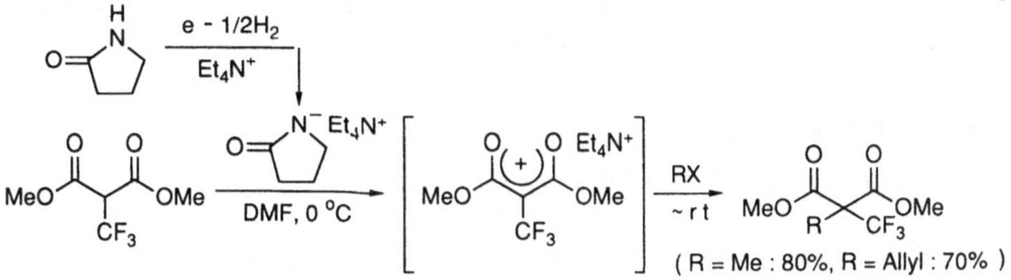

Scheme 4.1

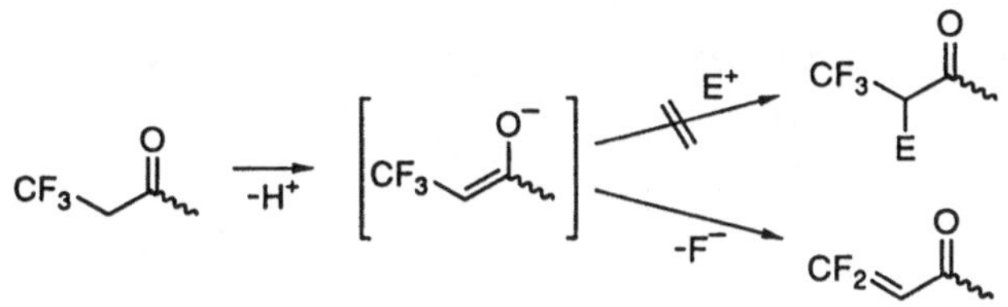

Scheme 4.2

A trifluoromethyl anion is also difficult to generate because it easily lose a fluoride ion to form difluoro carbene. Shono and Kashimura et al. have similarly achieved the generation of such a species from trifluoromethane using the electrogenerated α-pyrrolidone anion. The efficient trifluoromethylation of aldehydes and ketones in the presence of hexamethyldisilazane (HMDS) can thus be obtained (Scheme 4.3) [32].

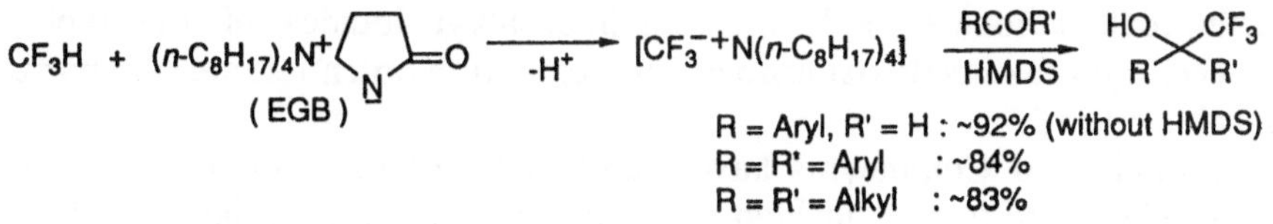

R = Aryl, R' = H : ~92% (without HMDS)
R = R' = Aryl : ~84%
R = R' = Alkyl : ~83%

Scheme 4.3

The electrogenerated base from pyrrolidone also promotes the introduction of a *gem*-difluoromethylen unit to aromatic aldehydes to give difluoromethyl carbinols in high yields (Scheme 4.4) [33].

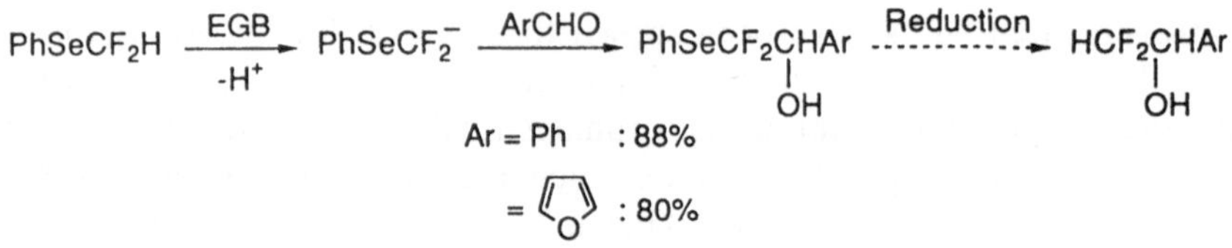

$$PhSeCF_2H \xrightarrow[-H^+]{EGB} PhSeCF_2^- \xrightarrow{ArCHO} PhSeCF_2\underset{\underset{OH}{|}}{C}HAr \xdashrightarrow{Reduction} HCF_2\underset{\underset{OH}{|}}{C}HAr$$

Ar = Ph : 88%

= ⟨O⟩ : 80%

Scheme 4.4

Quite recently, Troupel et al. have developed an effective synthesis of *gem*-difluoro-β-oxonitriles using an electrogenerated base derived from phenyl bromide and a sacrificial magnesium anode together with a nickel cathode coated with a small deposit of cadmium as shown in Scheme 4.5 [34].

At the anode : $Mg \longrightarrow Mg^{2+} + 2e$

At the cathode : $PhBr + 2e \longrightarrow Ph^- + Br^-$

(EGB)

$$Ph^- + RCH_2CN \longrightarrow PhH + R\bar{C}HCN$$

$$PhCF_2COOMe + R\bar{C}HCN \longrightarrow PhCF_2-\underset{\underset{Mg^{2+}\ldots O^-}{\overset{OMe}{|}}}{C}-CHRCN$$

$$\xrightarrow[-MeOH]{H^+} PhCF_2COCHRCN$$

R = H : 60%
= Ph: 83%

Scheme 4.5

5 Anodic Oxidation of Perfluoro and Polyfluoro Compounds

Perfluoroalkyl iodides R_fI and polyfluoroalkyl iodides of the type of $R_fCH_2CH_2I$ are directly oxidizable. Their oxidation potentials are summarized in Table 3 [35].

Germain and Commeyras have found that perfluoro sulfonic esters and fluorosulfates are formed in high yields by direct anodic oxidation of R_fI in perfluoroalkane sulfonic acids and fluorosulfuric acid (Eqs. 12 and 13) [36].

Table 3. Oxidation potentials (peak potentials, E_p) of perfluoro and polyfluoro iodides[a]

Compd.	E_p V vs Ag wire[b]
C_4F_9I	1.53
$C_6F_{13}I$	1.56
$C_8F_{17}I$	1.56
$C_{10}F_{21}I$	> 1.70[c]
$C_6F_{13}CH_2CH_2I$	1.55
$C_8F_{17}CH_2CH_2I$	1.39
$C_{10}F_{21}CH_2CH_2I$	2.15[c]

[a] In 0.1M Et_4NBF_4/MeCN; Pt anode; 0.1 V/s.
[b] + 1.5V vs SCE.
[c] In 0.1M Bu_4NBF_4/MeCN.

With diiodo compounds, the mono and diesters can be selectively obtained (Eq. 14). These are useful precursors to valuable perfluoro carboxylic acids.

$$R_fI \xrightarrow[R_{f'}SO_3H]{-e} R_{f'}SO_3R_f + 1/2I_2 \quad (12)$$

$$R_f = C_2F_5, R_{f'} = C_4F_9: 88\%$$
$$R_f = C_6F_{13}, R_{f'} = CF_3: 92\%$$

$$C_6F_{13}I \xrightarrow[FSO_3H]{-e} FSO_3C_6F_{13} + 1/2I_2 \quad (13)$$

$$85\%$$

$$(C_4F_9SO_3CF_2CF_2)_2 \xleftarrow[C_4F_9SO_3H]{-2e \text{ (galvanostatic)}} I(CF_2)_4I$$

$$\xrightarrow[C_4F_9SO_3H]{-e \text{ (potentiostatic)}} C_4F_9SO_3(CF_2)_4I \quad (14)$$

Since the reaction is shown to be a one-electron oxidation process, the reaction may proceed via a perfluoro carbocation intermediate as is seen with nonfluorinated iodides (Eq. 15).

$$R_fCF_2I \xrightarrow{-e} [R_fCF_2I]^{+\cdot} \xrightarrow{-I\cdot} [R_fCF_2]^+ \xrightarrow[\text{or } FSO_3^-]{R'_fSO_3^-} R'_fSO_3CF_2R_f \quad (15)$$
$$\text{or}$$
$$FSO_3CF_2R_f$$

Perfluoro iodide $C_6F_{13}CH_2CH_2I$ is also similarly oxidized to give the sulfonic ester quantitatively (Eq. 16).

13

$$C_6F_{13}CH_2CH_2I \xrightarrow[CF_3SO_3H]{-e} CF_3SO_3CH_2CH_2C_6F_{13} \qquad (16)$$

$$98\%$$

On the other hand, Becker et al. also have attempted the anodic oxidation of $RfCH_2CH_2I$ in acetonitrile and they have achieved the anodic transformation of $C_8F_{17}CH_2CH_2I$ to the corresponding acetamide, trifluoroacetate, and benzoate derivatives in good yields [35]. They propose a different reaction mechanism involving a hypervalent iodanyl radical intermediate as shown in Eq. 17.

$$C_8F_{17}CH_2CH_2I \xrightarrow{-e} [C_8F_{17}CH_2CH_2I]^{+\cdot}$$

$$\xrightarrow{Nu^-} [C_8F_{17}CH_2CH_2INu]^{\cdot}$$

$$\longrightarrow C_8F_{17}CH_2CH_2Nu + I\cdot(1/2I_2) \qquad (17)$$

$$(Nu^- = MeCN + OH^-, CF_3COO^-, PhCOO^-)$$

Recently, Germain et al. have also shown that the indirect anodic oxidation in fluorosulfuric acid of fluorocarbon derivatives of the type R_fCF_2X (X = H, COOH, SO_3H, CH_2OH, Br), which are not directly oxidizable, leads to fluorosulfates of the type $FSO_3CF_2R_f$ (Eqs. 18 and 19) [37, 38]. In these reactions, the peroxide $(FSO_3)_2$, partially dissociated in its free radicals, is the in-situ electrochemically produced reactive intermediate as shown in Scheme 5.1.

$$C_3F_7COOH \xrightarrow[FSO_3^-]{-2e} FSO_3C_3F_7 \qquad (18)$$

$$85\%$$

$$H(CF_2)_6CH_2OH \xrightarrow[FSO_3^-]{-2e} FSO_3(CF_2)_6O_3SF \qquad (19)$$

$$82\%$$

$$2FSO_3^- \xrightarrow{-2e} (FSO_3)_2 \rightleftharpoons 2FSO_3\cdot$$

$$FSO_3\cdot + R_fH \longrightarrow FSO_3H + R_f\cdot$$

$$FSO_3\cdot + R_f\cdot \longrightarrow FSO_3R_f \qquad R_f = C_2F_5 \ : 87\%$$
$$C_6F_{13} \ : 92\%$$

Scheme 5.1

These products are precursors of perfluoro carboxylic acids while the fluorosulfates derived from secondary and tertiary hydrofluorocarbons provide perfluoro ketones and, starting from the $1H$-perfluoro bicyclo[2.2.1]heptane, the corresponding perfluoro tertiary alcohol is obtained respectively (Eqs. 20 and 21) [39].

$$(R_f)_2CFH \xrightarrow[FSO_3^-]{-2e} FSO_3CF(R_f)_2 \xrightarrow[KF/CsF]{\Delta} (R_f)_2C=O \qquad (20)$$

$$(R_f)_3CH \xrightarrow[FSO_3^-]{-2e} FSO_3C(R_f)_3 \longrightarrow (R_f)_3COH \qquad (21)$$

6 Anodic Oxidation of Heteroatom Compounds Having Fluoroalkyl Groups

6.1 General Aspects

As mentioned in the introduction, partially fluorinated compounds are highly useful, however methods for their synthesis are strictly limited in many cases. For example, nucleophilic substitution occurs with difficulty at the position α to a trifluoromethyl group due to its strong electron-withdrawing effect, although sulfur and selenium nucleophiles undergo such a substitution rather efficiently (Scheme 6.1).

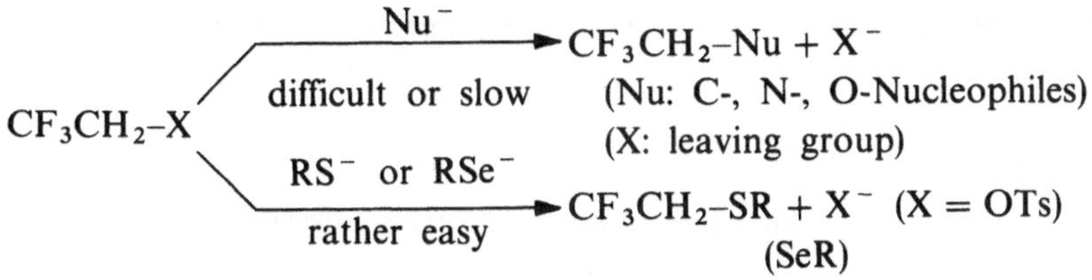

CF$_3$CH$_2$–X

$\xrightarrow{\text{Nu}^-}$ difficult or slow → CF$_3$CH$_2$–Nu + X$^-$
(Nu: C-, N-, O-Nucleophiles)
(X: leaving group)

$\xrightarrow{\text{RS}^- \text{ or RSe}^-}$ rather easy → CF$_3$CH$_2$–SR + X$^-$ (X = OTs)
(SeR)

Scheme 6.1

Therefore, achievement of this desired substitution, particularly the formation of a carbon-carbon bond at the α position is one of most important goals of modern organofluorine chemistry. Although anodic substitution is a characteristic of certain electrolytic reactions, no results pertaining to the electrolytic substitution of trifluoromethylated compounds have been reported. Recently, the use of the electrochemical technique has opened new avenues for the realization of such nucleophilic substitution [40–42] and construction of a carbon-carbon bond [43–45].

Here, regioselective anodic substitutions of heteroatom compounds having fluoroalkyl groups and the effects of fluorine atoms on both the synthetic behavior and the oxidation potentials will be mainly discussed.

6.2 Oxidation Potentials of Heteroatom Compounds Having Fluoroalkyl Groups

The oxidation potentials of various fluoroalkyl sulfides, selenides, and telluride have been comparatively investigated by cyclic voltammetry [42, 43, 46]. They show irreversible multiple anodic waves. Table 4 summarizes the first oxidation peak potentials E_p together with those of some nonfluorinated and other electronegatively substituted chalcogeno compounds as a comparison. In all cases, the fluorinated compounds are oxidized at a more positive potential than the corresponding nonfluorinated compounds due to the electron-withdrawing effect of the fluoroalkyl group. It should be noted that 2,2,2-trifluoroethyl and cyanomethyl selenides show almost equal oxidation potentials although the CF_3 group has a weaker electron-withdrawing effect than CN (Taft's σ^* of $CH_2CF_3 = 0.92$ vs 1.30 for CH_2CN). A similar trend is observed in the case of the corresponding amines (see Table 5).

A linear correlation of the oxidation potentials E_p of sulfides with Taft's σ^* values of fluoromethyl groups is obtained as shown in Fig. 1 [42]. This clearly indicates that the polar effect of the fluoroalkyl group plays a significant role in the electron-transfer step from the sulfides to anode. Namely, the oxidation potential increases linearly as the number of fluorine atoms of the fluoroalkyl group increases. However, interestingly the oxidation potential was not appreciably affected by the length of the perfluoroalkyl group (Table 4).

Oxidation potentials (half peak potentials, $E_{p1/2}$) of fluoroethylamines and related nonfluorinated amines are summarized in Table 5 [41, 47]. It is notable that chloromethyl groups such as CF_2Cl and $CHFCl$ shows slightly higher oxidation potentials than the corresponding fluoromethyl groups although a

Table 4. Oxidation potentials (peak potentials, E_p) of fluoroalkyl chalcogeno compounds[a]

[PhZCH₂Rf]

R_f	E_p V vs SCE		
	Z = S	Z = Se	Z = Te
CF_3	1.78	1.70	1.34
CF_3CF_2	1.82	1.71	—
$CF_3CF_2CF_2$	1.84	1.72	—
CHF_2	1.69	—	—
CH_2F	1.58	—	—
CH_2CF_3	1.63	1.50	—
CH_3	1.48	1.37	—
H	1.51	1.32	0.82
CN	1.84	1.70	—
COOEt	1.64	1.50	—

[a] In 0.1M Bu_4NBF_4/MeCN; 0.1 V/s; Pt anode.

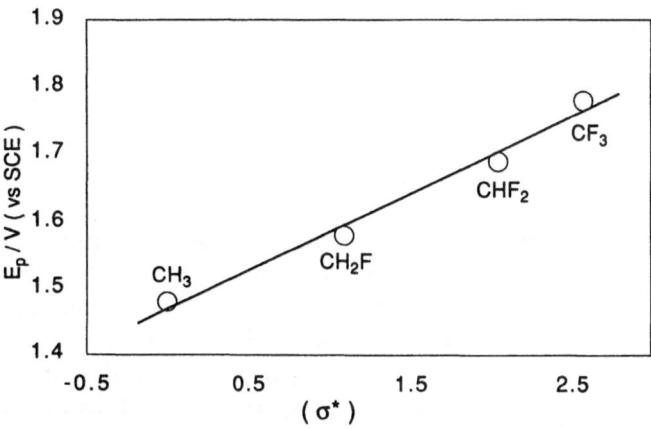

Fig. 1. Relationship between oxidation peak potentials (E_p) of fluoroalkyl sulfides and Taft's substituent constants (σ^*) (from Ref. 42).

Table 5. Oxidation potentials (half peak potentials, $E_{p1/2}$) of fluoro-ethylamines and related amines

$$\left[\begin{array}{c} R^1 \\ \diagdown \\ N-CH_2R_f \\ \diagup \\ R^2 \end{array} \right]$$

Amine		R_f	$E_{p1/2}$V vs SCE
R^1	R^2		
Ph	Me	CF_3	0.96[a]
Ph	Et	CF_3	0.96[a]
Ph	Et	CN	0.95[a]
Ph	Et	Me	0.64[a]
Ph	Et	CF_2Cl	1.00[a]
Ph	Et	CHFCl	0.92[a]
Ph	Et	CHF_2	0.88[a]
Ph	Et	CH_2F	0.76[a]
Bzl	Et	CF_3	1.24[b]
Bzl	Me	H	1.10[b]
n-Bu	n-Bu	CF_3	1.15[b]
Et	Et	Me	0.96[b]
$-(CH_2)_5-$		CF_3	1.18[b]
$-(CH_2)_5-$		H	1.10[b]

[a] In 0.1M Et_4NOTs/MeCN, 100 mV/s, glassy carbon anode.
[b] In 0.1M $NaClO_4$/MeCN, 100 mV/s, Pt anode.

chlorine atoms has a lower electronegativity. The oxidation potential of fluoro-ethylamines increases as the number of fluorine atoms of the fluoroalkyl group increases in a manner similar to the case of fluoroalkyl sulfides.

6.3 Anodic Substitutions of Fluoroalkyl Sulfides and Selenides

Anodic substitutions at some types of organic nitrogen and sulfur compounds are well-known to be useful key reactions for organic synthesis. Recently, Fuchigami et al. have found that aryl 2,2,2-trifluoroethyl sulfides, which are readily derived from cheap trifluoroethanol, undergo efficient anodic meth-oxylation and acetoxylation and they have successfully achieved introduction of oxygen nucleophiles to the α position to the CF_3 group as shown in Scheme 6.2 [40, 43]. In contrast, the anodic methoxylation of nonfluorinated sulfides does not occur. In addition, generally anodic acetoxylation of simple alkyl sulfides proceeds only when the concentration of both the substrate and the acetate ions of the supporting electrolyte is extremely high whereas the anodic acetoxylation of 2,2,2-trifluoroethyl sulfide takes place rather efficiently even at low concentrations. Similarly, the fluoroalkylated sulfide gives a much higher yield compared to nonfluorinated sulfide at high concentrations (Scheme 6.2).

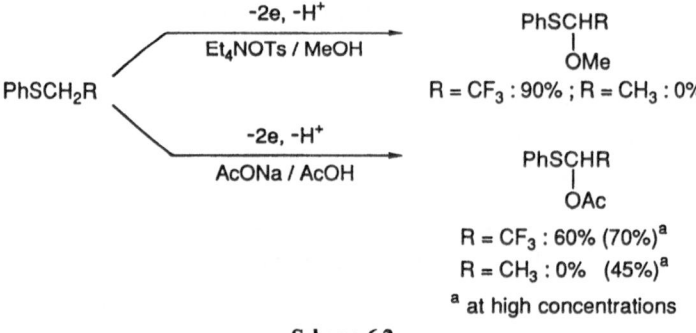

Scheme 6.2

α-Acetoxy sulfides are known to be obtained from sulfoxides by the Pum-merer reaction. However, the sulfoxide derived from **2** provides **3** in low yield (Scheme 6.3). In contrast, cyanomethyl phenyl sulfoxide is known to give α-acetoxy sulfide in good yield. Therefore, a CF_3 group interferes with the Pummerer reaction although its electron-withdrawing effect is similar to that of a cyano group.

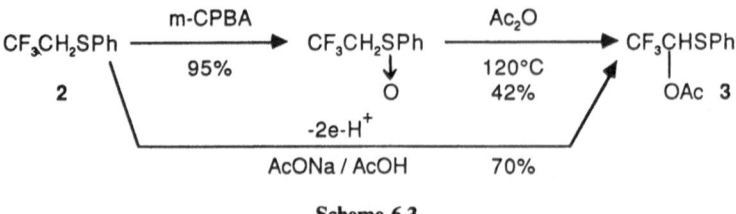

Scheme 6.3

Table 6. Efficiency of the anodic methoxylation and Hammett's σ and σ^+; values of substituents (X)

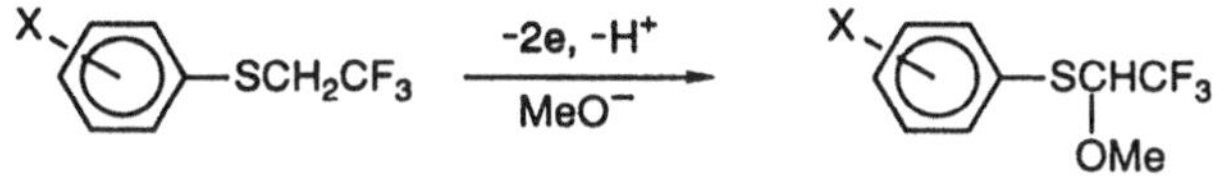

X	p-Cl	H	p-F	m-Me	p-Me	p-MeO
Yield/%	80	93	43	28	18	13
σ^+	+ 0.11	0	− 0.07	− 0.07	− 0.31	− 0.78
σ	+ 0.23	0	+ 0.66	− 0.07	− 0.17	− 0.27

These facts indicate that an α-trifluoromethyl group remarkably promotes anodic substitutions with oxygen nucleophiles. Since these are very few success-ful examples of anodic methoxylation of phenylalkyl sulfides are known, it is notable that an α-trifluoromethyl group facilitates the anodic methoxylation. Therefore, such anodic substitutions have been systematically investigated from both mechanistic and synthetic aspects [42].

It was found that the anodic methoxylation of aryl 2,2,2-trifluoroethyl sulfides is greatly affected by the substitutions at the benzene ring. Electron-withdrawing groups promote this anodic methoxylation while electron-dona-ting groups significantly interfere with the methoxylation (Table 6). As shown in Table 6, the product yields are better correlated to Hammet's σ^+ values than to the σ values, namely, as the σ^+ value becomes more negative, the yield decreases. Therefore, the reaction is not governed by the stability of the cationic inter-mediates **A** or **B** as illustrated in Scheme 6.4. Since the electron-withdrawing substituent (X = Cl) does not interfere with this reaction and promotes the methoxylation as observed in the case of non-substituted phenyl sulfide, this reaction is highly controlled by the ease of deprotonation of the radical cation intermediate **A** (step a).

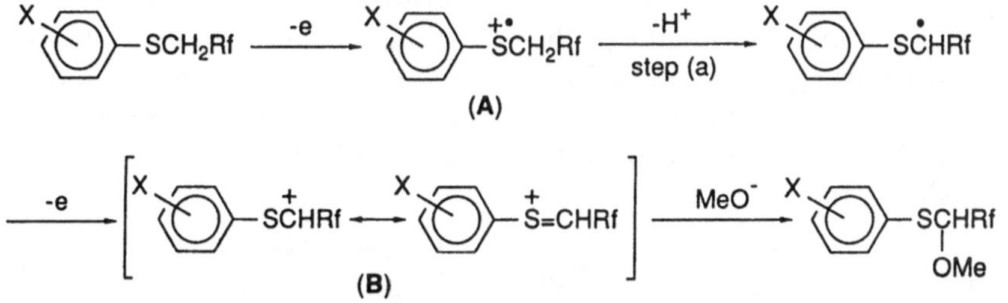

Scheme 6.4

The anodic methoxylation and acetoxylation are also significantly affected by the structure of the fluoroalkyl groups as shown in Table 7. Strong electron-withdrawing perfluoroalkyl groups promote the anodic methoxylation. Inter-estingly, the longer perfluoroalkyl groups exhibits less substitution when

Table 7. Anodic methoxylation and acetoxylation of fluoroalkyl phenyl sulfides

$$PhSCH_2R_f \xrightarrow[\text{YO}^-]{-2e, -H^+} \overset{\displaystyle PhSCHR_f}{\underset{\displaystyle OY}{|}}$$

R_f	Yield (%)						
Y	$CF_3CF_2CF_2$	CF_3CF_2	CF_3	CHF_2	CH_2F	$CHClF$	CH_3
Me	58	72	93	19	b	c	0
Ac[a]	43	46	60	28	20	16	0

[a] At low concentrations.
[b] Many complicated products were formed.
[c] Not detected.

compared with the trifluoromethyl group although these longer perfluoroalkyl groups show almost the same effect on the oxidation potentials of the sulfides as the trifluoromethyl group (see Table 4). In contrast, a difluoromethyl group causes a drastic decrease in the yield of the methoxylated product and mono-fluoromethyl and chlorofluoromethyl groups do not promote the methoxylation.

In contrast to the anodic methoxylation, the anodic acetoxylation takes place regardless of the fluoroalkyl group and the promotion effect on the acetoxylation is in the ordered $CF_3 > C_2F_5 > C_3F_7 \gg CHF_2 > CH_2F > CHClF$. This order is similar to that of the methoxylation. This order can be rationalized as being due to the ease of the deprotonation from **A** (kinetic acidity of **A**) in Scheme 6.4 (MeO→AcO). Thus, the anodic methoxylation and acetoxylation of aryl fluoroalkyl sulfides are greatly affected by both the substituents at the benzene ring and at the fluoroalkyl groups.

By the way, trifluoroacetaldehyde is a versatile fluoro building block. However the chemical or electrochemical oxidative transformation of trifluoro-ethanol to trifluoroacetaldehyde has been unsuccessful. Trifluoroacetaldehyde is therefore generally produced by the reduction of trifluoroacetic acid ester or acid chloride using an excess of LAH. The anodic substitution at fluoroalkyl phenyl sulfides is a useful alternative because it realizes the transformation of economical trifluoroethanol to highly valuable trifluoroacetaldehyde equivalents as shown in Scheme 6.5.

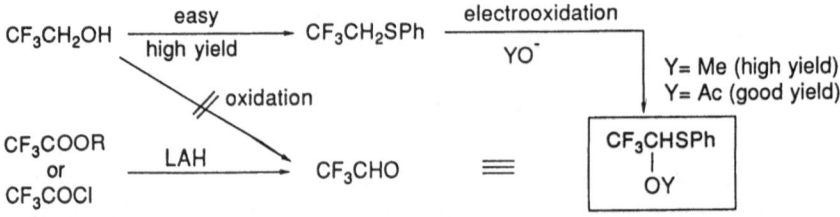

Scheme 6.5

In fact, **3** is easily converted into trifluoroacetaldehyde phenylhydrazone without any defluorination by alkali hydrolysis (Eq. 22).

$$CF_3CH(OAc)SPh \xrightarrow{\text{aq. } K_2CO_3} [CF_3CHO] \xrightarrow{\text{PhNHNH}_2} CF_3CH=NNHPh \qquad (22)$$

$$\textbf{3} \qquad\qquad\qquad\qquad\qquad\qquad 70\%$$

The α-methoxy sulfide is readily transformed into biologically interesting α-monofluoroalkanoic acids (Scheme 6.6).

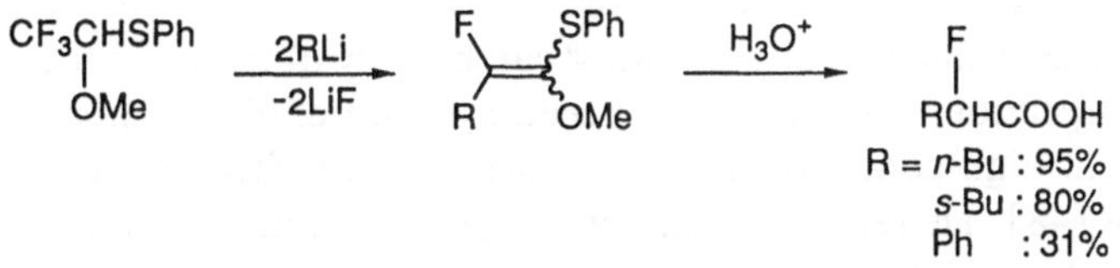

$$\begin{array}{ccc}
\underset{\underset{\text{OMe}}{|}}{CF_3CHSPh} & \xrightarrow[-2LiF]{2RLi} & \overset{F}{\underset{R}{\diagdown}}\underset{\text{OMe}}{\diagup}\overset{SPh}{} \xrightarrow{H_3O^+} & \underset{\underset{RCHCOOH}{}}{\overset{F}{|}}
\end{array}$$

R = n-Bu : 95%
s-Bu : 80%
Ph : 31%

Scheme 6.6

Since these methoxylated and acetoxylated sulfides have an acetal structure, it is expected that Lewis acid catalyzed demethoxylation should generate a carbocation intermediate which is stabilized by the neighboring sulfur atom. In fact, nucleophilic substitution with arenes has been successfully achieved as shown in Scheme 6.7 [43]. This procedure is useful for the preparation of trifluoroethyl aromatics. As already mentioned, generation of carbocations bearing an α-trifluoromethyl group is difficult due to the strong electron-withdrawing effect. Therefore, this carbon-carbon bond formation reaction is remarkable from both mechanistic and synthetic aspects.

$$CF_3CH(OMe)SAr \xrightarrow[-MeO^-]{AlCl_3} [CF_3\overset{+}{C}HSAr \rightleftharpoons CF_3CH=\overset{+}{S}Ar]$$

$$\xrightarrow{Ar'H} CF_3CH(Ar')SAr$$

$$Ar = Ar' = Ph: 83\%$$

$$CF_3CH(Ar')SAr \xrightarrow[-Bu_3SnSAr]{Bu_3SnH} CF_3CH_2Ar'$$
$$\text{quant.}$$

Scheme 6.7

Nucleophilic substitution at the β-position to a trifluoromethyl group is also generally difficult, except for sulfur nucleophiles owing to the predominant elimination to trifluoropropene as shown in Eq. 23.

$$CF_3CH_2CH_2X + Nu^- \xrightarrow[-X^-]{-NuH} CF_3CH=CH_2 \qquad (23)$$

Strongly basic carbon, oxygen, and nitrogen nucleophiles initiate E2 eliminations by deprotonation to give the corresponding olefins. In the case of E1 reactions, facile deprotonation from the β-position of the cationic intermediate takes place. The nucleophile may then be added to the double bond. It has been demonstrated that such nucleophilic substitution can be performed by anodic oxidation of polyfluoroalkyl iodides ($R_fCH_2CH_2I$) in the presence of some types of oxygen and nitrogen nucleophiles (see Sect. 5). On the other hand, β-trifluoromethylated O,S-acetals seem to be promising building blocks for this purpose. As described above, the Pummerer reaction is a well-known method for the preparation of α-acetoxy sulfides from sulfoxides. However, phenyl 1-(acetoxy)-3,3,3-trifluoropropyl sulfide **4** can be only obtained in low yield by the Pummerer reaction while the anodic acetoxylation of phenyl 3,3,3-trifluoropropyl sulfide at high concentrations provides **4** in satisfactory yield as shown in Scheme 6.8 [48]. Therefore, the electrochemical method is much superior to the conventional Pummerer reaction since the α-acetoxy sulfide was obtained in one step under mild conditions.

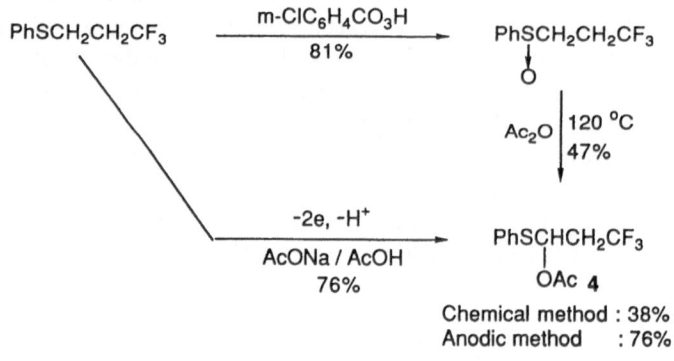

Scheme 6.8

The thus prepared β-trifluoromethylated O,S-acetal **4** allows us to make a carbon-carbon bond via a carbocation at the β-position to the CF_3 group as shown in Scheme 6.9 [48]. Interestingly, Lewis acid-mediated allylation and cyanation can be achieved efficiently only when electrogenerated acids (EGA) are employed.

Very recently, Fuchigami et al. have found that the CF_3 group also promotes the anodic fluorination of sulfides (Eq. 24) [49].

$$p\text{-}XC_6H_4SCH_2CF_3 \xrightarrow[\text{Et}_3\text{N}\cdot 3\text{HF/MeCN}]{-2e,\ -H^+} p\text{-}XC_6H_4SCHFCF_3 \qquad (24)$$

$$X = H \ : 62\%; \ X = MeO: 56\%$$
$$Me \ : 51\% \qquad Cl \ \ : 65\%$$

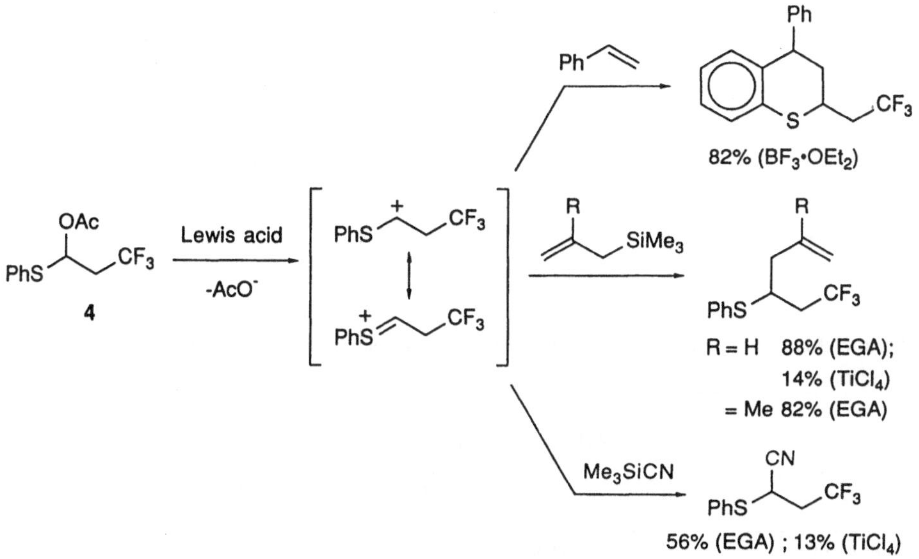

Scheme 6.9

Interestingly, not only a CF_3 group but also CHF_2 and CH_2F groups promote the anodic fluorination (Eq. 25) [50].

$$PhSCH_2R_f \xrightarrow[\text{Et}_3\text{N}\cdot 3\text{HF/MeCN}]{-2e, -H^+} PhSCHFR_f \tag{25}$$

$$R_f = CF_3 \ : 62\%$$
$$CHF_2: 53\%$$
$$CH_2F: 60\%$$

The anodic fluorination appears to proceed in a manner similar to that of the anodic methoxylation (Scheme 6.4). However, the effect of the substituents at the benzene ring and the fluoroalkyl group is quite different from that observed for the anodic methoxylation: Electron-donating groups on the benzene ring strongly retard the anodic methoxylation (Table 6), whereas electron-donating methyl and methoxy groups do not influence the anodic fluorination (Eq. 24). The electron-withdrawing ability of the flouroalkyl groups significantly affects the anodic methoxylation (Table 7), whereas it does not affect the anodic fluorination (Eq. 25).

When the anodic fluorination of 2-monofluoroethyl sulfide 5 is carried out in methanol containing $Et_3N \cdot 3HF$ instead of acetonitrile as a solvent, interestingly the α-methoxylated product 7 rather than the α-flourinated product 6 is obtained exclusively (Scheme 6.10) [51]. As described above, 7 is not obtained from 5 under conventional anodic methoxylation conditions. Therefore, that this novel fluoride ion promoted anodic methoxylation is remarkable. As shown

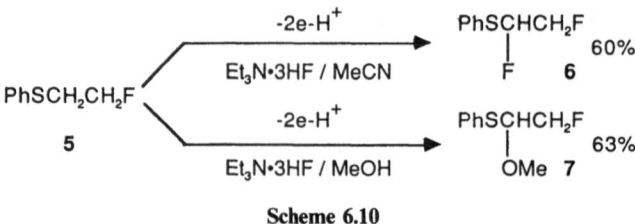

Scheme 6.10

Table 8. Fluoride ion promoted anodic methoxylation of 2-monofluoromethyl sulfide.

$$PhSCH_2CH_2F \xrightarrow[F^-/MeOH]{-2e, -H^+} PhSCHCH_2F$$
$$5 \qquad\qquad\qquad\qquad\qquad |$$
$$OMe$$

Supporting electrolyte[a]	Charge passed (F/mol)	Product yield, %
Et$_3$N·3HF	3.5	63
Et$_3$N·3HF (1 equiv)	3.5	42
Bu$_4$NF·3H$_2$O	3.5	50
NaF	3.5	trace
Et$_4$NBr	3.5	0
Bu$_4$NCl	7.2	0

[a] A 10:1 ratio with sulfide was used.

in Table 8, among the fluorides, Et$_3$N·3HF is most effective while sodium fluoride is not effective. It is notable that Bu$_4$NF·3H$_2$O is also effective and does not cause the formation of sulfoxide although it contains water.

This fluoride ion promoted substitution is widely applicable as shown in Table 9. Thus, the current efficiency and the yields of the desired methoxylated products are greatly improved by this method. Particularly, the product PhSCH(OMe)CH$_2$CF$_3$ is a useful building block as a CF$_3$CH$_2$CHO equivalent.

This novel anodic methoxylation may proceed via the fluorosulfonium ion **B** in a Pummerer-type mechanism as shown in Scheme 6.11. In this mechanism, the cation radical **A** of the sulfide is trapped by a fluoride ion, and this step should suppress side reactions from the cation radical **A** (such as dimerization and nucleophilic attack on an aromatic ring) even when deprotonation of **A** is slow due to the weak electron-withdrawing R$_f$ groups or electron-donating substituents on the benzene ring. Since fluoride ions are much weaker nucleophiles compared to methoxide, it is reasonable that the methoxylation predominates in methanol.

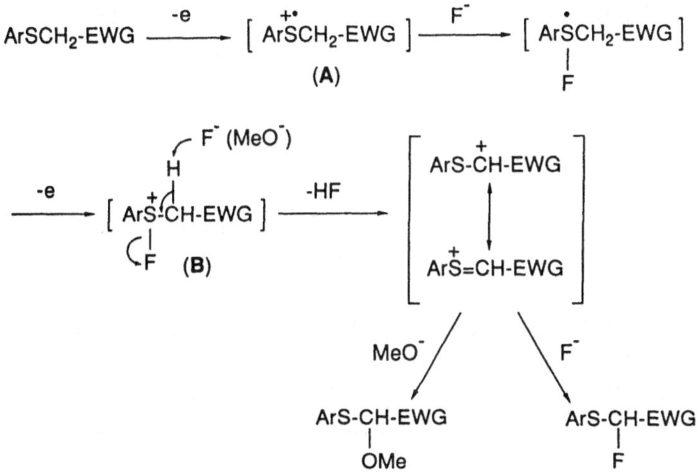

$$ArSCH_2\text{-EWG} \xrightarrow{-e} \left[ArSCH_2\text{-EWG} \right]^{+\bullet} \xrightarrow{F^-} \left[Ar\overset{\bullet}{S}CH_2\text{-EWG} \right]$$
$$\text{(A)} \qquad\qquad\qquad \underset{F}{|}$$

$$\xrightarrow{-e} \left[Ar\overset{+}{S}\overset{H}{\underset{F}{\overbrace{\cdot}}}CH\text{-EWG} \right] \xrightarrow[-HF]{F^- (MeO^-)} \left[\begin{array}{c} Ar S\text{-}\overset{+}{C}H\text{-EWG} \\ \updownarrow \\ Ar\overset{+}{S}=CH\text{-EWG} \end{array} \right]$$
$$\text{(B)}$$

$$\overset{MeO^-}{\swarrow} \qquad \overset{F^-}{\searrow}$$

$$\underset{OMe}{ArS\text{-}CH\text{-EWG}} \qquad\qquad \underset{F}{ArS\text{-}CH\text{-EWG}}$$

Scheme 6.11

Table 9. Fluoride ion promoted anodic methoxylation of fluoroalkyl sulfides

$$p\text{-}XC_6H_4SCH_2R_f \xrightarrow[\substack{Et_3N \cdot 3HF(A) \text{ or} \\ Et_4NOTs(B)/MeOH}]{-2e,\ -H^+} \underset{OMe}{p\text{-}XC_6H_4SCHR_f}$$

Sulfide		Supporting electrolyte	Charge passed	Product yield (%)
X	R_f			
H	CH_2F	A	3.5	63
H	CH_2F	B	3.5	0
H	CH_2F	A	5.1	98
H	CH_2F	B	10	19
MeO	CF_3	A	3.5	81
MeO	CF_3	B	10	13
H	CH_2CF_3	A	3.5	74
H	CH_2CF_3	B	10	trace

In contrast to these cases, it has been shown that bromosulfonium ions derived from sulfides and anodically generated positive bromonium ions can be used as so-called mediators for indirect anodic oxidation of alcohols [52]. Therefore, it is very interesting that the fluorosulfonium ions (**B**) noticeably promote the anodic methoxylation. This novel methoxylation can also be applied to fluoroalkyl selenides. Anodic α-substitution of organic selenium compounds has never been reported. As shown in Eq. 26, even perfluoroalkyl groups do not promote the anodic methoxylation of selenides efficiently under conventional methoxylation conditions using Et_4NOTs as a supporting electrolyte. This is quite different from that observed in the case of the corresponding

25

sulfides. (see Table 7). However, in the presence of fluoride ions, the anodic methoxylation can proceed in relatively good yields (Eq. 26) [53].

$$PhSeCH_2R_f \xrightarrow[Et_3N \cdot 3HF/MeOH]{-2e-H^+} PhSeCHR_f \qquad (26)$$
$$\underset{OMe}{|}$$

$$R_f = C_nF_{2n+1}(n = 1\text{--}3): 65\text{--}74\%$$
$$Et4NOTs/MeOH \qquad : 3\text{--}15\%$$

On the contrary, the anodic acetoxylation is markedly promoted by perfluoroalkyl groups (Eq. 27) [46]. It is notable that the promotion effect of a CF_3 group on the anodic methoxylation is more pronounced than that of a CN group although the CF_3 group has a smaller electron-withdrawing effect than CN.

$$PhSeCH_2R_f \xrightarrow[AcONa/AcOH]{-2e-H^+} PhSeCHR_f \qquad (27)$$
$$\underset{OAc}{|}$$

$$R_f = CF_3: 67\% \ (CH_3: 0\%, CN = 50\%)$$
$$C_2F_5, C_3F_7: \text{--}69\%$$

These methoxylated and acetoxylated selenides are α-perfluoroalkyl monoselenoacetals, which seem to be useful building blocks similar to those of the sulfur analogues described above. So far, only limited methods have been developed for the preparation of monoselenoacetals and they require rather complicated procedures or special reagents. In this regard, this electrochemical method has advantages since monoselenoacetals can be prepared in a one step reaction under mild conditions.

6.4 Anodic Oxidation of Fluoroalkyl Tellurides

In contrast to the cathodic reduction of organic tellurium compounds, few studies on their anodic oxidation have been performed. No paper has reported on the electrolytic reactions of fluorinated tellurides up to date, which is probably due to the difficulty of the preparation of the partially fluorinated tellurides as starting material. Quite recently, Fuchigami et al. have investigated the anodic behavior of 2,2,2-trifluoroethyl and difluoroethyl phenyl tellurides (8 and 9) [54]. The telluride 8 does not undergo an anodic α-substitution, which is totally different to the cases of the corresponding sulfide and selenide. Even in the presence of fluoride ions, the anodic methoxylation does not take place at all. Instead, a selective difluorination occurs at the tellurium atom effectively to provide the hypervalent tellurium derivative in good yield as shown in Scheme 6.12.

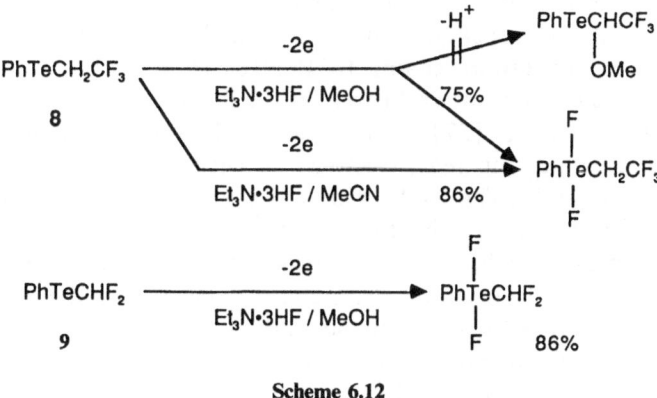

Scheme 6.12

6.5 Anodic Substitutions of Fluoroalkyl Amines

Recently, Fuchigami et al. have systematically studied the anodic α-methoxylation of various types of N-(fluoroethyl) amines, $ArRNCH_2R_f$ ($R_f = CF_3$, CHF_2, CH_2F etc.) at a graphite anode in alkaline methanol using an undivided cell [41, 55, 56]. A methoxy group is exclusively or preferentially introduced into the α-position toward the fluoromethyl (R_f) group depending on R_f and R groups as summarized in Table 10. N-Ethyl derivatives having a CF_3 group give only one regio isomer of the α-methoxylated product alone (runs 2 and 3).

Table 10. Anodic methoxylation of fluoroethylamines

$$\text{Ar} \diagdown \text{R} \diagup \text{NCH}_2\text{R}_f \xrightarrow[\text{MeONa/MeOH}]{-2e-H^+} \text{.Ar} \diagdown \text{R} \diagup \text{NCHR}_f \underset{\text{OMe}}{|}$$

Run	Amine			Charge passed (F/mol)	Yield (%)
	Ar	R	Rf		
1	Ph	Me	CF_3	3.8	71[a]
2	Ph	Et	CF_3	4.9	85
3	p-Tol	Et	CF_3	3.5	78
4	Ph	Ph	CF_3	7.8	81
5	Ph	Me	CHF_2	3.3	19[b]
6	Ph	Et	CHF_2	4.5	91
7	p-Tol	Et	CHF_2	3.5	87
8	Ph	Me	CH_2F	2.8	0[c]
9	Ph	Et	CH_2F	3.0	45

[a] $PhNHCH_2CF_3$ (29%) was formed.
[b] $PhNHCH_2CHF_2$ (48%) was formed.
[c] $PhN(CH_2OMe)CH_2CH_2F$ (84%) was formed.

27

It is quite interesting that the anodic methoxylation also occurs predominantly (71%) at the 2,2,2-trifluroethyl group of N-methylaniline (run 1) and only with 29% at the methyl group leading to PhNH-CH$_2$-CF$_3$. Such anodic methoxylation is known to occur exclusively at the methyl group of nonfluorinated N-ethyl-N-methylaniline [57]. Thus, it should be noted that the CF$_3$ group dramatically changes the regioselectivity of such anodic methoxylations as shown in Scheme 6.13. The CHF$_2$ group as well as the trifluoromethyl group markedly favors the anodic substitution at its α-position except for the N-methyl derivative (run 5), in which after methoxylation at the methyl group, carbon-nitrogen bond cleavage occurs preferentially. Notably, even the monofluoromethyl group promotes anodic α-methoxylation (run 9) although the yield is unsatisfactory. Consequently, the order of the promotion effect on the anodic methoxylation was found to be CF$_3$ > CHF$_2$ ≫ CH$_2$F [56].

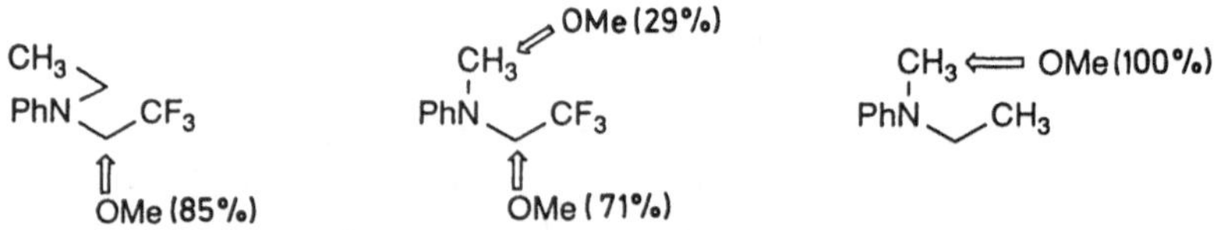

Scheme 6.13

The reaction proceeds via electrogenerated cationic species as its seen with the nonfluorinated amines, carbamates, and amides (Scheme 6.14). However, the regiochemistry of this anodic methoxylation is not governed by the stability of the cationic intermediates **B** and **B'** (thermodynamic control) since the main products are formed via the less stable intermediates **B**. Indeed, this remarkable promotion effect and unique regioselectivity can be explained mainly in terms of α-CH kinetic acidities of the cation radicals formed by one-electron oxidation of the amines since the stronger the acidity of the methylene hydrogen, the easier the deprotonation.

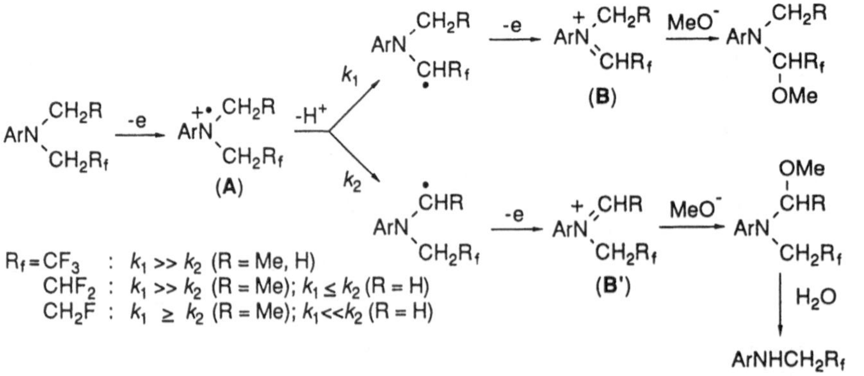

Scheme 6.14

The α-methoxylated products are highly useful building blocks for the construction of a carbon-carbon bond α to the trifluoromethyl and difluoromethyl groups, which is difficult to obtain by other methods, as shown in Scheme 6.15. Thus, α-tri and α-difluoromethylated α-aminonitriles, which are precursors to the corresponding fluorinated α-amino acids, have been prepared in good yields, and flourinated homoallyanilines have been also successfully prepared [44]. in addition, tri- and difluoromethylated tetra- and dihydroquinoline derivatives can be prepared by cationic polar cycloaddition in high yields [45].

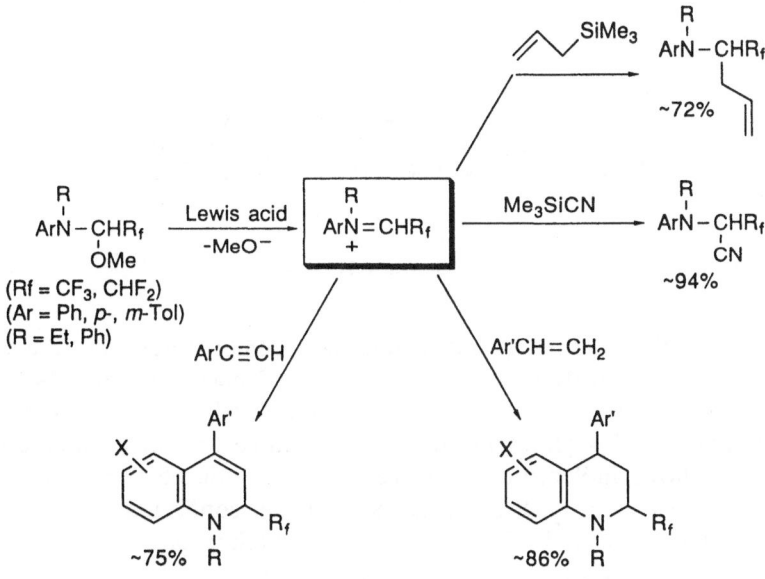

Scheme 6.15

Anodic cyanation of amines is a promising method for the preparation of α-amino nitriles as versatile synthetic precursors. In sharp contrast to the anodic methoxylation of N-(2,2,2-trifluoroethyl)amines, anodic cyanation does not occur at the position α to the CF$_3$ group but takes place at the carbon other than the trifluoroethyl group as shown in Scheme 6.16 [58]. Notably, even the amine **5**, which has no α-protons other than trifluoroethyl protons, does not undergo anodic α-cyanation at all as shown in Scheme 6.17. These facts indicate that the CF$_3$ group completely inhibits the cyanation at the α-position to the CF$_3$ group, regardless of the molecular structure of the amines. However, it is not clear why the regioselectivity of the anodic cyanation is different from that of the anodic methoxylation at present. The difference in the reaction medium or in the basicity and nucleophilicity of CN ions and MeO ions might cause such different regioselectivity.

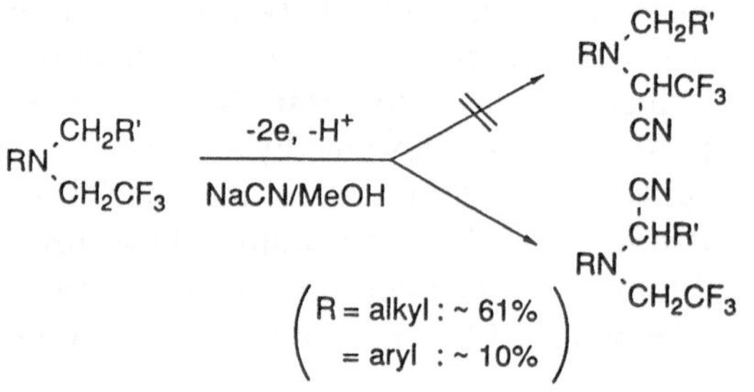

Scheme 6.16

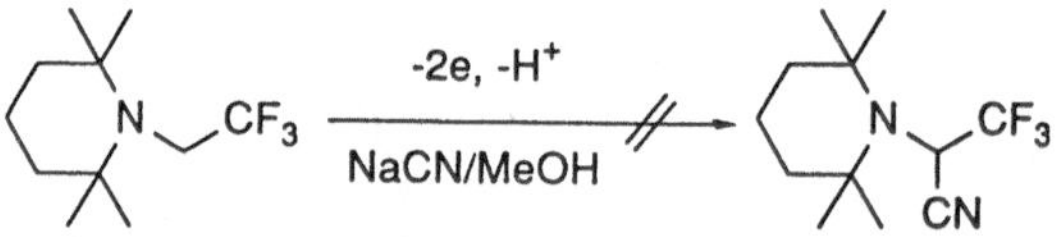

Scheme 6.17

Quite recently, Fuchigami and Konno have demonstrated that the photoaddition of N-(2,2,2-trifluoroethyl)-p-toluidine to 3-phenylcyclohex-2-en-1-one occurs at the α-position to the CF$_3$ group predominantly [59]. Interestingly, the regioselectivity of this photoaddition is comparable to that of anodic methoxylation as shown in Scheme 6.18. Since this photo induced electron transfer is a typical homogeneous reaction, it has been further confirmed that the unique regiochemistry observed in the anodic methoxylation of N-(2,2,2-trifluoroethyl)amines is not due to the heterogeneous system but mainly due to the control by the kinetic acidity.

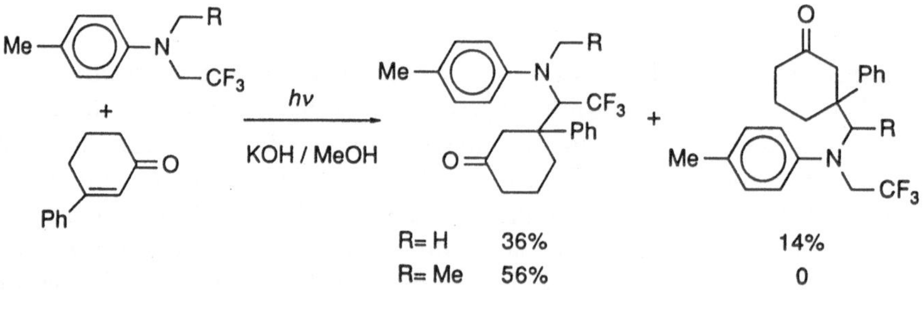

R= H	36%	14%
R= Me	56%	0

Scheme 6.18

Sulfenimines are versatile building blocks for the preparation of secondary and tertiary amines. Therefore, fluoroalkylated sulfenimines should be promising building blocks for the preparation of N-fluoroalkylamino compounds. In fact, very recently, trifluoromethylated sulfenimines have been easily prepared in one step by anodic oxidation of 2,2,2-trifluoroethylamine and diaryl disulfides in MeCN/Et$_4$NClO$_4$ using MgBr$_2$ as a redox mediator (Eq. 28) [60].

$$CF_3CH_2NH_2 + ArSSAr \xrightarrow[\text{MgBr}_2/\text{MeCN}]{-2e, -H^+} [CF_3CH_2NHSAr]$$

$$\xrightarrow{-2e, -H^+} CF_3CH = NSAr \qquad (28)$$

$$Ar = Ph : 72\%$$

$$= p\text{-Tol} : 58\%$$

The sulfenimines have been shown to be highly useful building blocks for the preparation of trifluoromethylated amines, aminoketones, and aminoalkanoates as illustrated in Scheme 6.19.

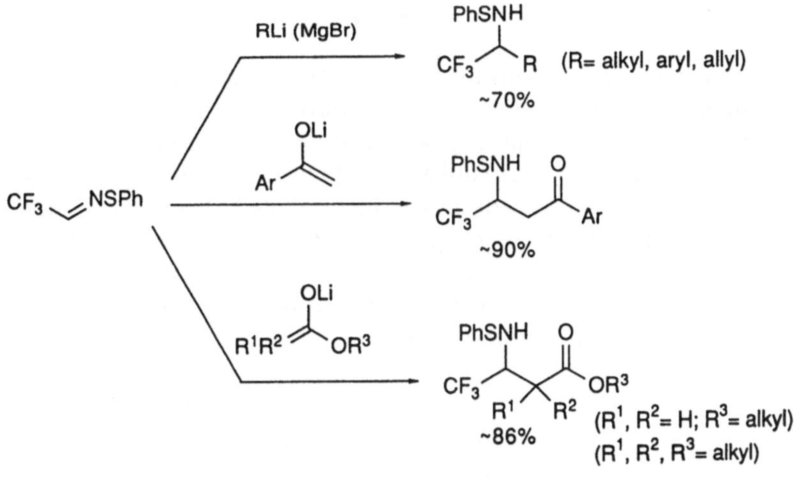

Scheme 6.19

7 Anodic Oxidation of Trifluoromethylated Carboxylic Acids

Electrooxidative generation of trifluoromethyl radicals (CF$_3$·) and their synthetic application have been developed since the early 1970s because trifluoroacetic acid (TFA) is readily available and one of the most economical starting materials for trifluoromethylation [61]. Heteroaromatics as well as olefins have been employed as substrates for the trifluoromethylation (Scheme 7.1) [62].

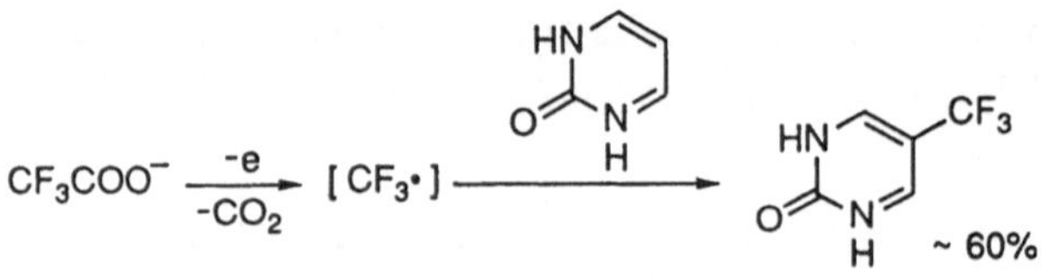

Scheme 7.1

However, the selectivity for such trifluoromethylation has been rather low in many cases. For example, anodic trifluoromethylation of olefins provides a mixture of several types of products in general as shown in Scheme 7.2 and the control of the product-selectivity has been difficult [63].

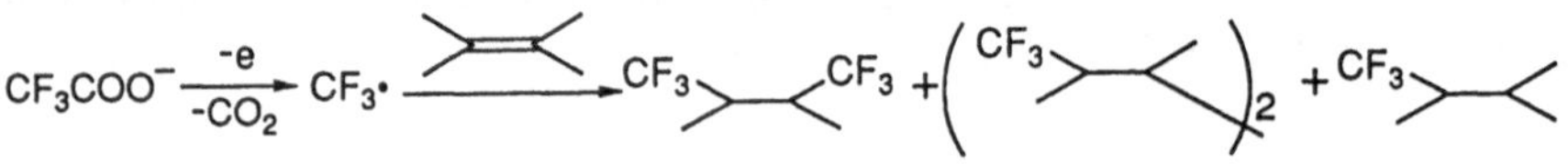

Scheme 7.2

Mullar has found that the formation of a dimer is suppressed in aq. acetone or aq. methanol while the mixture is mainly formed in aq. acetonitrile [64–67].

Recently, Uneyama et al. have systematically investigated the anodic generation of CF_3 radicals and their utilization (Scheme 7.3) [68–72]. They have clarified that trifluoromethyl radicals can be generated almost quantitatively in the oxidation of TFA at 0 °C in an aq. MeOH/Pt system using an undivided cell [70]. They have also found that the trifluoromethylation of electron-deficient olefins can be controlled by the current density, reaction temperature, and the substituents of the olefins. Interestingly, anodic trifluoromethylation of fumar-

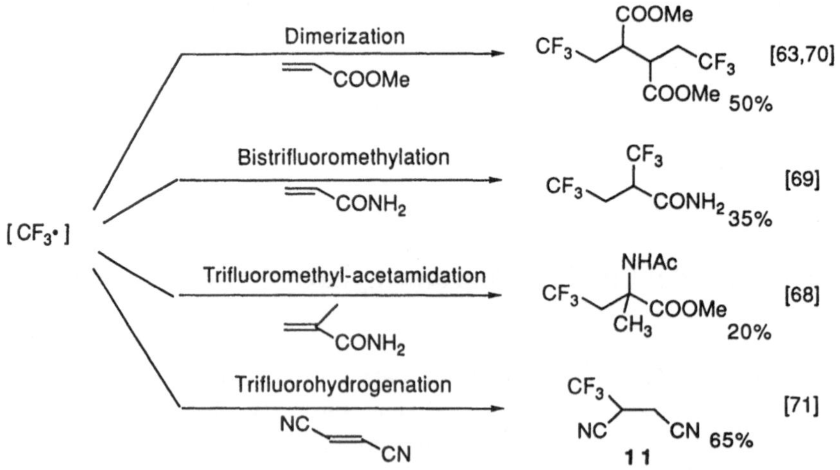

Scheme 7.3

onitrile leading to **11** (Scheme 7–3) is remarkably affected by the reaction temperature: The desired hydrotrifluoromethylation proceeds exclusively at approx. 55 °C while a simple hydrogenation of fumaronitrile predominates at around 0 °C. Both anodic and cathodic reactions are involved in this reaction as shown in Scheme 7.4 [71]. Furthermore, they have also achieved trifluoromethylation of enolizable active-methylene compounds and enolacetates [73]. In the latter case, the yields are good to moderate (Eq. 29).

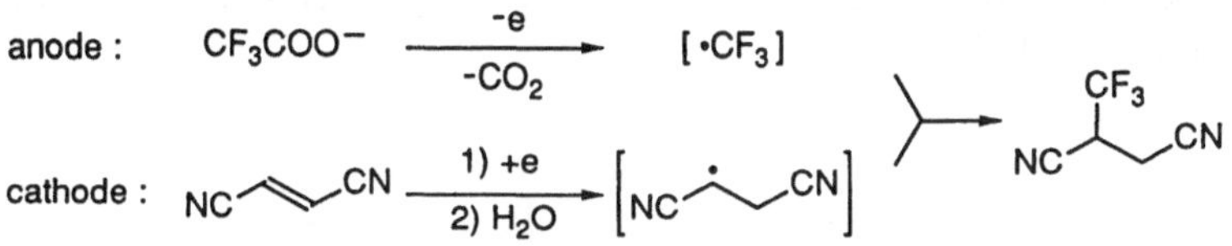

anode : $CF_3COO^- \xrightarrow[-CO_2]{-e} [\cdot CF_3]$

cathode : $NC\diagdown\diagup CN \xrightarrow[2) H_2O]{1) +e} \left[NC\diagdown\overset{\cdot}{\diagup} CN \right]$

$\longrightarrow NC-\overset{CF_3}{\underset{}{C}}\diagdown CN$

Scheme 7.4

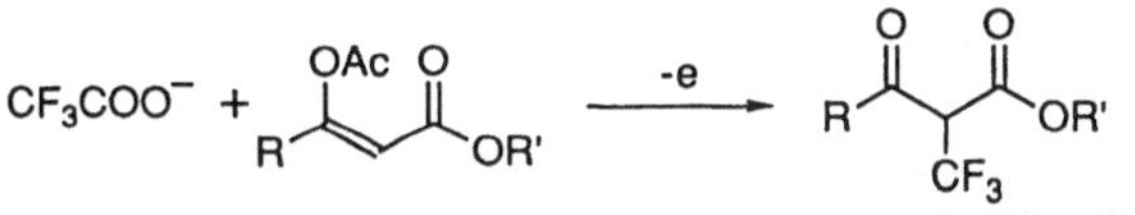

$$CF_3COO^- + \overset{OAc}{\underset{R}{\diagdown}}\overset{O}{\diagup}OR' \xrightarrow{-e} R-\overset{O}{\overset{\|}{C}}-\overset{CF_3}{\underset{}{C}}-\overset{O}{\overset{\|}{C}}-OR' \qquad (29)$$

$R = Me, R' = C_8H_{17} : 64\%$

On the other hand, trifluoromethylation through crossed Kolbe coupling is also known (Scheme 7.5) [74].

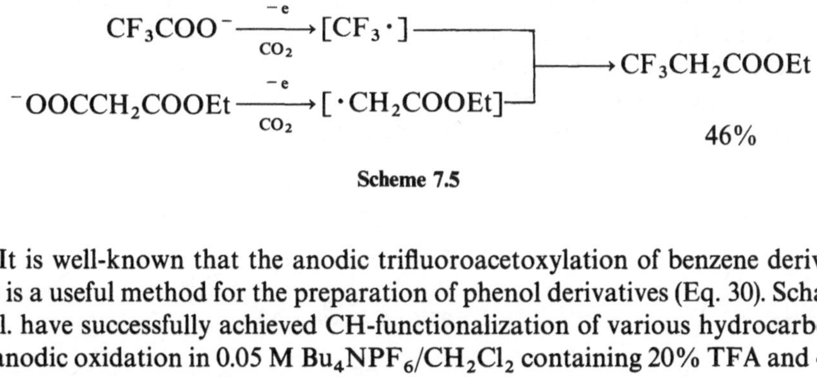

$$CF_3COO^- \xrightarrow[CO_2]{-e} [CF_3 \cdot]$$

$$^-OOCCH_2COOEt \xrightarrow[CO_2]{-e} [\cdot CH_2COOEt]$$

$$\longrightarrow CF_3CH_2COOEt$$

$$46\%$$

Scheme 7.5

It is well-known that the anodic trifluoroacetoxylation of benzene derivatives is a useful method for the preparation of phenol derivatives (Eq. 30). Schafer et al. have successfully achieved CH-functionalization of various hydrocarbons by anodic oxidation in 0.05 M Bu_4NPF_6/CH_2Cl_2 containing 20% TFA and 4% $(CF_3CO)_2O$ as shown in Eqs 31 and 32 [75].

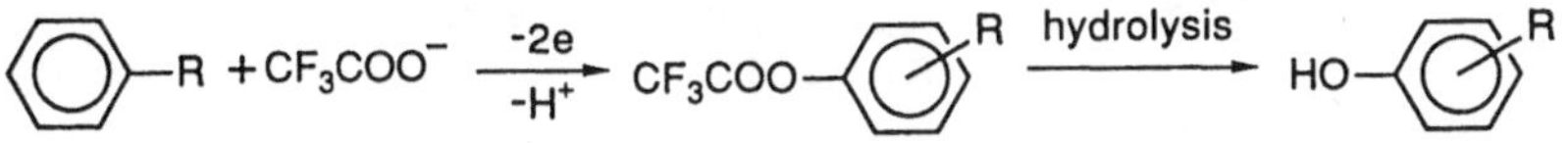

$$\langle\bigcirc\rangle-R + CF_3COO^- \xrightarrow[-H^+]{-2e} CF_3COO-\langle\bigcirc\rangle-R \xrightarrow{hydrolysis} HO-\langle\bigcirc\rangle-R \qquad (30)$$

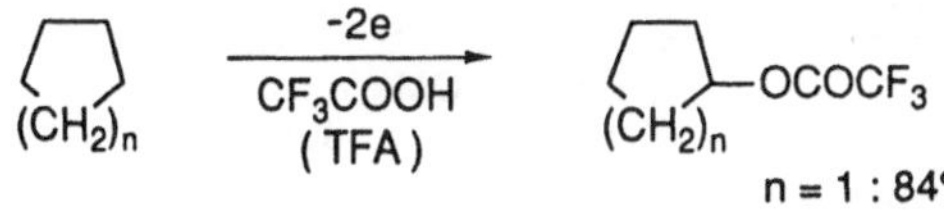

$$\underset{(CH_2)_n}{\square} \xrightarrow[CF_3COOH \\ (TFA)]{-2e} \underset{(CH_2)_n}{\square}-OCOCF_3 \qquad (31)$$

$n = 1 : 84\%$
$n = 2 : 92\%$

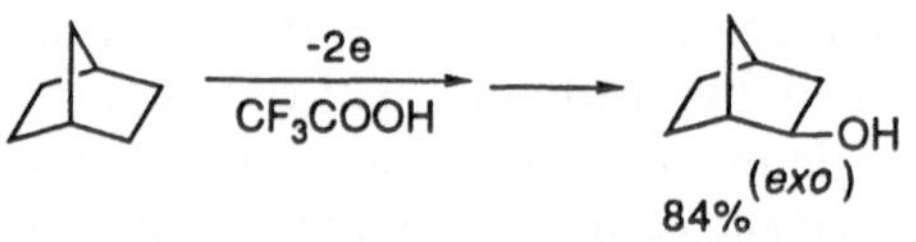

$$\text{(32)}$$

84% (exo)

Kolbe electrolysis of trilfuoromethylated carboxylic acids has been shown to be a versatile method for providing useful building blocks having a CF_3 group. Seebach and Renaud have prepared new types of trifluoromethylated chiral building blocks from enantiomerically pure 3-hydroxy-4,4,4-trifluorobutyric acid (Scheme 7.6) [76].

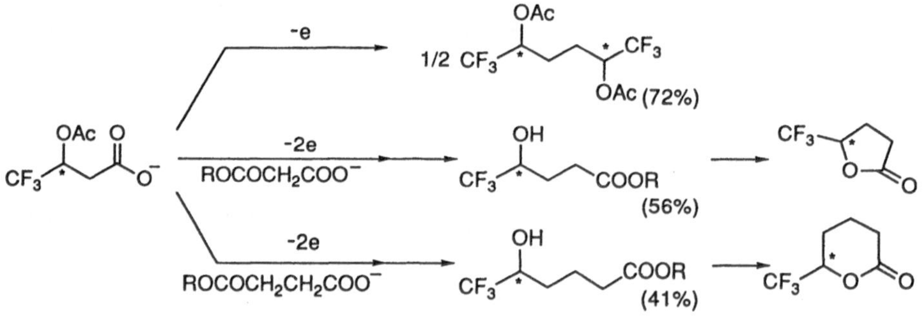

Scheme 7.6

These reactions are notable because α-branched carboxylic acids usually do not undergo efficient Kolbe coupling. Similarly, Kubota et al. have achieved highly efficient homo and crossed coupling reactions using trifluoromethylated carboxylic acids as shown in Scheme 7.7 [77, 78]. Notably, the protection of the hydroxy group of the acids **12** is not necessary.

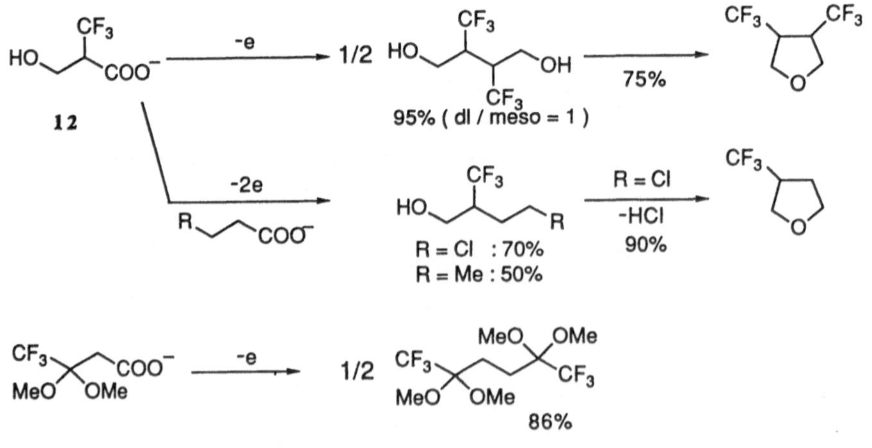

Scheme 7.7

8 Miscellaneous

Polyfluorobenzyl alcohols are one class of fluorinated starting materials for medicinal and agricultural usage. Practical electrochemical methods for the product-selective synthesis of polyfluorobenzyl alcohols have been developed [79, 80]. 2,3,4,5,6-Pentafluorobenzoic acid is reduced selectively to 2,3,4,5,6-pentafluorobenzyl alcohol at amalgamated lead, zinc and cadmium cathodes in aqueous sulfuric acid solutions, while 2,3,5,6-tetrafluorobenzyl alcohol can also be obtained selectively by using the non-amalgamated cathodes in solutions containing small amounts of a quarternary ammonium salt [79].

Conductive polymers have attracted increasing attention because they have wide applications. Recently, very stable poly(thiophenes) with polyfluorinated side chains have been electrochemically synthesized and characterized [81]. Furthermore, notably novel conductive materials have been prepared by cathodic electropolymerization of perfluoro cyclobutene and cyclopentene [82].

9 Conclusion

Organic electrochemistry itself is an inherently hybrid and interdisciplinary area. In spite of the recent increased interest in the chemistry of fluoro organic compounds, electrochemical reactions were not recognized as a powerful tool for fluoro organic synthesis until approx. 10 years ago. So far, great efforts have been made to solve the problems in fluoro organic synthesis and a number of new methodologies have been developed. It is hoped that this account has demonstrated that the electrochemical methodology is a novel and valuable tool for the molecular conversion and functionalization of fluoro organic molecules.

Hopefully, this review can stimulate not only organic electrochemists but also fluorine chemists to develop the new and attractive area of *"Electro Organic Fluorine Chemistry"*.

Acknowledgement: I wish to express my sincere thanks to my collaborators who are cited in the references. I am also grateful to Prof. Tsutomu Nonaka, Tokyo Institute of Technology, for his encouraging me to continue research in the new and attractive field of "Electro Organic Fluorine Chemistry". Finally but not least, I am thankful to Dr. Akinori Konno and Mr. Satoru Narizuka, who assisted me in preparing the manuscript.

10 References

1. Inesi A, Rampazzo NJ (1974) J Electroanal Chem 49: 85
2. Kopilov J, Evans DH (1990) J Electroanal Chem 280: 435
3. Coleman JP, Naser-ud-din, Gilde HG, Utley HP, Weedon CL, Eberson L (1973) J Chem Soc, Perkin Trans II 1903
4. Barker DJ, Brewis DM, Dahn RH (1978) Electrochim Acta 23: 1107
5. Marsella JA, Gilicinslci AG, Coughlin Am, Pez PG (1992) J Org Chem 57: 2856
6. Kariv-Miller E, Vajtner Z (1985) J Org Chem 50: 1394
7. Rozhkov IN, Igumnov SN, Pletnev SI, Borisov YA, Rempel GD (1989) In: 6th Japanese-Soviet Symposium on Fluorine Chemistry, 11–13 July 1989. Novosibirsk, USSR
8. Andrieux CP, Glis L, Medebielle M, Pinson J, Saveant JM (1990) J Am Chem Soc 112: 3509
9. Calas P, Moreau P, Commeyras A (1977) J Electroanal Chem 78: 271
10. Calas P, Moreau P, Commeyras A (1978) J Fluorine Chem 12: 67
11. Calas P, Moreau P, Commeyras A (1982) J Chem Soc, Chem Commun 433
12. Shono T, Kise N, Oka H (1991) Tetrahedron Lett 32: 6567
13. Fritz HP, Kornrumpf W (1979) J Electroanal Chem 100: 217
14. Andrieux CP, Gelis L, Saveant JM (1989) Tetrahedron Lett 30: 4961
15. Andrieux CP, Gelis L, Saveant JM (1990) J Am Chem Soc 112: 786
16. Medebielle M, Pinson J, Saveant JM (1990) Tetrahedron Lett 31: 1279
17. Medebielle M, Pinson J, Saveant JM (1991) J Am Chem Soc 113: 6872
18. Chaussard J, Folest JC, Nedelec JY, Perichon J, Sibille S, Troupel M (1990) Synthesis 369
19. Sibille S, d'Incan E, Leport L, Perichon J (1986) Tetrahedron Lett 27: 3129
20. Sibille S, Mcharec S, Perichon J (1989) Tetrahedron 45: 1423
21. Sibille S, Perichon J. Chaussard J (1989) Synth Commun 19: 2449
22. Saboureau C, Troupel M, Sibille S, Perichon J (1989) J Chem Soc, Chem Commun 1138
23. Folest JC, Nedelec JY, Perichon J (1988) Synth Commun 18: 1491
24. Nedelec JY, Mouloud HAH, Folest JC, Perichon J (1988) J Org Chem 53: 4720
25. Paratian JM, Sibille S, Perichon J (1992) J Chem Soc, Chem Commun 53
26. Mcharek S, Sibille S, Nedelek JY, Perichon J (1991) J Organomet Chem 401: 211
27. Meyer G, Rollin Y, Perichon J (1987) J Organomet Chem 333: 263
28. Rollin Y, Troupel M, Talec DG, Perichon J (1986) J Organomet Chem 303: 131
29. Saboureau C, Sibille S, Troupel M (1989) J Chem Soc, Chem Commun 885
30. Sibille S, Ratovelomanana V, Perichon J (1992) J Chem Soc, Chem Commun 283
31. Fuchigami T, Nakagawa Y (1987) J Org Chem 52: 5276
32. Shono T, Ishifune M, Okada T, Kashimura S (1991) J Org Chem 56: 2
33. Fuchigami T, Yoshiyama T (unpublished results)
34. Barhdadi R, Gal J, Heintz M, Troupel M (1992) J Chem Soc, Chem Commun 50
35. Becker JY, Smart BE, Fukunaga T (1988) J Org Chem 53: 5714
36. Germain A, Commeyras A (1981) Tetrahedron 37: 487
37. Germain A, Brunel D, Moreau P (1986) Bull Soc Chim Fr 895
38. Germain A, Brunel D, Moreau P (1989) J Fluorine Chem 43: 249
39. Brunel D, Germain A, Moreau P (1989) J Chem Soc, Perkin Trans I 2283
40. Fuchigami T, Nakagawa Y, Nonaka T (1986) Tetrahedron Lett 27: 3869
41. Fuchigami T, Nakagawa Y, Nonaka T (1987) J Org Chem 52: 5489
42. Fuchigami T, Yamamoto K, Konno A (1991) Tetrahedron 47: 625
43. Fuchigami T, Yamamoto K, Nakagawa Y (1991) J Org Chem 56: 137
44. Fuchigami T, Ichikawa S, Konno A (1990) Chem Lett 1987
45. Fuchigami T, Ichikawa S, Kandeel ZE, Konno A, Nonaka T (1990) Heterocycles 31: 415
46. Surowiec K, Fuchigami T (1992) J Org Chem 57: 5781
47. Fuchigami T, Ichikawa S: unpublished results
48. Fuchigami T, Yamamoto K, Yano H (1992) J Org Chem 57: 2946
49. Fuchigami T, Shimojo M, Konno A, Nakagawa K (1990) J Org Chem 55: 6074
50. Konno A, Nakagawa K, Fuchigami T (1991) J Chem Soc, Chem Commun 1027
51. Fuchigami T, Yano H, Konno A (1991) J Org Chem 56: 6732
52. Shono T, Matsumura Y, Hayashi J, Mizoguchi M (1980) Tetrahedron Lett 21: 1867
53. Surowiec K, Fuchigami T (1992) Tetrahedron Lett 33: 1065

54. Fuchigami T, Fujita T, Konno A: unpublished results
55. Fuchigami T, Fujita Y, Nonaka T (1990) J Electroanal Chem 284: 115
56. Fuchigami T, Ichikawa S: unpublished results
57. Shono T, Matsumura Y, Inoue K, Ohmizu H, Kashimura S (1982) J Am Chem Soc 104: 5753
58. Konno A, Fuchigami T, Fujita Y, Nonaka T (1990) J Org Chem 55: 1952
59. Konno A, Fuchigami T (1992) Chem Lett 2181
60. Fuchigami T, Ichikawa S, Konno A (1992) Chem Lett 2405
61. Schafer KJ (1990) Topics in Current Chemistry, vol. 152, p 91
62. Hein L, Cech D (1977) Z Chem 17: 415
63. Brookes C, Coe PL, Owen DM, Pedler AE, Tatlow JC (1974) J Chem Soc, Chem Commun 323
64. Muller N (1983) J Org Chem 48: 1370
65. Muller N (1984) J Org Chem 49: 2826
66. Muller N (1984) J Org Chem 49: 4559
67. Muller N (1986) J Org Chem 51: 263
68. Uneyama K, Nanbu H (1988) J Org Chem 53: 4598
69. Uneyama K, Morimoto O, Nanbu H (1989) Tetrahedron Lett 30: 109
70. Uneyama K, Makio S, Nanbu H (1989) J Org Chem 54: 872
71. Uneyama K, Watanabe S (1990) J Org Chem 55: 3909
72. Uneyama K (1991) Tetrahedron 47: 555
73. Uneyama K, Ueda K (1988) Chem Lett 853
74. Renaud RN, Champagne PJ (1975) Can J Chem 53: 529
75. Cramer E, Hembrock A, Schafer HJ (unpublished results)
76. Seebach D, Renaud P (1985) Helv Chim Acta 68: 2342
77. Kubota T, Aoyagi R, Sando H, Kawanishi M, Tanaka T (1987) Chem Lett 1435
78. Kubota T, Ishii T, Minamikawa H, Yamaguchi S, Tanaka T (1988) Chem Lett 1987
79. Iwasaki T, Yoshiyama A, Sato N, Fuchigami T, Nonaka T, Sasaki M (1987) J Electroanal Chem 238: 315
80. Iwasaki T, Sasaki M, Sato N, Yoshiyama A, Fuchigami T, Nonaka T (1990) Denki Kagaku 58: 83
81. Buchner W, Garreau R, Lemaire M (1990) J Electroanal Chem 277: 355
82. Briscoe MW, Chambers RD, Silvester MJ, Drakesmith FG (1988) Tetrahedron Lett 29: 1295

Electrochemical Reactions of Organosilicon Compounds

Jun-ichi Yoshida

Department of Material Science, Faculty of Science, Osaka City University, Sugimoto 3-3-138, Sumiyoshi, Osaka 558, Japan

Table of Contents

Topics in Current Chemistry Vol. 170
© Springer-Verlag Berlin Heidelberg 1994

Although the electrochemical reactivity of unfunctionalized tetraalkylsilanes is low, organosilicon compounds containing π-systems and heteroatoms exhibit unique electrochemical activity. Upon α-substitution, silyl groups have electron-accepting effects and upon β-substitution, silyl groups have electron-donating effects to π systems. Silyl groups also have electron-donating effects to β-situated heteroatoms such as oxygen, nitrogen, and sulfur. It is also noteworthy that silyl groups control the regiochemistry of the electrochemical reactions of these compounds. On the basis of such silicon effects, various types of electrochemical reactions of organosilicon compounds have been developed and some of them are utilized as useful tools in organic synthesis. The electrochemical behaviour of organosilicon compounds containing Si–Si bonds is also interesting. Si–Si bonds are susceptible to both anodic oxidation and cathodic reduction. Halosilanes are rather inert to electrochemical redox reactions and are often used as effective trapping agents of carbanion intermediates generated by cathodic reduction of organic halides. However, in the absence of other reducible substrates, halosilanes can be reduced cathodically to form Si–Si bonds.

1 Introduction

Organosilicon compounds are defined as those compounds which possess carbon–silicon bonds. Although silicon is the second most abundant element in the earth's crust, no organosilicon compound is known to occur naturally. A large variety of organosilicon compounds, however, have been synthesized so far, and some of them, such as silicone polymers, have been utilized as useful materials. Over the last twenty years, the application of organosilicon compounds as intermediates in organic synthesis has grown considerably and a number of chemical reactions of organosilicon compounds have been developed based upon the unique properties of silicon [1]. Electrochemical reactions of organosilicon compounds have also received significant research interest, and the activity in this area, from a view point of organic synthesis, has grown rapidly in the last five years. The aim of this article is to review the electrochemical properties and reactions of organosilicon compounds, and I have attempted to cover the recent exciting developments and to provide a future picture of the electrochemical aspects of organosilicon compounds.

2 Fundamental Electrochemical Properties of Organosilicon Compounds

This section will provide a brief outline of fundamental effects of silyl groups on the electrochemical properties of organic compounds, with special emphasis on the interpretation based upon molecular orbital (MO) theory.

2.1 Unfunctionalized Tetraalkylsilanes

Chemical reactivity of unfunctionalized organosilicon compounds, the tetra-alkylsilanes, are generally very low. There has been virtually no method for the selective transformation of unfunctionalized tetraalkylsilanes into other compounds under mild conditions. The electrochemical reactivity of tetraalkylsilanes is also very low. Kochi et al. have reported the oxidation potentials of tetraalkyl group-14-metal compounds determined by cyclic voltammetry [2]. The oxidation potential (E_p) increases in the order of $Pb < Sn < Ge < Si$ as shown in Table 1. The order of the oxidation potential is the same as that of the ionization potentials and the steric effect of the alkyl group is very small. Therefore, the electron transfer is suggested as proceeding by an outer-sphere process. However, it seems to be difficult to oxidize tetraalkylsilanes electrochemically in a practical sense because the oxidation potentials are outside the electrochemical windows of the usual supporting electrolyte/solvent systems (>2.5 V).

2.2 Organosilicon Compounds Containing π-Systems

It is well known that the silyl substituents have significant effects on the chemical reactivity of π-systems such as olefins and aromatic rings. On the basis of such silyl effects, a number of reactions of organosilicon compounds having π-systems, such as allylsilanes, benzylsilanes, arylsilanes, and vinylsilanes have been devised and they are widely utilized in organic synthesis. Silyl substituents have also significant effects on the redox properties of π-systems [3]. Upon α-substitution, silyl groups have electron-accepting effects and upon β-substitution, silyl groups have electron-donating effects. This dual donor-acceptor effects of silyl substituents on the redox properties of π-systems is one of the most intriguing aspects of the electrochemistry of organosilicon compounds.

Jun-ichi Yoshida

Table 1. Oxidation potential and ionization potentials of tetraalkyl group-14-metals [2]

Tetraalkylmetal	E_p (V vs. SCE)	I_p (eV)
Et_4Si	2.56	9.78
Et_4Ge	2.24	9.41
Et_4Sn	1.76	8.93
Et_4Pb	1.26	8.13

2.2.1 Organosilicon Compounds Having a π-System in the α-Position

Figure 1 shows the electron attachment energies (AE) and ionization potentials (IP) of silyl substituted π-systems and related compounds [4]. AE can be correlated with the energy level of the LUMO (lowest unoccupied molecular orbital) and IP can be correlated with the energy level of the HOMO (highest occupied molecular orbital). For α-substituted π-systems, the introduction of a silyl group produces a decrease in the π^*-(LUMO) level. This effect is attributed to the interaction between a low-lying silicon-based unoccupied orbital such as the empty d orbital of silicon and the π^* orbital (d_π–p_π interaction) as shown in Fig. 2. Recent investigations on these systems, however, indicate that d orbitals on silicon are not necessarily required for interpreting this effect; α-effects of SiR_3 can also be explained by the interaction between Si–R σ^* orbitals and the π-system.

Owing to such orbital interactions, α-silyl substitution causes the decrease in the LUMO level of the π-system and enhances the electron accepting ability of the π-system. Therefore, the reduction potentials of α-silyl-substituted π-systems are less negative than those of the parent π-systems, although the magnitude of this effect is not large.

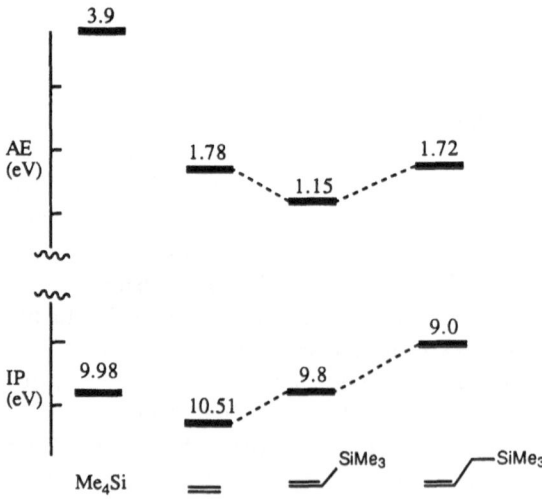

Fig. 1. Correlation diagram giving electron attachment energies (AE) and ionization potentials (IP) of α- and β-silyl substituted ethylenes [4].

42

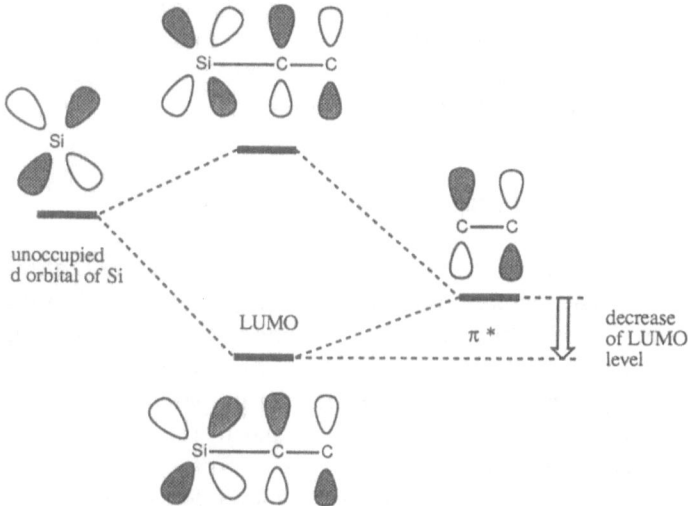

Fig. 2. The interaction of the d orbital of Si and the π^* orbital of the olefin (d_π–p_π interaction)

The effects of silyl groups on the chemical behavior of the anion radicals generated by cathodic reduction is also noteworthy. It is well known that silyl groups stabilize a negative charge at the α position. Therefore, it seems to be reasonable to consider that the anion radicals of π-systems are stabilized by α-silyl substitution. The interaction of the half-filled π^* orbital of the anion radical with the empty low-lying orbital of the silicon (such as d_π–p_π interaction) results in partial electron donation from the π-system to the silicon atom which eventually stabilizes the anion radical.

The effect of α-silyl groups in the anion radical is demonstrated by ESR spectroscopy [5]. The spectra of the anion radicals of arylsilanes generated by cathodic reduction consist of large couplings due to the *para* hydrogen and the hydrogen on silicon, and smaller couplings from the *ortho* and *meta* hydrogens. The spin density indicated by ESR spectroscopy can be explained in terms of molecular orbital theory [6]. In the anion radical, an additional electron must occupy an antibonding orbital of the parent compound. The two lowest lying antibonding orbitals, designated φ_4 and φ_5 are degenerate in benzene, but the degeneracy is lifted if a substituent is present. Electron-withdrawing substituents lower the energy of φ_5 because such substituents stabilize a negative charge at

φ_4 φ_5

43

Jun-ichi Yoshida

the adjacent carbon (*ipso* position). On the other hand, electron-releasing substituents raise the energy of φ_5. By virtue of the d_π–p_π interaction, silicon lowers the energy level of φ_5. Therefore, the unpaired electron occupies φ_5, which has large coefficients at the *ipso* and *para* positions.

2.2.2 Organosilicon Compounds Having a π-System in the β-Position

For β-substituted π-systems, silyl substitution causes the destabilization of the π-orbital (HOMO) [3, 4]. The increase of the HOMO level is attributed to the interaction between the C–Si σ orbital and the π orbital of olefins or aromatic systems (σ–π interaction) as shown in Fig. 3 [7]. The C–Si σ orbital is higher in energy than the C–C and C–H σ orbitals and the energy match of the C–Si orbital with the neighboring π orbital is better than that of the C–C or C–H bond. Therefore, considerable interaction between the C–Si orbital and the π orbital is attained to cause the increase of the HOMO level. Since the electrochemical oxidation proceeds by the initial electron-transfer from the HOMO of the molecule, the increase in the HOMO level facilitates the electron transfer. Thus, the introduction of a silyl substituents at the β-position results in the decrease of the oxidation potentials of the π-system. On the basis of this β-effect, anodic oxidation reactions of allylsilanes, benzylsilanes, and related compounds have been developed (Sect. 3.3).

The β-effect of silyl groups on π-systems depends on the geometry of the molecule, because the orbital interaction between the C–Si σ orbital and the π orbital reaches its maximum when they are in the same plane. For example, the

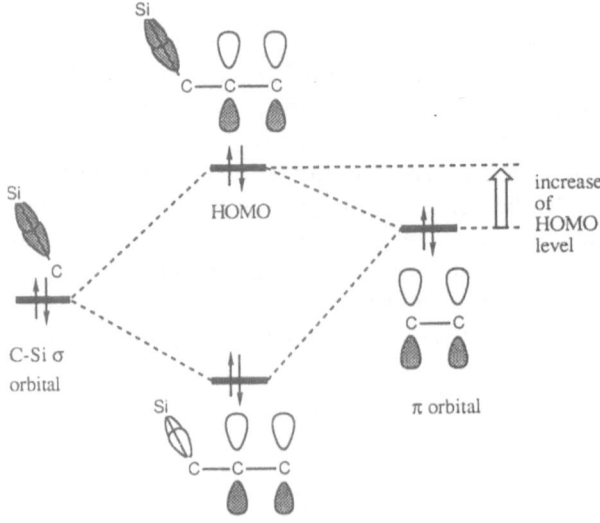

Fig. 3. The interaction of the C–Si σ orbital and the π orbital of an olefin (σ–π interaction)

I, IP = 8.13 eV   II, IP = 8.42 eV

C–Si bond can interact with the π-system quite effectively in compound I, which in turn lowers the ionization potential. But in the related silaindene, compound II, such interaction is weak because the C–Si bond and the π orbital are in perpendicular orientation [8].

It should be recognized that the stability of cation radicals generated by anodic oxidation is also affected by β-silyl substitution. Stabilization of carbocations by a silyl group situated at the β-position is well known as the "β effect". The interaction of the C–Si σ orbital with the empty p orbital of the carbon stabilizes the carbocation. Therefore, we can expect similar effects of silicon for cation radical species. The interaction of the filled C–Si σ orbital with the half-filled orbital of the carbon may stabilize the cation radical.

Bock et al. studied the cation radicals of various organosilanes having a π-system in the β position extensively and found that the charge is delocalized considerably into the silyl groups [9, 10]. Kira and Sakurai et al. also carried out an ESR study of cation radicals of allylsilanes and revealed the large polarization of the SOMO (singly occupied molecular orbital) as a consequence of the σ–π interaction [11].

2.3 Organosilicon Compounds Containing Heteroatoms

The introduction of silyl substituents also causes significant effects on the redox properties of compounds containing heteroatoms such as oxygen, nitrogen, and sulfur.

2.3.1 Organosilicon Compounds Having a Heteroatom in the α-Position

It has been known for some time that the basicities of a heteroatom decrease upon α-silyl substitution [12]. For example, alkyl silyl ethers (R_3Si–O–R′) are less basic than dialkly ethers. Silylamines are weak bases compared to alkylamines. This electron-withdrawing effect of silyl groups has been explained in terms of the interaction between low lying vacant orbitals such as 3d orbitals of silicon or σ* orbitals with the nonbonding p orbitals (lone pairs) of the heteroatom (Fig. 4). This interaction decreases the HOMO level which in turn lowers the basicity of the heteroatom. Such effect may also cause the increase of the oxidation potentials, but little study has been reported on the electrochemical properties of this type of compounds.

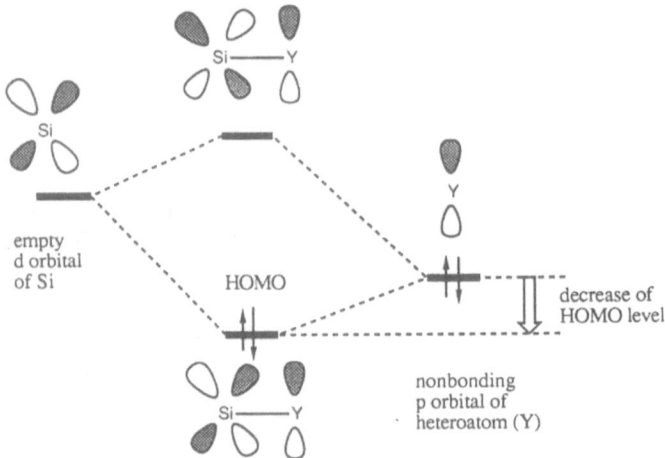

Fig. 4. The interaction of the empty *d* orbital of Si and the nonbonding *p* orbital of a heteroatom (Y)

2.3.2 Organosilicon Compounds Having a Heteroatom in the *β*-Position

Although little information has been available for the effect of α-silyl substitution on the electrochemical properties of heteroatom compounds, extensive
studies have been carried out on the effect of *β*-silyl substitution [10, 13]. For the
β-substituted heteroatom compounds (substitution at the α carbon), the introduction of a silyl group results in a significant decrease of the oxidation
potentials, although the magnitude depends upon the nature of the heteroatom.
This effect is explained in terms of the interaction between the C–Si σ orbital and
the nonbonding *p* orbital of the heteroatom (Fig. 5). This interaction raises the
HOMO level which in turn favours the electron transfer.

Since the magnitude of the orbital interactions depends upon the relative
orientations of the two orbitals, the effect of silicon depends upon the geometry
of the molecule. In Fig. 6 the HOMO level of SiH_3CH_2OH, which is a model
compound of silyl-substituted heteroatom compounds, obtained by ab initio
calculations, are plotted against the torsion angle of Si–C–O–H [13]. The
HOMO energy increases with the torsion angle and becomes the maximum
when the torsion angle is about 90°. In this conformation the C–Si bond is
almost in the same plane as the nonbonding *p* orbital of the oxygen so that they
can interact with each other effectively. Further increase in the torsion angle
results in decrease in the HOMO energy. At 180° the HOMO level becomes the
minimum. In this conformation the C–Si bond and the *p* orbital of the oxygen
are perpendicular, and therefore they cannot interact with each other. The
difference in the HOMO energies between the maximum and the minimum is
about 0.9 eV.

The results of MO calculations are consistent with the oxidation potentials
of silyl-substituted ethers in which the rotation around the C–O bond is

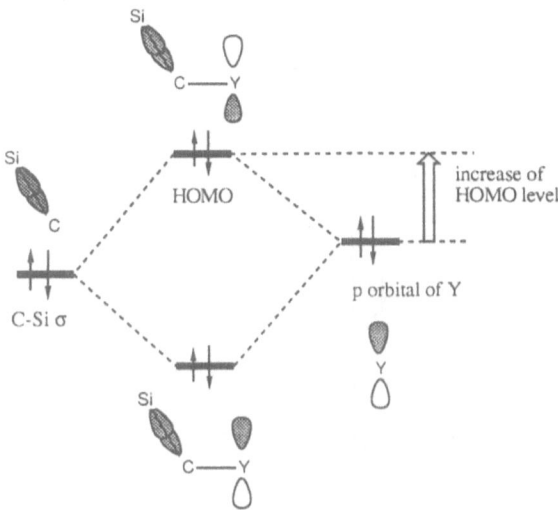

Fig. 5. The interaction of the Si–C σ orbital and the p orbital of the heteroatom (Y)

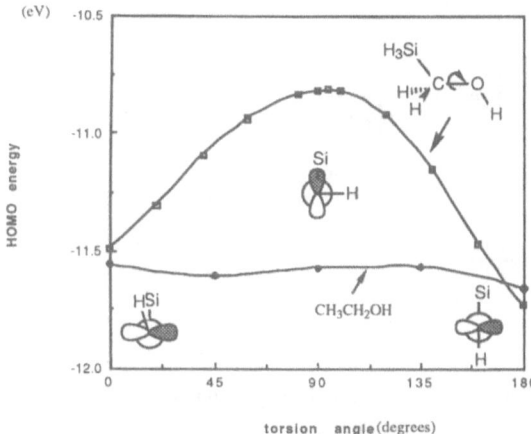

Fig. 6. Plots of the HOMO level of SiH_3CH_2OH and that of CH_3CH_2OH vs. the torsion angle of Si–C–O–H or C–C–O–H [13]

restricted (Fig. 7) [13]. The fact that the oxidation potentials vary dramatically with the torsion angle indicates that one can control the oxidation potential simply by changing the geometry of the molecule. This idea seems to provide a useful concept for designing organosilicon compounds of specific oxidation potentials, and various applications will hopefully appear in the near future.

The introduction of an additional silyl-substituent in the β-position to the heteroatom causes a further decrease in the oxidation potential of heteroatom-substituted organosilanes. The introduction of an additional silyl group results in the rise of the HOMO level. Increase in the population of the favorable conformers for the electron transfer also seems to be important.

V vs Ag/AgCl

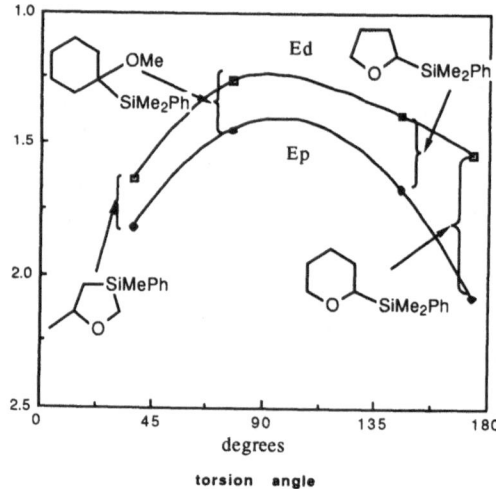

Fig. 7. Plots of the peak potentials (E_p, ●) and decomposition potentials (Ed, □) for the oxidation of silyl substituted ethers in which rotation around Si–C bond is restricted vs. torsion angle of Si–C–O–C [13]

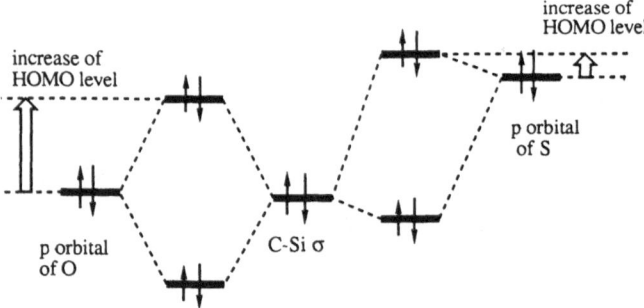

Fig. 8. The interaction of the Si–C σ orbital and the p orbital of oxygen and sulfur

It is worth noting that the effect of silyl groups also depends upon the nature of the heteroatom (Sect. 3.5, 3.6). Silyl substitution causes a small decrease in oxidation potentials of sulfides whereas a dramatic silicon effect is observed for ethers. Presumably, the energy level of the p orbitals of sulfur is much higher than the C–Si bond, and they do not interact with each other effectively (Fig. 8). The effect of silyl groups on nitrogen compounds such as carbamates is greater than that for sulfur compounds but smaller than that for oxygen compounds.

The cation radical at the heteroatom generated by one-electron oxidation is stabilized by a silyl group situated at the β-position. The C–Si σ orbital interacts with the SOMO to stabilize the cation radical when they are in the same plane.

In fact, Kira and Sakurai revealed in an ESR study that the SOMO of the cation radical of $Me_3SiCH_2OCH_3$ is no longer regarded as an oxygen nonbonding orbital but as one including a large contribution from the C–Si σ orbital [14]. However, ab initio MO calculations indicated that the stabilization energy of heteroatom cation radicals by β-silyl substitution is much smaller than that of carbocations [13].

2.3.3 Acylsilanes

Acylsilanes are a class of compounds in which a silyl group is directly bound to the carbonyl carbon, and they have received considerable research interest from the point of view of both physical organic and synthetic organic chemistry [15]. Acylsilanes have a structure quite similar to the structure of α-silyl-substituted ethers; a silyl group is attached to the carbon adjacent to the oxygen atom, although the nature of the C–O bond is different. Therefore, one can expect β-silicon effects in the electron-transfer reactions of acylsilanes.

R^1 $\diagdown$ OR^2
$\diagup$
SiR^3_3

R^1 $\diagdown$ O
$\diagup$
SiR^2_3

α-silyl-substituted ether acylsilane

Although oxidation potentials of aldehydes and ketones are generally very high, silyl substitution at the carbonyl carbon results in a significant decrease in the oxidation potential [16]. The decrease in the oxidation potentials is attributed to the rise of the HOMO level by the interaction of the C–Si σ-bond and the nonbonding p orbital (lone pair) of the carbonyl oxygen (Fig. 9). In the case of α-silyl-substituted ethers, the rotation around the C–O bond is free and,

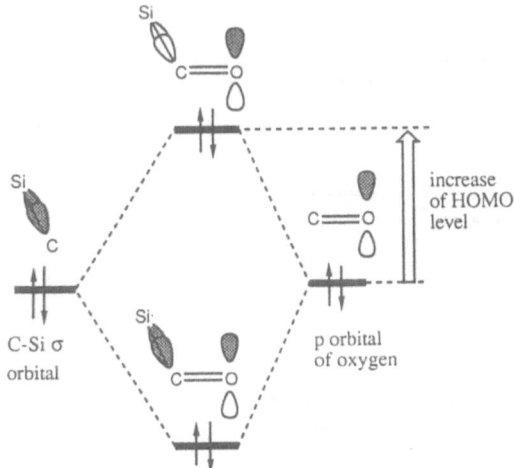

Fig. 9. The interaction of the C–Si σ orbital and the p orbital of oxygen in acylsilane

therefore, the C–Si bond and the *p* orbital of the oxygen atom do not always overlap effectively. In the case of acylsilanes, however, they are fixed in the same plane, and therefore, they can always interact with each other quite effectively. Thus, acylsilanes seem to be more susceptible to oxidation than α-silyl-substituted ethers.

2.4 Compounds Containing Si–Si Bonds

The special chemical and electrochemical properties of Si–Si bonds are noteworthy [17, 18]. For example, disilanes undergo facile reactions with electrophiles such as halogens. Therefore, Si–Si bonds resemble carbon–carbon double bonds. It is also worth noting that linear polysilanes absorb ultraviolet light, the wave length increasing with increasing chain length, implying a sort of conjugation of Si–Si bonds. The energy levels of Si–Si σ-bonding orbitals, which are the HOMOs of disilanes and polysilanes, are much higher than those of C–C σ-bonding orbitals and therefore are responsible for the high reactivity of Si–Si bonds. As shown in Fig. 10, the HOMO levels of polysilanes increase with increasing chain length. Thus, it is easy to understand that Si–Si bonds are electrochemically quite active and the oxidation potentials decrease with the increasing number of Si in polysilanes.

Another interesting feature of Si–Si bonds is the low energy level of the LUMO. The LUMO levels decrease with increasing chain length of polysilanes (Fig. 10). Because of the low lying LUMO, polysilanes can be reduced by both

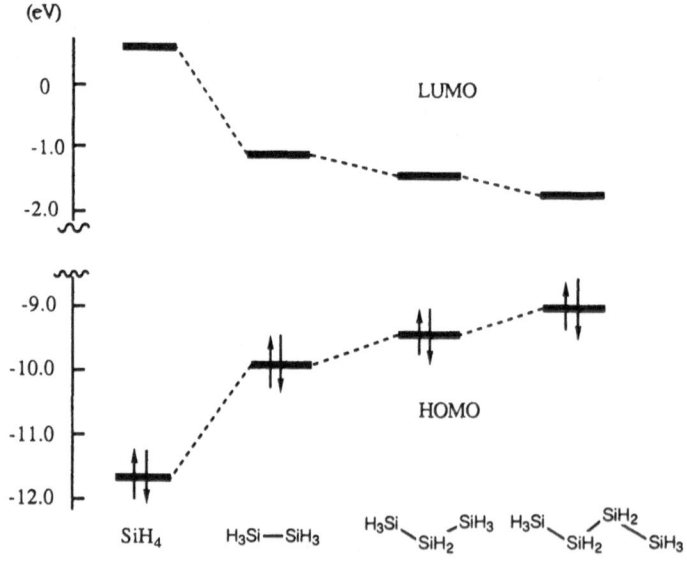

Fig. 10. The energy level diagram of HOMO and LUMO of polysilanes (PM3)

chemical and electrochemical means to generate the corresponding anion radicals.

The redox properties of cyclic polysilanes are interesting because they resemble those of aromatic hydrocarbons. For instance, cyclic polysilanes can be reduced to anion radicals or oxidized to cation radicals. ESR spectra for both the cation and anion radicals indicate that the unpaired electron is fully delocalized over the ring [17, 19, 20]. The aromatic properties of the cyclic polysilanes are ascribed to a high energy delocalized HOMO and a relatively low energy LUMO. Because the HOMO and LUMO levels lie at similar level to those of benzene, cyclic polysilanes can serve either as electron donors or electron acceptors.

2.5 Halosilanes

Although halosilanes undergo smooth nucleophilic substitution at the silicon, they are rather inert to redox reactions. Therefore, chlorosilanes are usually used as trapping agents of anionic intermediates generated by electroreduction of organic compounds. However, in the absence of other reactive substrates halosilanes are reduced electrochemically to form Si–Si bonds. Indeed, there are a number of reports in the literature of the cathodic reduction of chlorosilanes (Sect. 4.2).

3 Anodic Oxidation of Organosilicon Compounds

3.1 Tetraalkylsilanes

Although it is difficult to oxidize unfunctionalized tetraalkylsilanes (Sect. 2.1), in the presence of fluoride ions they can be oxidized electrochemically because the fluoride ions decreases the oxidation potential of tetraalkylsilanes. For example, the oxidation potential of phenyltrimethylsilanes becomes less positive (2.3 V vs $Ag/AgNO_3$) in the presence of $Et_4NH_3F_4$ than that in the absence of fluoride ion (2.6 V) [21]. Since the oxidation potential of fluoride ions (2.55 V) is also more positive than that of phenyltrimethylsilane in the presence of $Et_4NH_3F_4$, a mechanism involving fluoride ions assisted one-electron transfer from tetraalkylsilanes has been suggested (vide infra). This is consistent with facile electron transfer from extracoordinate fluoride containing organosilicon compounds as in organofluorosilicates ($RSiF_5^{2-}$) [22].

The constant potential electrolysis of tetraalkylsilanes in the presence of $Et_4NH_3F_4$ at 2.3 V results in the cleavage of the C–Si bond and the formation of the corresponding fluorosilanes [21]. Presumably, the first step involves the formation of the pentacoordinate anionic species ($RR_3'SiF^-$) which is oxidized

$$\text{R---SiR'}_3 \xrightarrow[\text{Et}_4\text{NH}_3\text{F}_4 \text{ / CH}_3\text{CN}]{\text{-e (Pt) 2.3 V vs Ag/AgNO}_3} \left[\begin{array}{c} \text{R} \\ \diagdown \\ \text{F} \diagup \end{array} \text{SiR'}_3 \right]^{\bullet}$$

$$\xrightarrow{\hspace{1cm}} \quad \text{F---SiR'}_3 \quad + \quad \text{R} \cdot$$

$$\downarrow$$

$$\text{R---R} , \quad \text{R-H}, \quad \text{R(-H)} \qquad \textbf{Scheme 1.}$$

Table 2. Anodic oxidation of organosilanes in the presence of fluoride ions [21]

Substrate	Products (% yield)
PhSi(CH$_3$)$_3$	PhSi(CH$_3$)$_2$F (80), CH$_4$, C$_2$H$_6$
(CH$_3$)$_3$SiCH$_2$Ph	(CH$_3$)$_3$SiF (50), PhCHO
(C$_2$H$_5$)$_3$SiCH$_2$Ph	(C$_2$H$_5$)$_3$SiF (70), PhCHO
CH$_3$Si(C$_2$H$_5$)$_3$	CH$_3$Si(C$_2$H$_5$)$_2$F (45), C$_2$H$_5$F, CH$_2$=CH$_2$, FSi(C$_2$H$_5$)$_3$ (5), CH$_4$, C$_2$H$_6$

on the surface of the anode to generate the pentacoordinate radical (Scheme 1). The most stable alkyl radical is eliminated to give the corresponding fluoro-silane. The alkyl radical intermediate undergoes either radical reactions (radical coupling or hydrogen abstraction) or electrochemical oxidation to give olefinic products. A major drawback of this process, from a synthetic point of view, is the low selectivity of the products derived from the eliminated alkyl radical.

3.2 Arylsilanes

Replacement of an alkyl group by an aromatic group slightly decreases the oxidation potential of silanes. For example, tetraphenylsilane exhibits an oxidation wave at the peak potential of 2.1 V (vs Ag/Ag$^+$ in LiClO$_4$/MeCN) [23]. The oxidation potential of phenyltrimethylsilane is also less positive than those of tetraalkylsilanes. The cation radical species of aromatic silanes generated on the surface of the anode are relatively stable in anhydrous medium (in the presence of activated alumina) and can be detected by ESR [23]. With residual water, however, the electrolysis leads to a black deposit on the anode. These coatings are mainly conjugated aromatic polymers which are very rich in phenyl rings.

3.3 Allylsilanes and Benzylsilanes

Allylsilanes and benzylsilanes are more susceptible to anodic oxidation than tetraalklsilanes and arylsilanes. It should also be noted that the oxidation potentials of allylsilanes and benzyl silanes are much less positive than those of the parent compounds without silyl substituents (σ–π interaction, Sect. 2.2.2). For example, the peak potential for the oxidation of geranyltrimethylsilane is much less positive than those of trisubstituted olefins such as 2-methyl-2-butene as shown in Table 3 [24].

The anodic oxidation of allylsilanes was investigated by three groups independently [24, 25, 26]. The electrolysis of allylsilanes in an undivided cell results in the facile cleavage of the C–Si bond and the introduction of nucleophiles such as alcohols onto the carbon. It is noteworthy that in the case of geranyltrimethylsilane, only the carbon–carbon double bond of the allylsilane moiety is oxidized without affecting the other carbon–carbon double bond (Scheme 2). This fact is in accord with the electron-donating ability of the β-silyl group. Another important feature of this reaction is that the C–Si bond is cleaved selectively without affecting the carbon-allylic hydrogen bonds. Various nucleophiles including oxygen nucleophiles such as alcohols, water, and acetic acid and nitrogen nucleophiles such as carbamates and sulfonamides can be introduced onto the allylic carbon. Since allylsilanes normally undergo reactions with various electrophiles, the electrochemical process provides an efficient method for "umpolung" of allylsilanes [27].

The fact that the anodic oxidation of allylsilanes usually gives a mixture of two regioisomers suggests a mechanism involving the allyl cation intermediate (Scheme 3). The initial one-electron transfer from the allylsilane produces the cation radical intermediate [9]. Although in the case of anodic oxidation of simple olefins the carbon–allylic hydrogen bond is cleaved [28], in this case the

Table 3. Oxidation potentials of allylsilanes and olefins

Compound	Ep (V vs. Ag/AgCl)
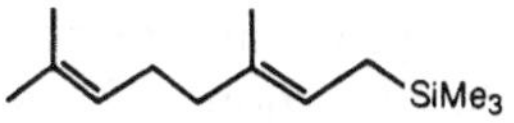 SiMe3	1.30
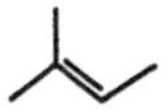	1.85

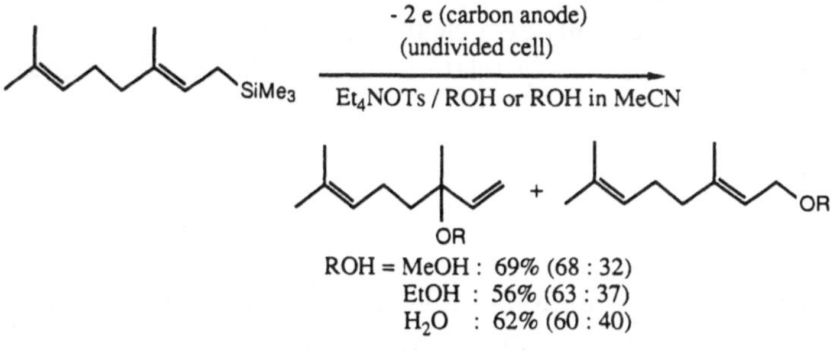
SiMe3
- 2 e (carbon anode)
(undivided cell)
Et4NOTs / ROH or ROH in MeCN
+
OR
OR
ROH = MeOH : 69% (68 : 32)
EtOH : 56% (63 : 37)
H2O : 62% (60 : 40)

Scheme 2.

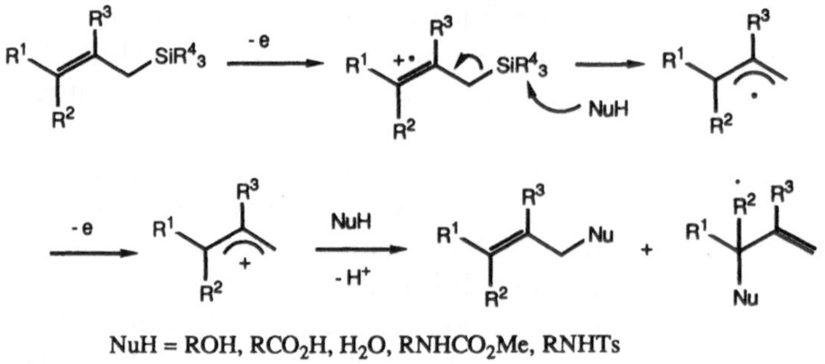
-e
-e
NuH
-H+
NuH = ROH, RCO2H, H2O, RNHCO2Me, RNHTs

Scheme 3.

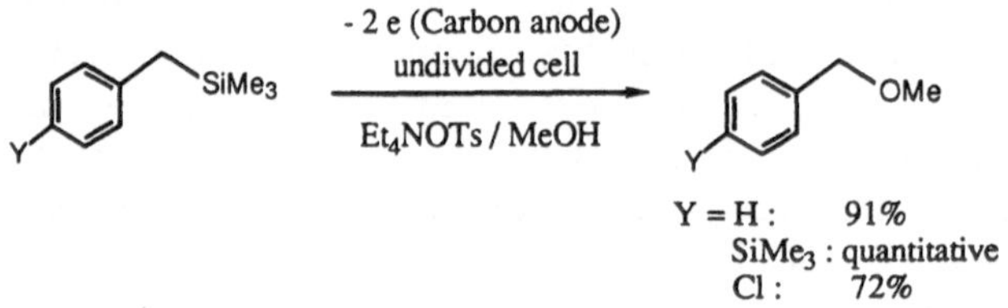
- 2 e (Carbon anode)
undivided cell
Et4NOTs / MeOH
SiMe3
OMe
Y = H: 91%
SiMe3 : quantitative
Cl : 72%

Scheme 4.

silyl group acts as a "super proton" and the C–Si bond is cleaved selectively to give the allyl radical. The allyl radical is spontaneously oxidized on the surface of the anode to give the allyl cation intermediate which reacts with the nucleophile to give the substitution product as a mixture of two regioisomers.

Benzylsilanes are also oxidized under similar conditions with selective cleavage of the C–Si bond (Scheme 4) [24, 25]. Nucleophiles are introduced onto the benzylic carbon exclusively. Nuclear substitution products are not formed. The high chemoselectivity observed for the reaction of *p*-(trimethylsilyl)benzyltrimethylsilane is interesting. Only the benzyl C–Si bond is cleaved without affecting the aromatic C–Si bond.

Closely related reactions have been accomplished by photoelectron-transfer reactions of allylsilanes and benzylsilanes, and a similar mechanism involving the cation radical intermediate is suggested [29]. Chemical oxidation of allylsilanes [27] and ferrocenylsilanes [30] also cleaves the C–Si bond and mechanism of these reactions seem to closely relate to that of the electrochemical process.

3.4 1-Silyl-1,3-dienes

It is well known that the anodic oxidation of 1,3-dienes in nucelophilic solvents such as methanol and acetic acid gives mainly 1,4-addition products together with a small amount of 1,2-addition products [31]. If the 1,3-dienes substituted by a silyl group at the terminal position are electrolyzed, the 1,4-adducts thus formed have an allylsilane structure and therefore are expected to be further oxidized under the electrochemical condition giving rise to the facile cleavage of the C–Si bond and the introduction of nucleophiles onto the carbon.

The anodic oxidation reactions of 1-silyl-1,3-dienes in methanol take place exactly as expected to give the corresponding 1,1,4-trimethoxy-2-butene derivatives (Scheme 5) [32]. It is worth noting that the third methoxy group is introduced exclusively at the terminal position of the original diene, whereas regioselectivity of the anodic oxidation of allylsilanes is usually relatively low. A cation-stabilizing effect of the methoxy group which has been introduced at the terminal position by the first two-electron oxidation seems to play a major role.

1,1,4-Trimethoxy-2-butene derivatives thus obtained can be hydrolyzed by the treatment with dilute acid to afford the corresponding γ-methoxy-α,β-unsaturated aldehydes. Since 1-trimethylsilyl-1,3-dienes are readily prepared by the reaction of the anion of 1,3-bis(trimethylsilyl)propene with aldehydes or ketones [33], 1,3-bis(trimethylsilyl)propene offers a β-formylvinyl anion equivalent for the reaction with carbonyl compounds (Scheme 6) [34].

The diastereoselectivity of the anodic oxidation of 1-silyl-1,3-dienes is interesting [32]. The anodic oxidation of 5-phenyl-1-trimethylsilyl-1,3-hexa-diene in methanol gives a mixture of two diastereomers in 4:1 ratio. On the basis

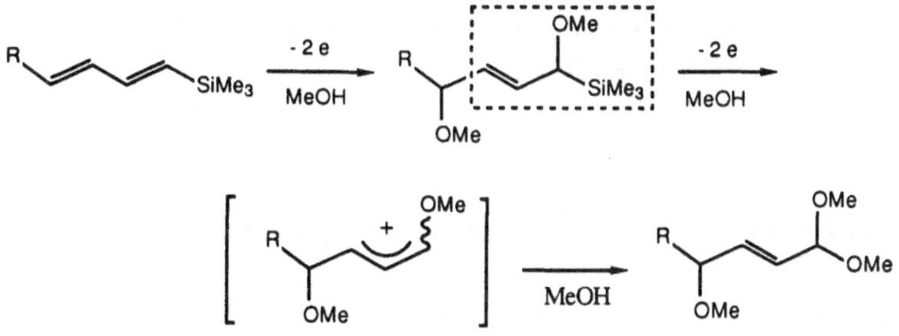

Scheme 5.

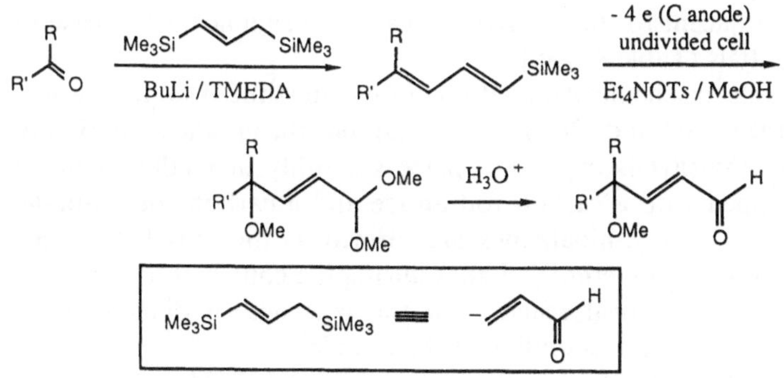

Scheme 6.

55

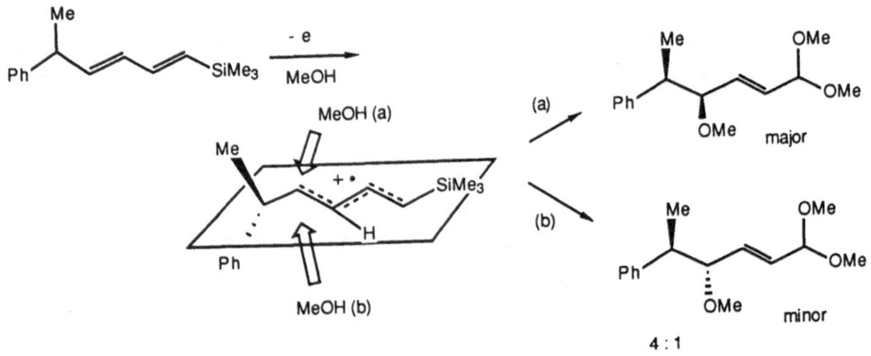

Scheme 7.

of the concept of allylic strain, diastereoselectivity seems to be determined by the attack of methanol to the predominant conformation of the cation radical intermediate (Scheme 7). The attack from the less hindered face rationalizes the observed selectivity. However, if the cation radical is adsorbed on the electrode surface at the less hindered face, methanol should attack the cation radical from the opposite face. But this is not the case. The adsorption at the more hindered face by the attractive interaction between the phenyl group and the electrode surface can also account for the results.

3.5 Organosilicon Compounds Containing Sulfur, Nitrogen, and Phosphorus

The electrooxidation of organosilicon compounds containing heteroatoms has been investigated extensively and various synthetic applications have been developed. Cooper and Owen studied the oxidation potentials of a series of silyl-substituted amines, phosphines, and sulfides, and observed that silyl substitution at the carbon adjacent to the heteroatom caused a significant decrease in the oxidation potentials (Table 4) [35].

The preparative electrochemical oxidation of silyl-substituted sulfides results in the cleavage of the C–Si bond [36–38]. For example, the anodic oxidation of 1-phenylthio-1-trimethylsilylalkanes takes place smoothly in methanol in an undivided cell equipped with a carbon rod anode and a carbon rod cathode. Although 1-methoxy-1-phenylthioalkanes are formed as the initial products, they are converted into 1,1-dimethoxyalkanes during the course of the reaction (Scheme 8). The electrochemical reaction in the presence of diols such as ethylene glycol affords the corresponding cyclic acetals.

The mechanism shown in Scheme 9 has been suggested. The initial one-electron oxidation produces the cation radical species. Nucleophilic attack of

Table 4. Oxidation potentials of silicon-substituted nitrogen, phosphorous, and sulfur compounds [35]

Compound	$E_{p/2}$ (V vs. SCE)	Compound	$E_{p/2}$ (V vs. SCE)
Me_3SiCH_2NHPh	0.44	Me_3SiCH_2SPh	1.15
Me_3CCH_2NHPh	0.60	$Me_3Si(CH_2)_2SPh$	1.26
$Me_3Si(CH_2)_2NHPh$	0.60	$Me_3Si(CH_2)_3SPh$	1.28
$Me_3SiCH_2PPh_2$	0.63		
$Me_3Si(CH_2)_2PPh_2$	0.88		

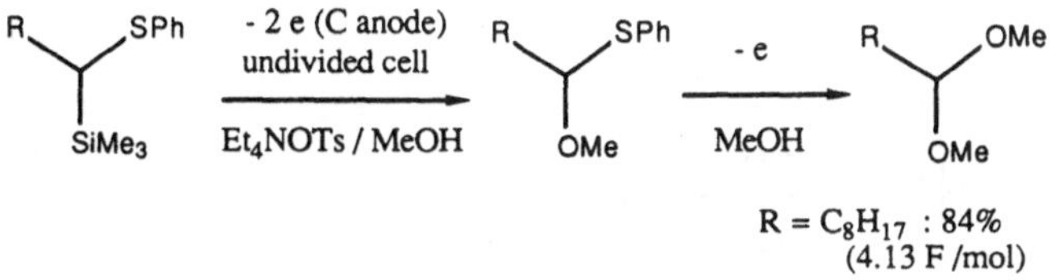

Scheme 8.

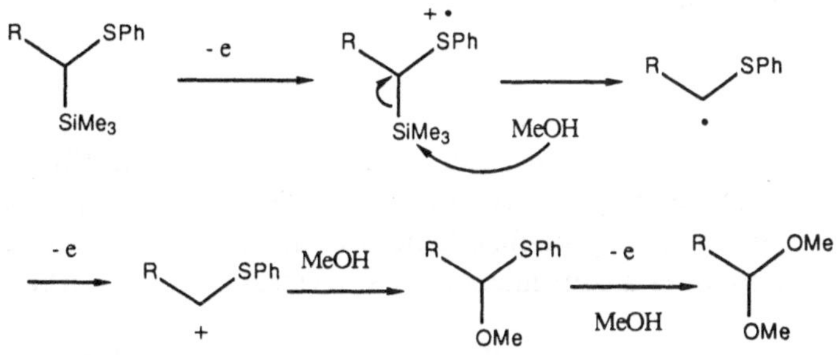

Scheme 9.

methanol on silicon cleaves the C–Si bond to generate the carbon radical species adjacent to sulfur, which is further oxidized on the anode to afford the carbocation intermediate. This carbocation reacts with methanol to give 1-methoxy-1-phenylthioalkanes. 1-Methoxy-1-phenylthioalkanes suffer the cleavage of the C–S bond on the electrochemical oxidation [39] and the introduction of methanol onto the carbon.

1-Phenylthio-1-trimethylsilylalkanes are easily prepared by the alkylation of (phenylthio)(trimethylsilyl)methane as shown in Scheme 10 [40]. The treatment of (phenylthio)(trimethylsilyl)methane with butyllithium/tetramethylethylene-diamine (TMEDA) in hexane followed by the addition of alkyl halides or epoxides produces alkylation products which can be oxidized electrochemically to yield the acetals. Since acetals are readily hydrolyzed to aldehydes, (phenylthio)(trimethylsilyl)methane provides a synthon of the formyl anion. This is an alternative to the oxidative transformation of α-thiosilanes to aldehydes via Sila-Pummerer rearrangement under application of MCPBA as oxidant [40, 41].

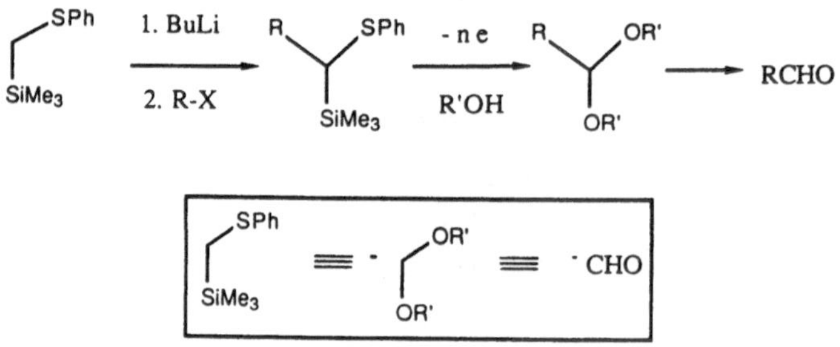

Scheme 10.

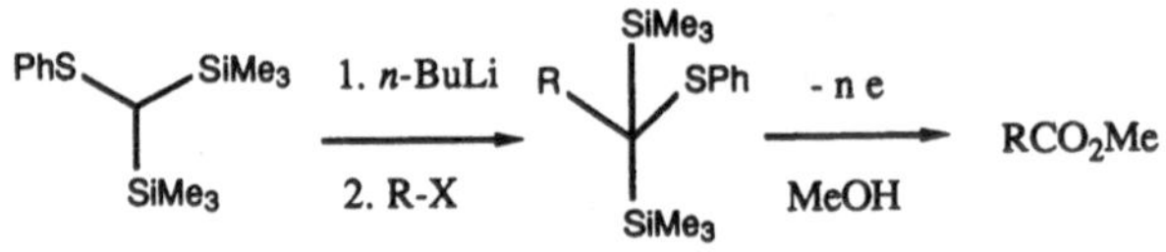

R = C$_{12}$H$_{25}$: 81% (11.5 F/mol)
R = C$_8$H$_{17}$: 72% (9.83 F/mol)

Scheme 11.

Sulfides having two silyl groups are also oxidized electrochemically in methanol to give the corresponding methyl esters (Scheme 11) [36, 37]. The alkylation of (phenylthio)bis(trimethylsilyl)methane with alkyl halides followed by the anodic oxidation provides a convenient access to esters with one-carbon homologation.

Suda and coworkers described the anodic oxidation of 2-silyl-1,3-dithianes which have two sulfur atoms on the carbon adjacent to silicon [42]. In this case, however, the C–Si bond is not cleaved, but the C–S bonds are cleaved to give the corresponding acylsilanes (Scheme 12). Although the detailed mechanism has not been clarified as yet, the difference in the anode material seems to be responsible for the different pathway of the reaction. In fact, a platinum plate anode is used in this reaction, although a carbon anode is usually used for the oxidative cleavage of the C–Si bond. In the anodic oxidation of 2-silyl-1,3-dithianes the use of a carbon anode results in a significant decrease in the yield of acylsilanes. The effects of the nature of the solvent and the supporting electrolyte may also be important for the fate of the initially formed cation radical intermediate. Since various 2-alkyl-2-silyl-1,3-dithianes can be readily synthesized, this reaction provides a convenient route to acylsilanes.

The electrochemical behaviour of silyl-substituted nitrogen compounds is also interesting. The introduction of a silyl group at the carbon adjacent to the nitrogen of carbamates causes a significant decrease in the oxidation potentials, although such effect is much smaller for amines. Preparative electrochemical oxidation of silyl-substituted carbamates in methanol results in smooth and selective cleavage of the C–Si bond and introduction of methanol at the α-

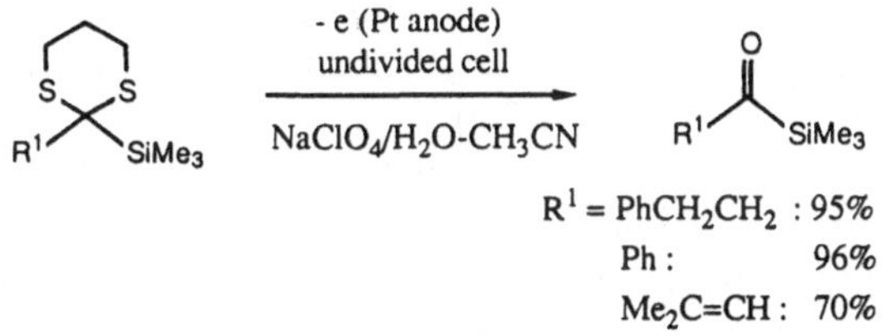
- e (Pt anode)
undivided cell
NaClO4/H2O-CH3CN
R1 SiMe3
R1 = PhCH2CH2 : 95%
Ph : 96%
Me2C=CH : 70%

Scheme 12.

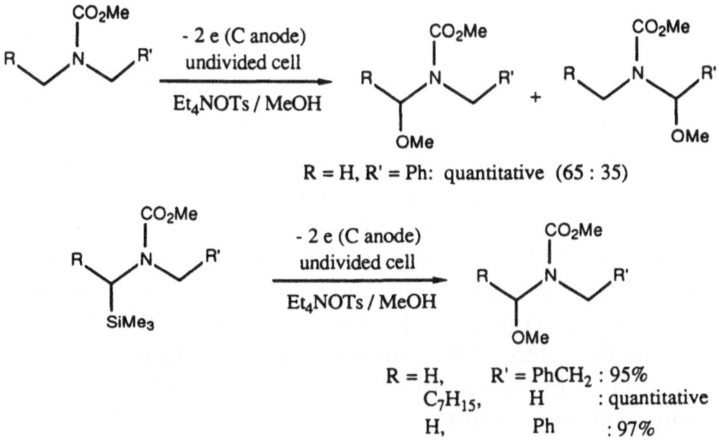
- 2 e (C anode)
undivided cell
Et4NOTs / MeOH
R = H, R' = Ph: quantitative (65 : 35)
- 2 e (C anode)
undivided cell
Et4NOTs / MeOH
R = H, R' = PhCH2 : 95%
C7H15, H : quantitative
H, Ph : 97%

Scheme 13.

carbon (Scheme 13) [43]. Shono and coworkers carried out an extensive study of the electrochemical oxidation of carbamates. Carbamates which do not have a silyl group also give the methoxylated compounds, but when unsymmetrical carbamates are used, usually a mixture of two regioisomers is obtained [44]. However, by the pre-introduction of a silyl group at one of the α-carbons, the regiochemistry of the methoxylated product can be controlled completely. Disilyl-substituted carbamates can be readily prepared regioselectively by the reaction of *N*-alkylcarbamates with trimethylsilylmethyl iodide or by the sequence consisting of mesylation of α-silyl alcohols, reaction with amines, and *N*-methoxycarbonylation with methyl chloroformate.

Since the methoxyl group attached to the carbon adjacent to nitrogen can be readily replaced by various carbon nucleophiles, α-methoxylated carbamates are useful intermediates for the syntheses of nitrogen-containing compounds. The electrochemical oxidation of silyl-substituted carbamates provides a highly regioselective route to these useful compounds.

3.6 Organosilicon Compounds Containing Oxygen

The electrochemical behaviour of silyl-substituted ethers have been investigated extensively. Generally, it is difficult to oxidize aliphatic ethers because of their high oxidation potentials (>2.5 V). However, the ethers having a silyl group at the carbon adjacent to the oxygen exhibit an oxidation wave at a peak potential of 1.6 to 1.7 V (Table 5), indicating that silyl-substitution lowers the oxidation potential by at least 0.8 V [45] (Sect. 2.3.2).

Constant current electrolysis of such silyl-substituted ethers takes place smoothly using a carbon anode in methanol, giving rise to facile cleavage of the C–Si bond and introduction of the methoxy group at the carbon. Both the phenyldimethylsilyl group and the trimethylsilyl group are effective as the activating groups. The anodic oxidation of silyl substituted allyl ethers affords the corresponding mixed acetals in good yields without affecting the carbon–carbon double bond.

Studies on the electrochemical oxidation of silyl-substituted ethers have uncovered a rich variety of synthetic application in recent years. Since acetals, the products of the anodic oxidation in the presence of alcohols, are readily hydrolyzed to carbonyl compounds, silyl-substituted ethers can be utilized as efficient precursors of carbonyl compounds. If we consider the synthetic application of the electrooxidation of silyl-substituted ethers, the first question which must be solved is how we synthesize ethers having a silyl group at the carbon adjacent to the oxygen. We can consider either the formation of the C–C bond (Scheme 15a) or the formation of the C–O bond (Scheme 15b). The formation of the C–Si bond is also effective, but this method does not seem to be useful from a view point of organic synthesis because the required starting materials are carbonyl compounds.

The carbon–carbon bond formation is accomplished by the reaction of the silicon-stabilized carbanions with electrophiles. Magnus and Roy have reported that methoxy(trimethylsilyl)methane is deprotonated with sec-butyllithium in

Table 5. Oxidation potentials of silyl-substituted ethers and related compounds

Compound	Ep (V vs. Ag/AgCl)	Compound	Ep (V vs. Ag/AgCl)
$C_7H_{15}CHOCH_3$ $\quad$| $\quad SiMe_2Ph$	1.60	Et_4Si	2.5^a
		$C_8H_{17}SiMe_2Ph$	2.25
$C_7H_{15}CHOCH_3$ $\quad$| $\quad SiMe_3$	1.72	$C_8H_{17}OCH_3$	> 2.5

^{a} Ref 2 (vs. SCE)

$$\begin{array}{ccc} R\text{-}OR' & \xrightarrow[\text{Et}_4\text{NOTs / MeOH}]{\substack{-2\,e\ (\text{C anode}) \\ \text{undivided cell}}} & R\text{-}OR' \\ | & & | \\ SiR''_3 & & OMe \end{array}$$

$R = C_9H_{19}$, $R' = Me$: 95%

C_9H_{19}, $CH_2CH=CH_2$: 87%　　**Scheme 14.**

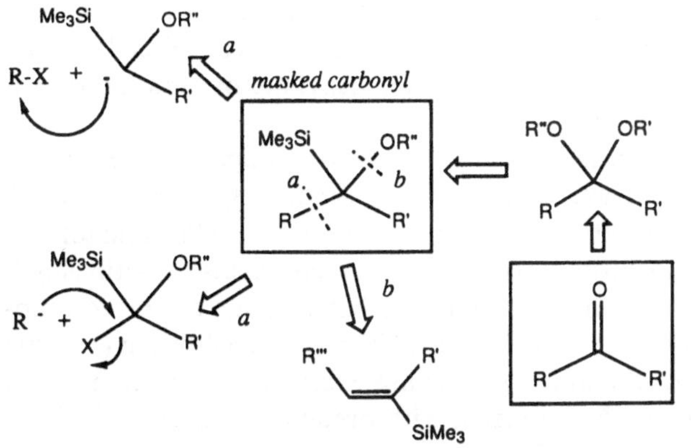

Scheme 15.

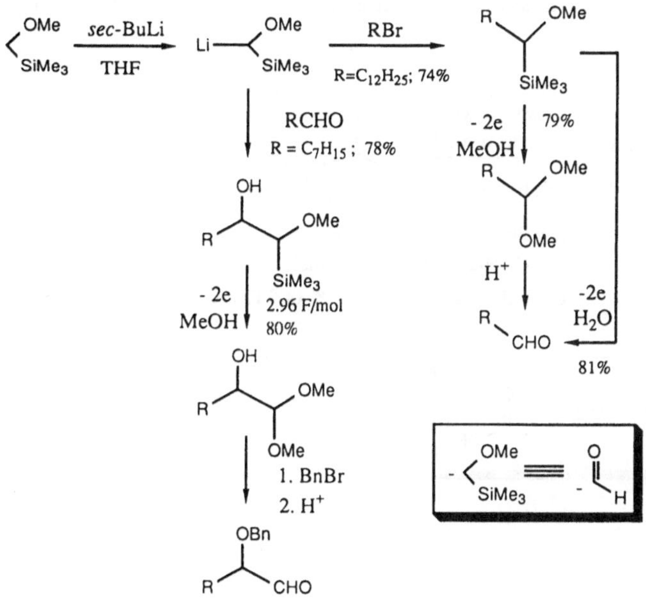

Scheme 16.

THF to give the corresponding carbanion (Scheme 16) [46]. The anion of methoxy(trimethylsilyl)methane readily reacts with carbonyl compounds to give β-hydroxy-α-silylethers [37, 47]. The electrochemical oxidation in methanol affords the corresponding dimethyl acetal without affecting the free hydroxyl group. Protection of the hydroxyl group, followed by acid-catalyzed hydrolysis gives the α-benzyloxyaldehyde. The anion of methoxy(trimethylsilyl)methane can also be alkylated with organic halides. The electrochemical oxidation of the alkylated product in methanol gives the dimethyl acetal which can be converted into the aldehyde by acid-catalyzed hydrolysis. Electrochemical oxidation in the presence of water in dichloromethane produces the aldehyde directly under neutral conditions. Therefore, the methoxy(trimethylsilyl)methyl anion provides a synthon of the formyl anion [37].

The anodic oxidation of ethers having two silyl groups on the carbon adjacent to the oxygen atom are also useful in organic synthesis [37, 47]. Deprotonation of methoxy(trimethylsilyl)methane with *sec*-butyllithium follow- ed by the treatment with chlorotrimethylsilane gives methoxybis(trimethylsilyl)- methane. Deprotonation of methoxybis(trimethylsilyl) methane can be accomp- lished with butyllithium, which is a weaker base than *sec*-butyllithium (Scheme 17). This is probably due to the anion-stabilizing effect of the additional silyl group. The resulting anion can be alkylated with organic halides to afford the corresponding alkylation products. The electrochemical oxidation in methanol gives 1,1-dimethoxy-1-silylalkanes which are converted into the corresponding methyl ester by further electrolysis. Thus, the anion of methoxybis(tri- methylsilyl)methane provides a synthon of the anion of the alkoxycarbonyl group.

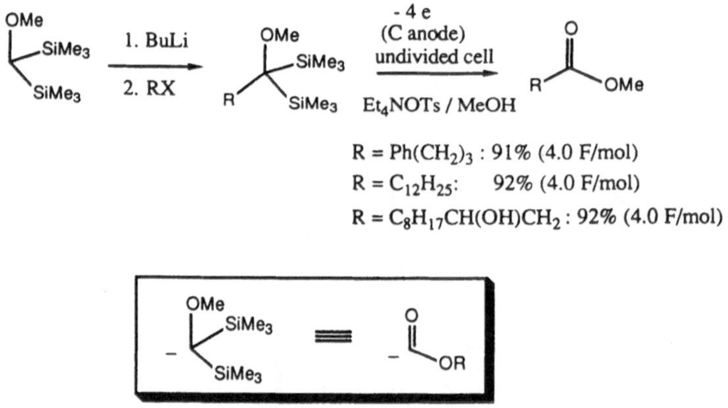

$R = Ph(CH_2)_3$: 91% (4.0 F/mol)
$R = C_{12}H_{25}$: 92% (4.0 F/mol)
$R = C_8H_{17}CH(OH)CH_2$: 92% (4.0 F/mol)

Scheme 17.

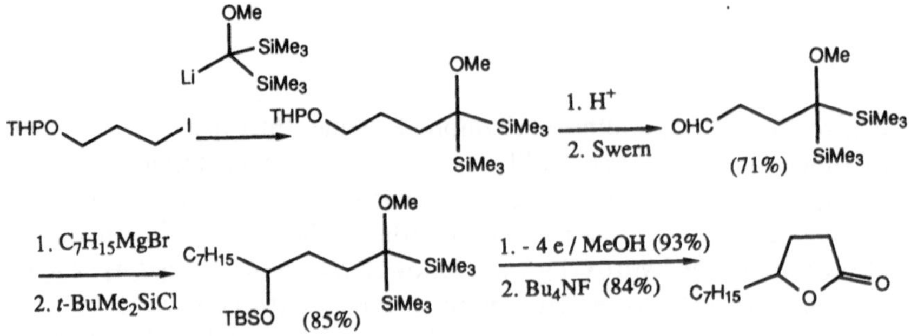

Scheme 18.

The potentiality of the present methodology is demonstrated by the synthesis of γ-undecalactone, as shown in Scheme 18 [37, 47]. The treatment of the THP-protected ω-hydroxyalkyl iodide with the anion of methoxybis(trimethylsilyl) methane gave the corresponding alkylation product. Acidic deprotection of the hydroxyl group followed by Swern oxidation produced the aldehyde. The aldehyde was allowed to react with heptylmagnesium bromide, and the resulting alcohol was protected as *tert*-butyldimethylsilyl ether. The electrochemical oxidation in methanol followed by the treatment with fluoride ion afforded the γ-undecalactone.

It is worth noting that α-silyl-substituted ethers are compatible with various conditions, including acidic conditions as necessary for the removal of THP, chemical oxidations such as Swern oxidation, and basic conditions as in Grignard reactions. Such compatibility, together with their thermal stability, allows synthetic reactions employing silyl-substituted ethers to be run under a wide range of conditions. The mild reaction conditions of the electrochemical method and the easy separation from the volatile silicon-containing by-products are also advantageous. Thus, ethers having a silyl group at the carbon adjacent to the oxygen provide a powerful building block in organic syntheses.

The silyl-substituted ethers can also be synthesized from vinylsilanes. Vinylsilanes are known to be efficient precursors of carbonyl compounds which can be unmasked by epoxidation followed by acid-catalyzed rearrangement [48]. However, by using the electrochemical method, vinylsilanes can be directly converted into carbonyl compounds having a hydroxyl group at the α-position. Thus, BF_3-catalyzed ring opening of epoxysilanes in methanol [49] gives the β-hydroxy-α-silylethers, which are readily converted into α-benzyloxyaldehydes by anodic oxidation in methanol followed by hydrolysis (Scheme 19) [50].

On the basis of the electrochemical oxidation of silyl-substituted ethers, a general and iterative route to optically active polyols has been developed (Scheme 20) [51]. The key intermediates of this iterative process are β-hydroxy-α-methoxysilanes, a protected form of α-hydroxyaldehydes. The electrochemical

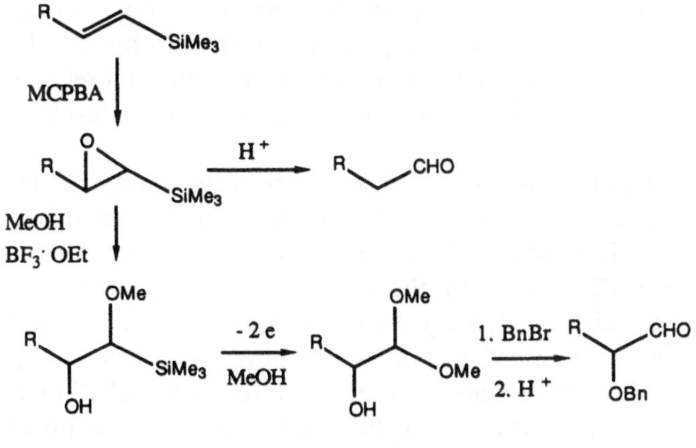

Scheme 19.

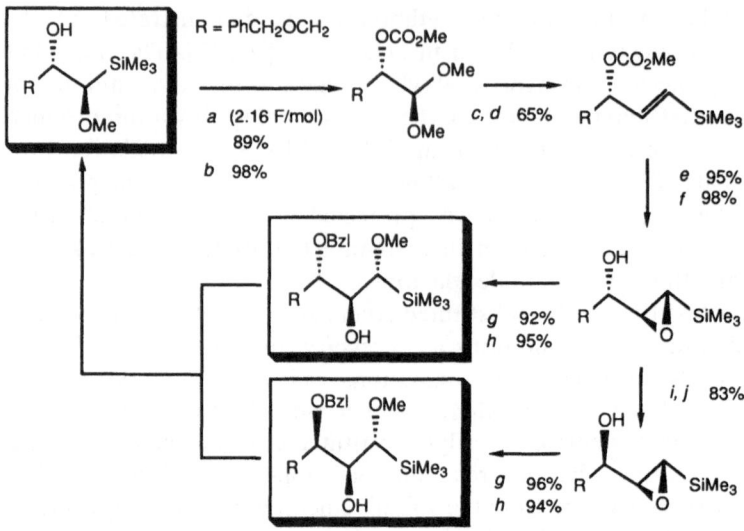

a: anodic oxidation Et₄NOTs/MeOH; b: ClCO₂Me, pyridine; c: H⁺; d: CrCl₂/Br₂CHSiMe₃;
e: LiAlH₄;; f: L-(+)-DIPT, Ti(OPrⁱ)₄, TBHP; g: NaH, PhCH₂Br; h: BF₃·OEt₂/ MeOH;
i: Ph3P, diethylazodicarbonate, PhCO2H; j: NaOH/MeOH

Scheme 20.

oxidation in methanol followed by the protection of the hydroxyl group gives the corresponding acetal. The acid-catalyzed hydrolysis of the acetal to give the aldehyde followed by chromium-promoted olefination with dibromotrimethylsilylmethane [52] affords alkenylsilanes. Deprotection of the carbonate and Sharpless epoxidation [53] gives the epoxy alcohol. BF₃-catalyzed ring opening affords the homologated β-hydroxy-α-methoxysilanes which is a protected form of *anti*-2,3,-dihydroxyaldehydes. Using this product as starting material for the next cycle, the whole process can be repeated without any difficulty.

The corresponding *syn*-compound can also be synthesized by simply inverting the stereochemistry of the hydroxyl group of the epoxy alcohol by the Mitsunobu reaction [54]. Therefore, this method provides a simple and reliable method for the synthesis of any enantiomers and diastereomers of straight-chain 1,2-polyols.

Nitrogen compounds are also effective as nucleophiles in the anodic oxidation of silyl-substituted ethers. The electrochemical oxidation in the presence of a carbamate or a sulfonamide in dry THF or dichloromethane results in the selective cleavage of the C–Si bond and the introduction of the nitrogen nucleophile at the carbon (Scheme 21) [55]. Since α-methoxycarbamates are useful intermediates in the synthesis of nitrogen-containing compounds [44], this reaction provides useful access to such compounds. Cyclic silyl-substituted ethers such as 2-silyltetrahydrofurans are also effective for the introduction of nitrogen nucleophiles. The anodic oxidation in the presence of a carbamate or a

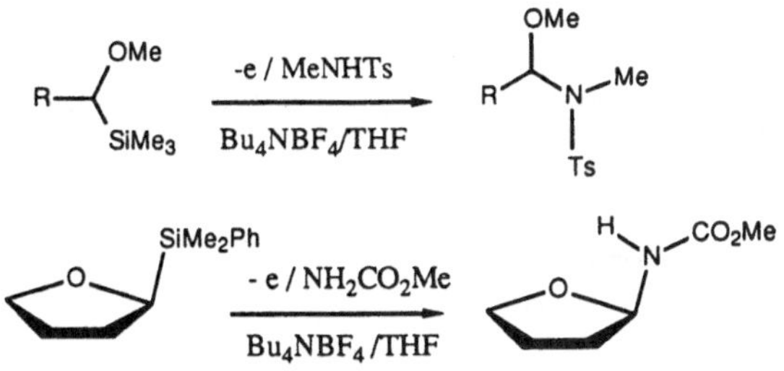

Scheme 21.

sulfonamide affords the corresponding nitrogen-containing products, providing a potentially useful method for the synthesis of the nucleotide.

3.7 Acylsilanes

Although the oxidation potentials of aldehydes and ketones are generally very high, silyl substitution at the carbonyl carbon results in a significant decrease in the oxidation potential (Table 6) [16] (Sect. 2.3.3). For example, undecanoyl-trimethylsilane exhibits an oxidation wave at the peak potential of 1.45 V. The effect of the silyl-substitution in the oxidation potential of aryl-substituted carbonyl compounds, however, is much smaller [56].

Preparative electrochemical oxidations of acylsilanes proceed smoothly in methanol in an undivided cell equipped with carbon rod electrodes to give the corresponding methyl esters. The C–Si bond is cleaved and methanol is introduced at the carbonyl carbon (Scheme 22) [16].

The mechanism shown in Scheme 23 has been suggested; The first step involves the transfer of an electron from the acylsilane to produce the cation-radical intermediate. Attack of methanol at the silicon cleaves the C–Si bond to give the acyl radical intermediate, although there is no direct evidence for the acyl radical intermediate. The acyl radical is then oxidized anodically to the acyl cation, which reacts with methanol to give the corresponding methyl ester.

Various nucleophiles other than methanol can be introduced onto the carbonyl carbon. Anodic oxidation of acylsilanes in the presence of allyl alcohol, 2-methyl-2-propanol, water, and methyl N-methylcarbamate in dichloromethane affords the corresponding esters, carboxylic acid, and amide derivatives (Scheme 24) [16]. Therefore, anodic oxidation provides a useful method for the synthesis of esters and amides under neutral conditions.

Recently, acylsilanes have been utilized as useful intermediates in organic synthesis [57]. For example, treatment of acylsilanes with the fluoride ion generates the corresponding acyl anions which react with electrophiles. On the other hand, by using the electrochemical method, acylsilanes serve as acyl cation equivalents because nucleophiles are introduced at the carbonyl carbon. Chemical oxidation of acylsilanes with hydrogen peroxide which affords the corresponding carboxylic acids has been reported [58]. However, the anodic oxidation provides a versatile method for the introduction of various nucleophiles

Table 6. Oxidation potentials of acylsilane and related compounds

Substrate	E_p (V vs. Ag/AgCl)	Substrate	E_p (V vs. Ag/AgCl)
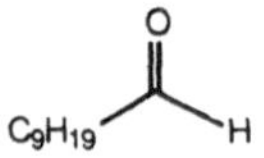 (structure: C_9H_{19}–C(=O)–H)	> 2.5	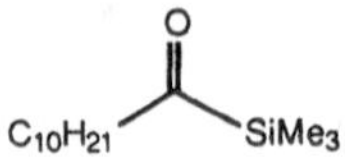 (structure: $C_{10}H_{21}$–C(=O)–SiMe$_3$)	1.45
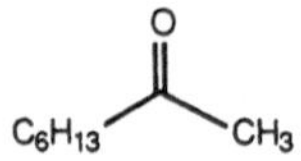 (structure: C_6H_{13}–C(=O)–CH$_3$)	> 2.5	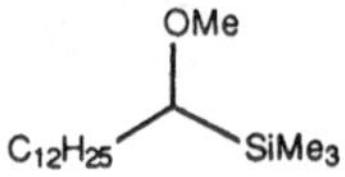 (structure: $C_{12}H_{25}$–CH(OMe)–SiMe$_3$)	1.70

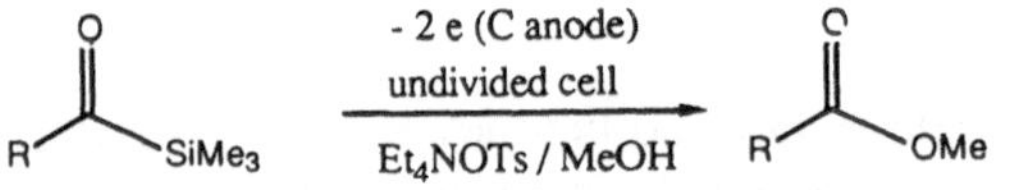

$$R = C_{10}H_{21} : 90\%$$
$$PhCH_2 : 83\%$$

Scheme 22.

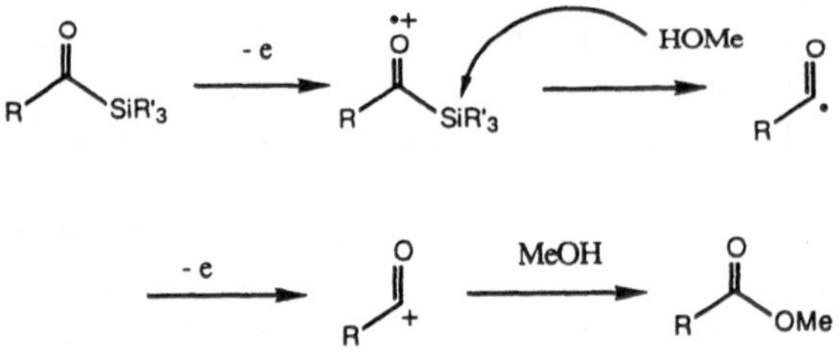

Scheme 23.

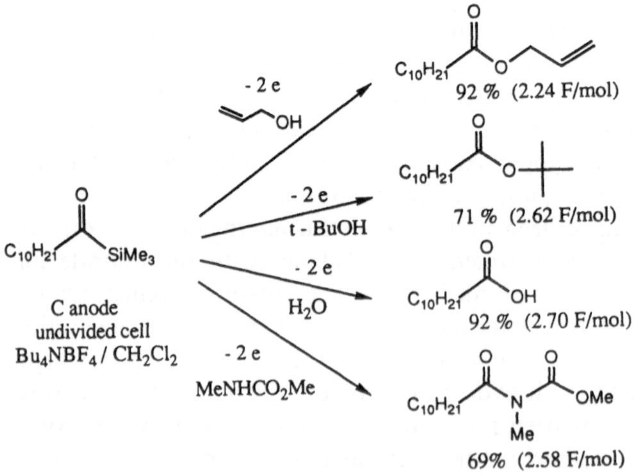

$C_{10}H_{21}$–C(=O)–O–CH$_2$CH=CH$_2$
92 % (2.24 F/mol)

$C_{10}H_{21}$–C(=O)–O–C(CH$_3$)$_3$
71 % (2.62 F/mol)

$C_{10}H_{21}$–C(=O)–OH
92 % (2.70 F/mol)

$C_{10}H_{21}$–C(=O)–N(Me)–C(=O)–OMe
69% (2.58 F/mol)

Scheme 24.

onto the carbonyl carbon. Mild reaction conditions of the electrochemical method are also advantageous from a synthetic stand-point.

3.8 Silyl Enol Ethers and Related Compounds

Silyl enol ethers are powerful intermediates in organic synthesis. Reactions of silyl enol ethers with various electrophiles provide effective methods for the synthesis of various carbonyl compounds. In this section we will briefly touch on the electrochemical reactions of silyl enol ethers and related compounds. The electrochemical behaviour of silyl enol ethers is expected to be closely related to that of allylsilanes and benzylsilanes because silyl enol ethers also have a silyl group β to the π-system.

Schäfer reported that the electrochemical oxidation of silyl enol ethers results in the homo-coupling products, 1,4-diketones (Scheme 25) [59]. A mechanism involving the dimerization of initially formed cation radical species seems to be reasonable. Another possible mechanism involves the decomposition of the cation radical by Si–O bond cleavage to give the radical species which dimerizes to form the 1,4-diketone. In the case of the anodic oxidation of allylsilanes and benzylsilanes, the radical intermediate is immediately oxidized to give the cationic species, because oxidation potentials of allyl radicals and benzyl radicals are relatively low. But in the case of α-oxoalkyl radicals, the oxidation to the cationic species seems to be retarded. Presumably, the oxidation potential of such radicals becomes more positive because of the electron-withdrawing effect of the carbonyl group. Therefore, the dimerization seems to take place preferentially.

Miller and coworkers investigated the electrochemical oxidation of hydroquinone disilylethers. Anodic oxidation using platinum or graphite anodes in Bu_4NBF_4/CH_3CN gives the corresponding quinones 80–90% yields (Scheme 26) [60]. The reaction is also effective in $LiClO_4/CH_3CN$ and Bu_4NClO_4/CH_2Cl_2. On the basis of the cyclic voltammetry a mechanism involving initial one-electron oxidation to generate a cation radical which decomposes by Si–O bond cleavage to eventually form quinone has been

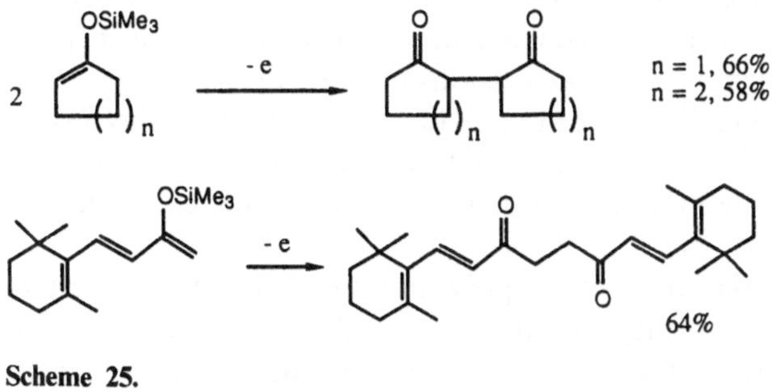

Scheme 25.

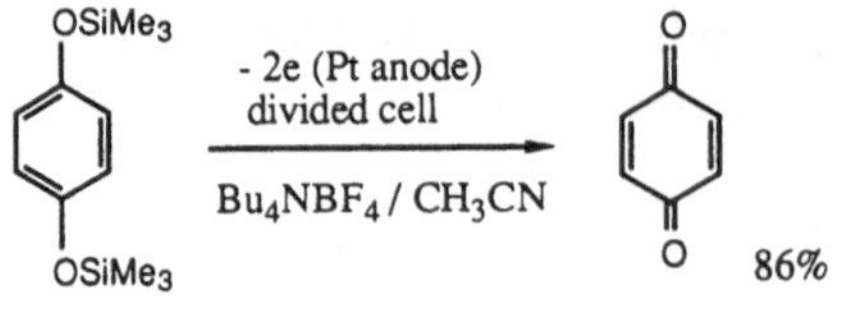

OSiMe₃ ... Scheme 26.

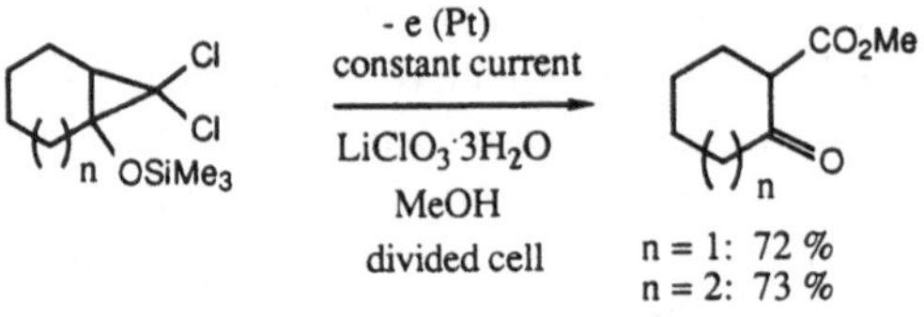

Scheme 26.

- 2e (Pt anode)
divided cell

Bu₄NBF₄ / CH₃CN

86%

Scheme 26.

- e (Pt)
constant current

LiClO₃·3H₂O
MeOH
divided cell

CO₂Me

n = 1: 72 %
n = 2: 73 %

Scheme 27.

suggested. The relative rates for decomposition of the cation radicals depends upon the nature of the silyl group (Me_3Si: 10, Et_3Si: 5, t-$BuMe_2Si$: 1 at $-60\,^\circ$C).

Since the chemical reactivity of a cyclopropane ring is sometimes compared with that of a carbon–carbon double bond, it is interesting to study the electrochemical behaviour of silyloxycyclopropanes. Torii et al. described the anodic oxidation of silyloxycyclopropanes in the presence of $Fe(NO_3)_3$ which resulted in the opening of the cyclopropane ring as shown in Scheme 27 [61]. The reaction temperature has a profound effect on the yields of the products; the best resuls are obtained at -13 to $-10\,^\circ$C.

3.9 Organosilicon Compounds Containing Si–Si Bonds

The chemistry of compounds containing Si–Si bond(s) is an intriguing subject in the field of organosilicon chemistry because Si–Si bonds have unique physical and chemical properties. The reactivities of Si–Si bonds is often compared with those of carbon–carbon double bonds. The current interest in polysilanes in material science stems from the fact that they exhibit unusual properties implying considerable electron delocalization in the polymer chain [62]. This section concerns with the unique elecrochemical properties of compounds containing Si–Si bonds (Sect. 2.4).

Boberski and Allred reported that oxidation potentials of permethylpoly-silanes determined by a.c. polarography decrease with increasing chain length, and that the oxidation potentials are correlated almost linearly with the energies of the HOMO as determined by MO calculations (Table 7) [63].

The electrochemical oxidation of hexamethyldisilane in acetonitrile results in the cleavage of the Si–Si bond and the formation of a short lived intermediate shown in Scheme 28. The number of electrons transferred is calculated as being 2.

The electrochemical oxidation of cyclic polysilanes has also been investigated [64]. Cyclic polysilanes display at least two anodic waves separated by 0.2–0.4 V. The first oxidation potential is 1.1 to 1.4 V (vs SCE) which depends

Table 7. Lowest oxidation potentials of permethylpolysilanes [63]

Compound	E_{ac}(V vs. SCE)
Me(SiMe$_2$)$_2$Me	1.88
Me(SiMe$_2$)$_3$Me	1.52
Me(SiMe$_2$)$_4$Me	1.33
Me(SiMe$_2$)$_5$Me	1.18
Me(SiMe$_2$)$_6$Me	1.08

$$Me_3SiSiMe_3 \xrightarrow[CH_3CN]{-2\,e} 2\,Me_3Si^+(NCCH_3)_n$$

Scheme 28.   short-lived intermediate

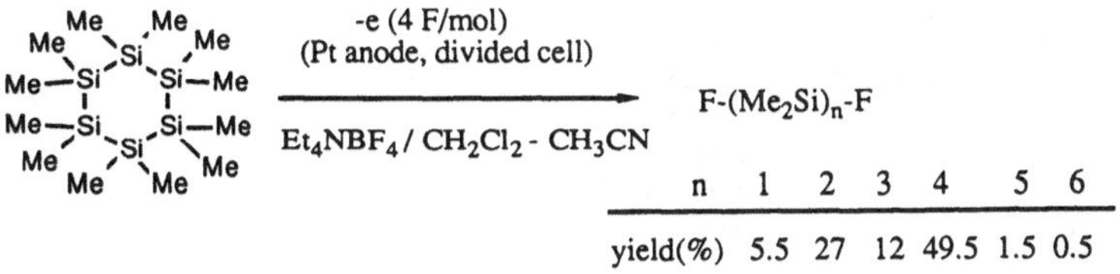

Scheme 29.

both upon the ring size and on the nature of the substitutents on silicon. The controlled potential electrolysis of cyclic polysilanes $(R_2Si)_n$ in Et_4NBF_4/CH_2Cl_2–CH_3CN results in ring opening to form α,ω-difluorosilanes as major products (Scheme 29).

The following mechanism has been suggested (Scheme 30). The initial one-electron oxidation produces a highly reactive cyclic cation radical intermediate which may undergo Si–Si bond cleavage to generate a fluoro-substituted silyl radical. Further oxidation to the cationic species followed by the reaction with BF_4^- leads to the formation of the α,ω-difluoropolysilane. Further oxidative cleavage of the initially formed difluoropolysilane generates their lower homologues.

The cyclic voltammetry of polysilanes adsorbed on the electrode surface has also been investigated [65]. The oxidation potentials depend upon the nature of the organic groups on silicon. The electrochemical oxidation is irreversible to give soluble products which are liberated from the surface of the anode.

The electrochemical behaviour of the compounds containing bonds between silicon and other group-14-metals is also interesting. Mochida et al. reported the electrochemical oxidation potentials of group-14-dimetals [66]. As shown in Table 8, there is a good correlation between the oxidation potentials and the ionization potentials which decrease in the order: Si–Si > Si–Ge > Ge–Ge > Si–Sn > Ge–Sn > Sn–Sn in accord with the metal–metal ionic bond dissociation energy.

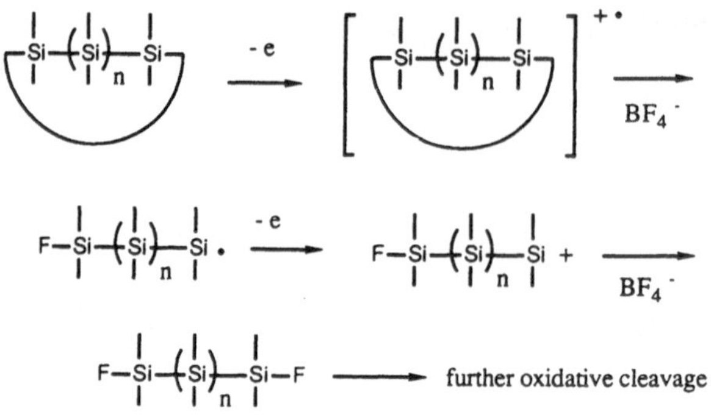

Scheme 30.

Table 8. Oxidation and ionization potentials of group 14 dimetals [66]

Compound	E_p (V vs. Ag/AgCl)[a]	IP (eV)[b]
$Me_3SiSiMe_3$	1.76	8.68
$Me_3SiGeMe_3$	1.76	8.62
$Me_3GeGeMe_3$	1.70	8.60
$Me_3SiSnMe_3$	1.60	8.39
$Me_3GeSnMe_3$	1.44	8.36
$Me_3SnSnMe_3$	1.28	8.20

[a] Determined in Et_4NClO_4/CH_3CN with glassy carbon, anode. [b] Determined with photoelectron spectroscopy.

3.10 β-Silyl Carboxylic Acids

Shono et al. reported that anodic oxidation of β-silyl-substituted carboxylic acids results in decarboxylation and desilylation to form the olefinic products (Scheme 31) [67]. The first step of this reaction involves the oxidation of the carboxylate with the elimination of CO_2 to give the carbocation intermediate having a silyl group at the β-position. In the second step the silyl group is eliminated spontaneously to give the olefin. It should be noted that in the absence of the silyl group a mixture of two regioisomeric olefins is obtained. Therefore, the silyl group controls the regiochemistry of the reaction.

β-Silyl carboxylic acids are readily synthesized by the alkylation of silylmethylmalonic esters followed by decarboxylation. Therefore, the silylmethyl malonic esters can be considered to be a vinyl anion equivalent (Scheme 32).

Schäfer and coworkers devised a useful synthetic procedure based on this reaction [26]. Thus, the Diels-Alder reaction of β-silylacrylic acid with cyclopentadiene gave the adduct which was oxidized anodically with the elimination of the carboxyl and the silyl groups. Successful formation of norbonadiene indicates that β-silylacrylic acid can be used an a synthon of acetylene in Diels-Alder reactions with dienes (Scheme 33).

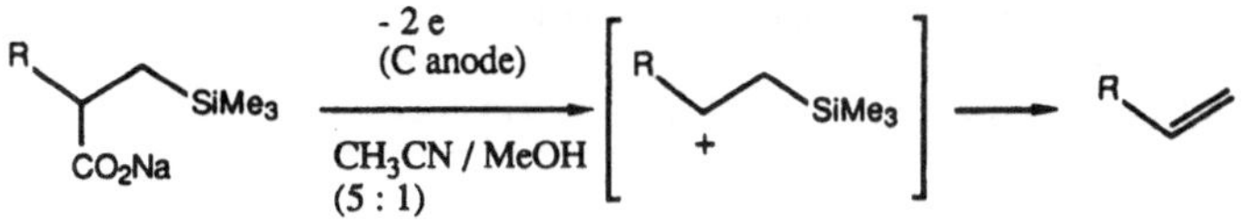

Scheme 31.

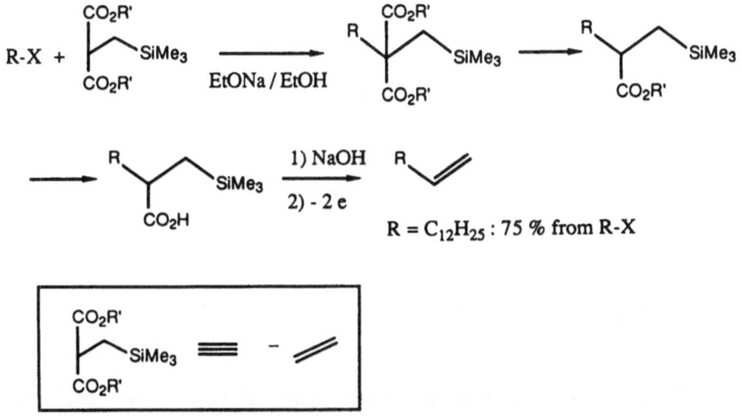

$R = C_{12}H_{25}$: 75 % from R-X

Scheme 32.

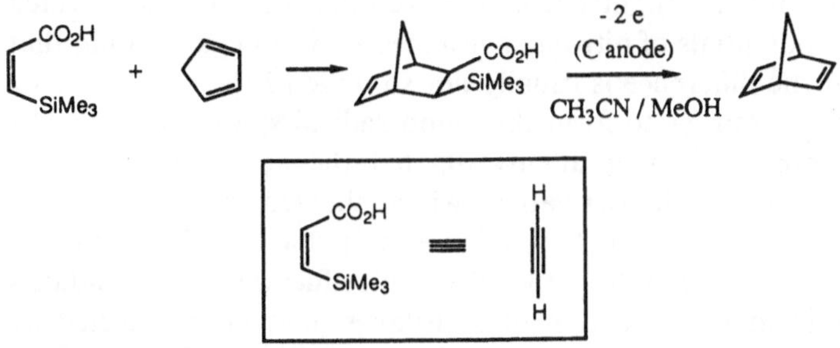

Scheme 33.

4 Cathodic Reduction of Organosilicon Compounds

4.1 Arylsilanes

We know that reduction potentials of arylsilanes are less negative than those of the corresponding aromatic compounds without silyl substituents (Sect. 2.2.1). The effect of silyl groups to facilitate the electron transfer to the neigbouring aromatic group is explained in terms of $d_\pi - p_\pi$ interaction. For example, half wave reduction potentials of naphthylsilanes are less negative than that of

Table 9. Reduction potentials of silyl-substituted naphthalenes [68]

Compound	$E_{1/2}$ (V vs. Ag/AgCl)	$E_{1/2}$ (V vs. Hg pool)
(naphthalene structure)	− 2.58	− 1.98
(1-SiMe₃ naphthalene structure)	− 2.52	− 1.93
(2-SiMe₃ naphthalene structure)	− 2.57	− 1.97
(1,4-bis-SiMe₃ naphthalene structure)	− 2.38	− 1.83

The polarography was carried out in Bu_4NI/DMF

naphthalene as shown in Table 9, although the difference is very small [68]. An additional silyl substitution causes further shift of the reduction potential.

Reduction potentials of silyl-substituted biphenyls have also been investigated [69]. The half wave potentials of 4-trimethylsilylbiphenyl and 4,4'-bis(trimethylsilyl)biphenyl are slightly less negative than that of unsubstituted biphenyl. Reduction potentials of nitrophenylsilanes are less negative than that of nitrobenzene, but the difference is usually not so large [70].

The initial electron transfer to form the anion radical species seems to be reversible. For example, Allred et al. investigated the ac polarography of bis(trimethylsilyl)benzene and its derivatives which showed two waves in dimethylformamide solutions [71]; the first one is a reversible one-electron wave, and the second one corresponds to a two-electron reduction. Anion radicals generated by electrochemical reduction of arylsilanes have been detected by ESR. The cathodic reduction of phenylsilane derivatives in THF or DME at − 76° C gives ESR signals due to the corresponding anion radicals [5] (See Sect. 2.2.1).

Preparative electrochemical reduction of aryltrimethylsilanes in methylamine in the presence of LiCl gives the Birch-type products, 1,4-cyclohexandienes (Scheme 34) [6]. A mechanism involving the electrochemical formation of lithium metal which chemically reduces the substrate has been suggested. The hydrogen atom is introduced on the carbon adjacent to the silicon preferentially. This regioselectivity is consistent with the spin density of the anion radical determined by ESR spectroscopy (Sect. 2.2.1).

Paquette and coworkers reported the electrochemical behaviour of silyl-substituted cyclooctatetraene [72]. Two waves are observed with polarography in hexamethylphosphoramide (HMPA) solutions. The silyl substitution causes a

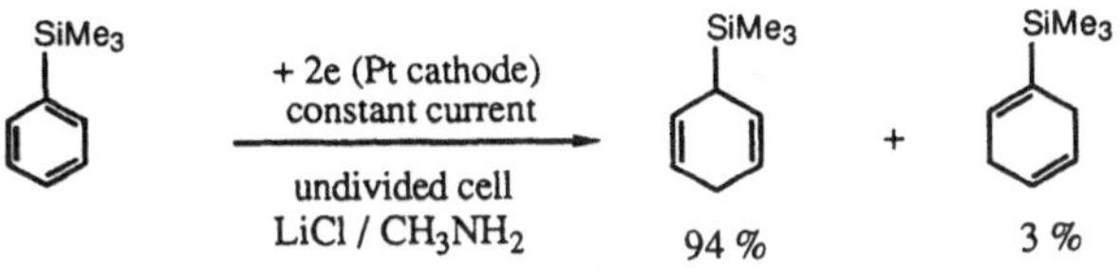

Scheme 34.

Table 10. Reduction potentials of cyclooctatetraene derivatives [72]

Compound	$E^1_{1/2}$	$E^2_{1/2}$
	− 1.606	− 1.921
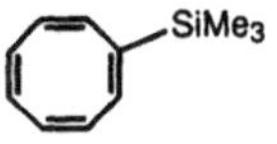	− 1.640	− 1.804
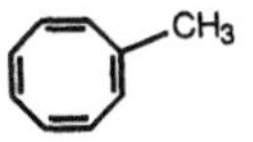	− 1.730	− 1.955

Determined by linear sweep voltammetry in HMPA. The values are correlated to SCE.

slight decrease in the first reduction potential and an increase in the second reduction potential. The second effect is probably responsible for the stabilization of the dianion by $d_\pi-p_\pi$ interaction between the aromatic 10π-system and the silicon. The faster rate of the decomposition of the dianion due to the Si–C cleavage may also play a role.

4.2 Halosilanes

Although it is well known that organic halides which have carbon–halogen (C–X) bonds, undergo cathodic reduction, it is rather difficult to reduce halosilanes which have silicon–halogen (Si–X) bonds. As a matter of fact, when organic halides such as allyl and benzyl halides are electrolyzed in the presence of a chlorosilane, organic halides are reduced preferentially to generate the corresponding carbanion which is trapped by the chlorosilane to give organosilicon compounds [73, 74]. This reaction provides a convenient and efficient method for the formation of carbon–silicon (C–Si) bonds.

The use of reactive metal electrodes are also effective for the silylation of various organic halides and simple arenes [75]. For instance, Dunoguès et al. reported that electrolysis of aryl chlorides in the presence of excess Me_3SiCl in a one-compartment cell equipped with a sacrificial aluminum anode in 80:20 THF/HMPA gave the corresponding aryltrimethylsilanes (Scheme 36). When

73

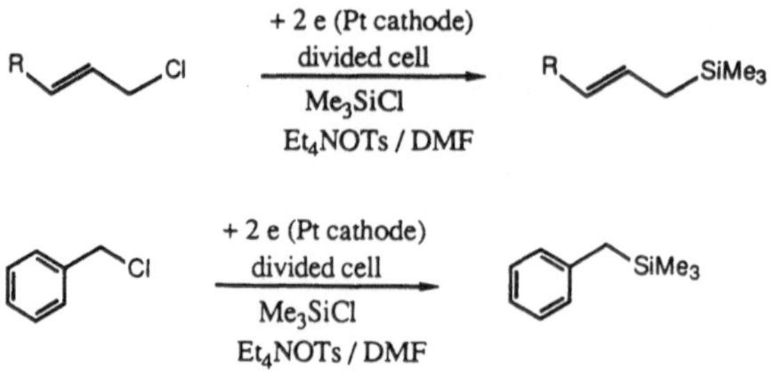

Scheme 35.

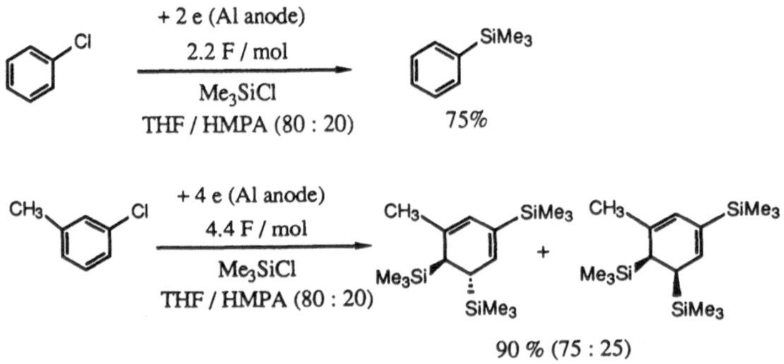

Scheme 36.

excess electricity (up to 4.4 F/mol) was passed, a mixture of *cis*- and *trans*-tris(trimethylsilyl)cyclohexadienes is produced. This is consistent with the electrochemical reduction of arylsilanes described in Sect. 4.1, although the regioselectivity is different. As a matter of fact, aryltrimethylsilanes can be reduced under similar conditions to give the disilylated products, tris-(trimethylsilyl)cyclohexadienes.

Selective silylation of polychloromethanes using reactive metal electrodes such as zinc and magnesium has also been reported as shown in Scheme 37 [76, 77]. The electroreduction of carbon tetrachloride and chloroform in the presence of chlorotrimethylsilane affords the monosilylated and disilylated products. The product selectivity seems to depend upon the electrode material.

In the cathodic reduction of activated olefins, chlorosilanes also act as trapping agents of anionic intermediates. Nishiguchi and coworkers described the electrochemical reduction of α,β-unsaturated esters, nitriles, and ketones in the presence of Me_3SiCl using a reactive metal anode (Mg, Zn, Al) in an undivided cell to afford the silylated compounds [78]. This reaction provides a valuable method for the introduction of a silyl group into activated olefins.

In the absence of other substrates which are easily reduced, halosilanes can be reduced by cathodic reaction. However, it is rather difficult to determine the reduction potentials of halosilanes, because halosilanes are readily hydrolyzed during voltammetric measurements. Although early reports stated that the reduction potential of Me_3SiCl is not very negative, extensive studies by Corriu

$$CCl_4 \; + \; Me_3SiCl \; \xrightarrow[\text{Et}_4\text{NBF}_4\,/\,\text{DMF}]{\substack{+\,2\,\text{e}\;(2.2\;\text{F/mol}) \\ (\text{Zn anode})}} \; \underset{94\%}{Me_3SiCCl_3} \; + \; \underset{6\%}{(Me_3Si)_2CCl_2}$$

$$CCl_4 \; + \; Me_3SiCl \; \xrightarrow[\text{Et}_4\text{NBF}_4\,/\,\text{DMF}]{\substack{+\,4\,\text{e}\;(4.4\;\text{F/mol}) \\ (\text{Mg anode})}} \; \underset{68\%}{(Me_3Si)_2CCl_2}$$

$$HCCl_3 \; + \; Me_3SiCl \; \xrightarrow[\text{Et}_4\text{NBF}_4\,/\,\text{DMF}]{\substack{+\,2\,\text{e}\;(2.2\;\text{F/mol}) \\ (\text{Zn anode})}} \; \underset{94\%}{Me_3SiCHCl_2}$$

Scheme 37.

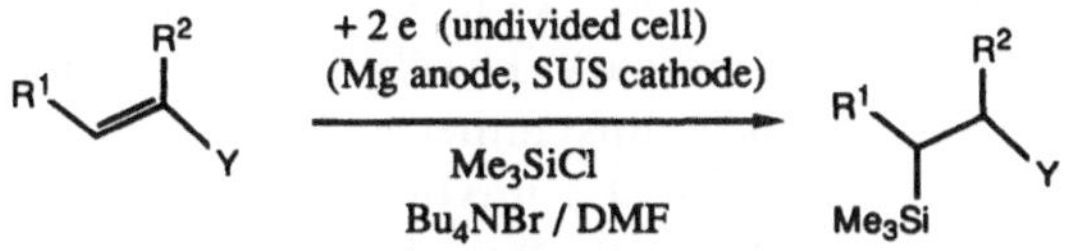

Scheme 38. $Y = CO_2Me,\; CN,\; COCH_3$

$$Me_3SiCl \; \xrightarrow[\text{Bu}_4\text{NClO}_4\,/\,\text{DME}]{+\,\text{e}\;(\text{Pb cathode, Hg anode})} \; \underset{95\%\;\text{current efficiency}}{Me_3Si\text{-}SiMe_3}$$

Scheme 39.

et al. [79] revealed that the reduction peaks usually observed at about -0.4 V vs SCE actually refers to the reduction of HCl or HBr formed by hydrolysis arising from traces of water in the solvent. Therefore, the reduction potential of Me_3SiCl seems to be quite negative. Indeed, Ph_3SiCl which is less sensitive towards hydrolysis shows a single reduction wave in the low cathodic range (about -2.4 V vs SCE) in anhydrous solvents.

There is a rich literature on preparative electrochemical reductions of halosilanes. Dessy et al. reported that Ph_3SiH was formed by the electrochemical reduction of Ph_3SiCl in 1,2 dimethoxyethane (DME) using a dropping mercury cathode at -3.1 V vs. $AgClO_4/Ag$ [80]. Although the ultraviolet spectra of the solutions after electrolysis were identical with the spectra of an authentic sample of Ph_3SiH, the product was not isolated.

Hengge et al. reported the formation of Si–Si bond in the electrochemical reduction of chlorosilanes in DME without control of the applied potential. For the formation of the Si–Si bond the choice of electrode materials seems to be of great importance [81]. Lead and mercury have proved suitable as cathode materials.

Allred et al. studied the electrochemical reduction of Me_3SiCl in acetonitrile using a platinum cathode [82]. They reported that the choice of the supporting electrolyte is important for the formation of the disilane. When Bu_4NClO_4 was

Jun-ichi Yoshida

used as the supporting electrolyte, the cathodic reduction of Me₃SiCl in acetonitrile using a platinum cathode gave Me₃SiOSiMe₃ as the only silicon-containing product. When Bu₄NCl was used as the supporting electrolyte, Me₃SiSiMe₃ was the only silicon-containing product. They also reported the attachment of permethylpolysilane groups to the platinum electrode by cathodic reduction of chloropermethylpolysilane. Corriu et al. also demonstrated that chlorotriphenylsilane could be electrochemically coupled, but large quantities of disiloxanes are formed [83].

Very recently the use of reactive metal anodes such as Mg, Al, Cu, Hg and Ag electrodes in an undivided cell was found to be quite effective for the formation of Si–Si bonds.

Kunai and Ishikawa et al. have reported that electrolysis of monochloro-silanes in 1,2-dimethoxyethane using a platinum cathode and a mercury anode gives disilanes in high yield (Scheme 40) [84]. Silver can also be used as an excellent anode material in place of mercury. The electrolysis of a mixture of two different monochlorosilanes produces unsymmetrical disilanes. Trisilanes can also be synthesized by the electrolysis of a mixture of monochlorosilanes and dichlorosilanes. They also reported that the use of copper electrodes is effective for the synthesis of disilanes, trisilanes, tetrasilanes, and pentasilanes [85].

Shono and Kashimura et al. have reported the elegant electrochemical synthesis of disilanes from chlorosilanes by using magnesium for both the anode and the cathode and by alternating the direction of the current at some interval [86]. The application of this method to dichlorosilanes results in the efficient formation of polysilanes (Scheme 41). Sonication results in a marked increase in the yields of polysilanes. The electrolysis of dichlorodisilanes also gives the corresponding polysilanes of higher molecular weight although the yields are low.

Magnesium electrodes are also effective for the formation of Si–Ge bonds [87]. Thus, the electroreduction of a mixture of PhMe₂SiCl and Me₃GeCl affords PhMe₂SiGeMe₃ in 60% yield together with Me₃GeGeMe₃. Silane–germane copolymers are also obtained by the electroreduction of a mixture of R₂SiCl₂ and R₂GeCl₂.

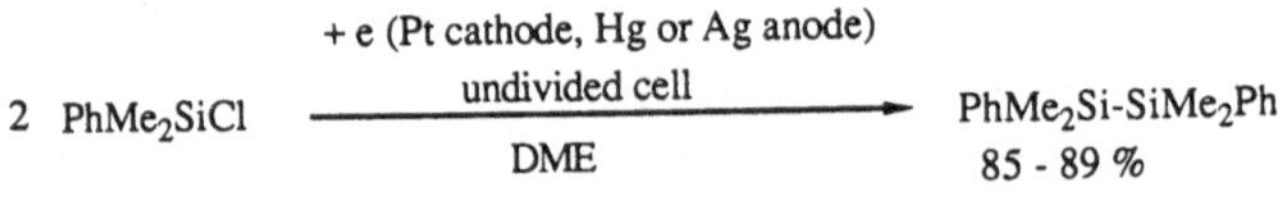

$$2 \ PhMe_2SiCl \xrightarrow[\text{DME}]{\substack{+ e \ (Pt \ cathode, \ Hg \ or \ Ag \ anode) \\ \text{undivided cell}}} PhMe_2Si\text{-}SiMe_2Ph$$
85 - 89 %

Scheme 40.

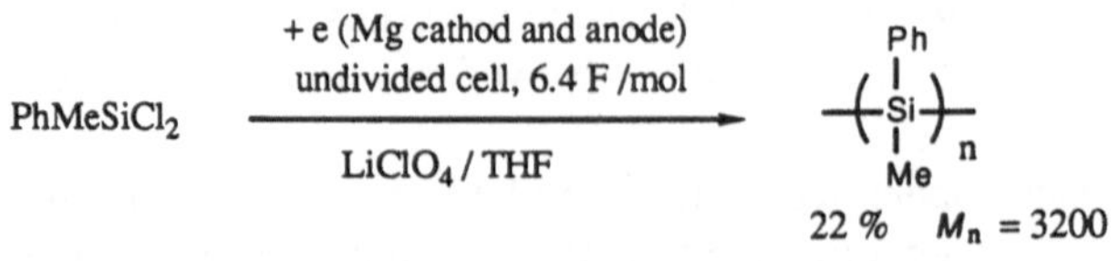

$$PhMeSiCl_2 \xrightarrow[\text{LiClO}_4 / THF]{\substack{+ e \ (Mg \ cathod \ and \ anode) \\ \text{undivided cell, 6.4 F /mol}}} \left(\begin{array}{c} Ph \\ | \\ Si \\ | \\ Me \end{array}\right)_n$$
22 % $M_n = 3200$

Scheme 41.

76

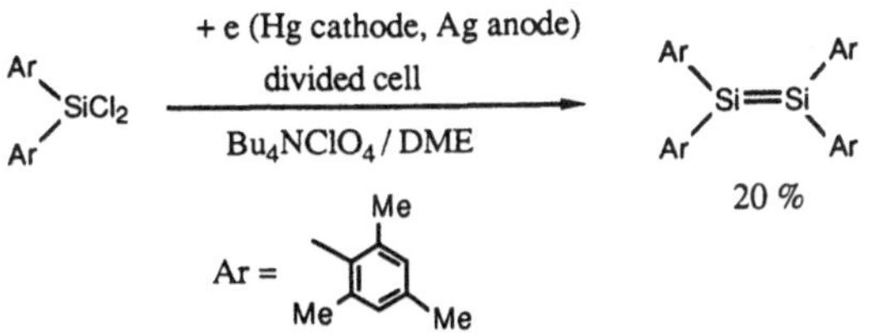

$$\underset{Ar}{\overset{Ar}{\diagdown}}SiCl_2 \quad \xrightarrow[\substack{\text{divided cell} \\ Bu_4NClO_4 \, / \, DME}]{+ \, e \, \text{(Hg cathode, Ag anode)}} \quad \underset{Ar}{\overset{Ar}{\diagdown}}Si{=}Si\underset{Ar}{\overset{Ar}{\diagup}}$$

20 %

$$Ar = \underset{Me}{\overset{Me}{\diagdown}}\underset{Me}{}$$

Scheme 42.

The use of sacrificial aluminum electrodes is also effective for the electrore-ductive synthesis of disilanes and polysilanes from monochlorosilanes and dichlorosilanes, respectively as reported by Nonaka et al. and Dunoguès et al. independently [88].

Dunoguès and coworkers also devised the electrochemical reduction of dimethyldichlorosilane without solvent [89]. Cathodic reduction of dimethyl-dichlorosilane in the presence of a small amount of complexing agent such as hexamethylphosphoric triamide (HMPA) or tris(3,6-dioxaheptyl)amine (TDA-1) in an undivided cell fitted with a sacrificial aluminum anode and a stainless steel cathode gave polydimethylsilane in very high faradaic yield. In the absence of the complexing agent, the polysilane is contaminated with metallic aluminum resulting from reduction of $AlCl_3$.

In some cases, the cathodic reduction of dichlorosilane gives the correspond-ing disilene. For example, the electrolysis of dimesityldichlorosilane in a divided cell equipped with a mercury pool cathode and silver anode under controlled potential conditions (-3.2V vs Ag/Ag$^+$) affords tetramesityldisilene in 20% yield (Scheme 42) [90].

4.3 Halomethylsilanes

Generally, the two-electron reduction of organic halides produces carbanion species. In fact, cathodic reduction of organic halides under certain conditions gives the product derived from the corresponding carbanion intermediates. Silicon is known to stabilize the carbanion at the α position by $d_\pi - p_\pi$ interaction. Therefore, we can expect that silicon promotes the electron transfer from carbon–halogen bonds and the formation of the carbanion at the α position.

The polarographic study of (idomethyl)trialkylsilanes has revealed that the introduction of a trialkylsilyl group into iodoalkanes decreases the value of the half wave potential ($E_{1/2}$), i.e. favours the electron-transfer (Table 11) [91].

Although the preparative cathodic reduction of halomethylsilanes has not been investigated extensively, Dunoguès and coworkers revealed that the electrochemical reduction of (chloromethyl)dimethylchlorosilane with an alumi-num cathode afforded polycarbosilanes [92]. 1,1.3,3-Tetramethyl-1,3-disilacyclobutane is also formed under these condition (Scheme 43).

It is also possible to produce $Cl(CH_2SiMe_2)_2Cl$ rather selectively (43%) by choosing the appropriate solvent system. The electrolysis of (chloro-methyl)dimethylchlorosilane in the presence of dimethyldichlorosilane provides

Table 11. Reduction potentials of (iodomethyl) trialkylsil-
anes

Compound	$E_{1/2}'$	$E_{1/2}''$
Et_3SiCH_2I	− 1.54	− 1.72
$CF_3CH_2CH_2SiMe_2CH_2I$	− 1.33	− 1.54
$PhSiMe_2CH_2I$	− 1.43	− 1.56
$p\text{-}FC_6H_4SiMe_2CH_2I$	− 1.40	− 1.53

Determined by polarography with the dropping Hg electrode in 0.09 N KCl in 57% alcohol.

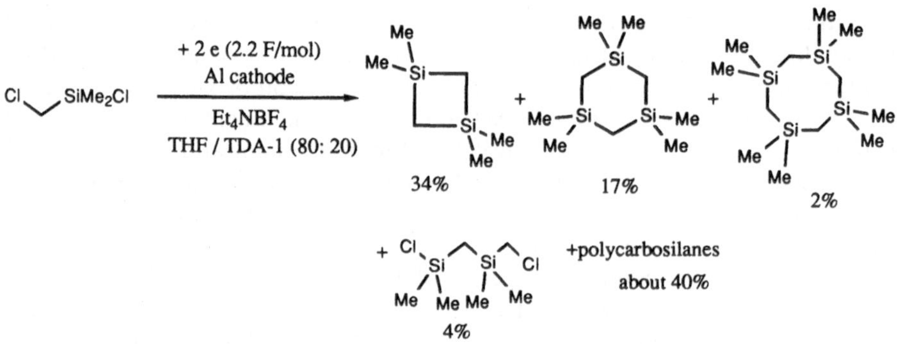

Scheme 43.

bis(dimethylchlorosilyl)methane (60%) which is a useful precursor of polycarbosilanes.

5 Conclusions

Silicon is no longer an exotic element in electroorganic chemistry. The electrochemical behaviour of organosilicon compounds is easily understood, because
the following concepts help us to appreciate their unique properties.

1) Silicon activates organic molecules;
   (a) Silyl substitution α to π systems lowers the LUMO of the molecule and
   facilitates cathodic reduction.
   (b) Silyl substitution β to π systems and heteroatoms raises the HOMO of
   the molecule and facilitates anodic oxidation.
2) Silicon controls the regiochemistry of the reaction;
   (a) In anion radical intermediates silicon facilitates the formation of a
   negative charge at the α position.
   (b) In cation radical or cation intermediates the β-silyl group acts as "super
   proton" and is eliminated spontaneously without affecting other protons.

On the basis of these concepts a number of electrochemical reactions of organosilicon compounds have been developed. Although a rich variety of synthetic applications of the anodic oxidation of organosilicon compounds has been made in recent years, the application of the cathodic reduction of such compounds has been less studied and will hopefully be uncovered in the near future.

It should be also recognized that enormous advances have been made in the study of the electrochemistry of halosilanes. In this field, reactive metal electrodes provide powerful tools for the formation of Si–Si bonds. The electrochemistry of polysilanes is also a fascinating area of research because Si–Si bonds serve as unique electron pools.

It is hoped that the examples and the accompanying mechanistic discussions of electrochemical reactions of organosilicon compounds shown in this review will provide a guide to the potential utility of such reactions in organic synthesis and to the development of new electroorganic chemistry based on the unique properties of silicon.

6 References

1. Weber WP (1983) Silicon reagents for organic synthesis. Springer, Berlin Heidelberg New York
   Colvin EW (1981) Silicon in organic synthesis. Butterworths, London
2. Klingler RJ, Kochi JK (1980) J Am Chem Soc 102: 4790
3. Bock H (1989) Angew Chem Int Ed Engl 28: 1627
4. Giordan JC (1983) J Am Chem Soc 105: 6544
5. Wan Y-P, O'Brien DH, Smentowski FJ (1972) J Am Chem Soc 94: 7680
6. Eaborn C, Jackson RA, Pearce R (1974) J Chem Soc Perkin I: 2055
7. Pitt CG (1973) J Organomet Chem 61: 49
8. West R, Barton TJ (1980) J Chem Education 57: 165
9. Bock H, Kaim W (1980) J Am Chem Soc 102: 4429; Kira M, Nakazawa H, Sakurai H (1985) Chem Lett: 1845
10. Bock H, Kaim W (1982) Acc Chem Res 5: 9
11. Kira M, Nakazawa H, Sakurai H (1985) Chem Lett 1845
12. Bock H, Mollère P, Becker G, Fritz G (1973) J Organomet Chem 61: 113
13. Yoshida J, Maekawa T, Murata T, Matsunaga S, Isoe S (1990) J Am Chem Soc 112: 1962
14. Kira M, Nakazawa H, Sakurai H (1986) Chem Lett: 497
15. Bock H, Alt H, Seidl H (1969) J Am Chem Soc 91: 355; Ramsey BG, Brook A, Bassindale AR, Bock H (1974) J Organomet Chem 74: C41
16. Yoshida J, Matsunaga S, Isoe S (1989) Tetrahedron Lett 30: 5293; Yoshida J, Itoh M, Matsunaga S, Isoe S (1992) J Org Chem 57: 4877
17. West R (1982) In: Wilkinson G, Stone FGA, Abel EW (eds) Comprehensive organometallic chemistry, vol 2, Pergamon Press, Oxford, p 365
18. Pitt CG, Carey RN, Toren Jr EC (1972) J Am Chem Soc 94: 3806
19. Carberry E, West R, Glass GE (1969) J Am Chem Soc 91: 5446
20. Bock H, Kaim W, Kira M, West R, J Am Chem Soc 101: 7667
21. Alyev IY, Rozhkov IN, Knunyants IL (1976) Tetrahedron Lett: 2469
22. Yoshida J, Tamao K, Kumada M, Kawamura T (1980) J Am Chem Soc 102: 3269 and ref cited therein
23. Genies EM, Omar FEl (1983) Electrochimica Acta 28: 547
24. Yoshida J, Murata T, Isoe S (1986) Tetrahedron Lett 27: 3373

25. Koizumi T, Fuchigami T, Nonaka T (1986) Chem Express 1: 355; Koizumi T, Fuchigami T, Nonaka T (1989) Bull Chem Soc Jpn 62: 219
26. Schäfer H, Hermeling D, Lange KH (1984) Spring Meeting of the Electrochemical Society, Cincinnati, Ohio, Extend Abstracts: 441
27. Ochiai M, Arimoto M, Fujita E (1981) Tetrahedron Lett 22: 4491; Ochiai M, Fujita E (1983) Tetrahedron Lett 24: 777; Ochiai M, Fujita E, Arimoto M, Yamaguchi H (1985) Chem Pharm Bull 33: 989
28. Shono T, Ikeda A (1972) J Am Chem Soc 94: 7892; Yoshida K, Kanbe T, Fueno T (1977) J Org Chem 42: 2313
29. Ohga K, Mariano PS (1982) J Am Chem Soc 104: 617; Kavash RW, Mariano PS (1989) Tetrahedron Lett 30: 4185; Mizuno K, Terasaka K, Ikeda M, Otsuji Y (1985) Tetrahedron Lett 26: 5819; Mizuno K, Ikeda M, Otsuji Y (1985) Tetrahedron Lett 26: 461; Mizuno K, Terasaka K., Yasueda M, Otsuji Y (1988) Chem Lett: 145; Mizuno K, Yasueda M, Otsuji Y (1988) Chem Lett: 229; Baciocchi E, Giacco TD, Rol C, Sebastiani GV (1989) Tetrahedron Lett 30: 3573
30. Kondo T, Yamamoto K, Kumada M (1973) J Organomet Chem 60: 303
31. Shono T, Ikeda A (1976) Chem Lett: 311; Shono T, Nishiguchi I, Okawa M (1976) Chem Lett: 573; Baltes H, Stork L, Schäfer HJ (1977) Angew Chem 89: 425; Baltes H, Steckhan E, Schäfer HJ (1978) Chem Ber 111: 1294; Baltes H, Stork L, Schäfer HJ (1979) Ann: 318; Baltes H, Stork L, Schäfer HJ (1979) Chem Ber 112: 807
32. Yoshida J, Murata T, Isoe S (1987) Tetrahedron Lett 28: 211
33. Corriu R, Escudie N, Guerin C (1984) J Organomet Chem 264: 207
34. Meyers AI, Spohn RF (1985) J Org Chem 50: 4872; Julia M, Badet B (1975) Bull Soc Chim France; 1363; Kondo K, Tunemoto D (1975) Tetrahedron Lett: 1007; Corey EJ, Erickson BW, Noyori R (1971) J Am Chem Soc 93: 1724
35. Cooper BE, Owen WJ (1971) J Organomet Chem 29: 33
36. Yoshida J, Isoe S (1987) Chem Lett: 631
37. Yoshida J, Matsunaga S, Murata T, Isoe S (1991) Tetrahedron 47: 615
38. Koizumi T, Fuchigami T, Nonaka T (1987) Chem Lett: 1095
39. Uneyama K, Torii S (1971) Tetrahedron Lett: 329
40. Ager DJ, Cookson RC (1980) Tetrahedron Lett 21: 1677; Kocienski PJ (1980) Tetrahedron Lett 21: 1559
41. Ogura F, Otsubo T, Ohira N (1983) Synthesis: 1006
42. Suda K, Watanabe J, Takanami T (1992) Tetrahedron Lett 33: 1355
43. Yoshida J, Isoe S (1987) Tetrahedron Lett 28: 6621
44. Shono T, Hamaguchi H, Matsumura Y (1975) J Am Chem Soc 97: 4264
45. Yoshida J, Murata T, Isoe S (1988) J Organomet Chem 345: C23
46. Magnus P, Roy G (1982) Organometallics 1: 553
47. Yoshida J, Matsunaga S, Isoe S (1989) Tetrahedron Lett 30: 219
48. Stork G, Colvin E (1971) J Am Chem Soc 93: 2080
49. Hudrlik PF, Arcoleo JP, Schwartz RH, Misra RN, Rona RJ (1977) Tetrahedron Lett: 591; Davis AP, Hughes GJ, Lowndes PR, Robbins CM, Thomas EJ, Whitham GH (1981) J Chem Soc Perkin I: 1934
50. Yoshida J, Maekawa T, Isoe S (unpublished results)
51. Yoshida J, Maekawa T, Morita Y, Isoe S (1992) J Org Chem 57: 1321
52. Takai K, Kataoka Y, Okazoe T, Utimoto K (1987) Tetrahedron Lett 28: 1443
53. Sharpless KB, Behrens CH, Katsuki T, Lee AWM, Martin VS, Takatani M, Viti M, Walker FJ, Woodward SS (1983) Pure & Appl Chem 55: 589; Kitano Y, Matsumoto T, Sato F (1988) Tetrahedron 44: 4073
54. Mistunobu O (1981) Synthesis: 1
55. Yoshida J, Shiozawa S, Matsunaga S, Isoe S (unpublished results)
56. Mochida K, Okui S, Ichikawa K, Kanakubo O, Tsuchiya T, Yamamoto K (1986) Chem Lett 805
57. Ricci A, Degl'innocenti A (1989) Synthesis: 647
58. Miller JA, Zweifel G (1981) Synthesis: 288; Miller JA, Zweifel G (1981) J Am Chem Soc 103: 6217; Mandai T (unpublished results, personally communicated)
59. Schäfer HJ (1981) Angew Chem Int Ed Engl 20: 911
60. Stewart RF, Miller LL (1980) J Am Chem Soc 102: 4999
61. Torii S, Okamoto T, Ueno N (1978) J Chem Soc Chem Commun: 293
62. West R (1986) J Organomet Chem 300: 327
63. Boberski WG, Allred AL (1975) J Organomet Chem 88: 65

64. Becker JY, Shakkour E, West R (1992) Tetrahedron Lett 33: 5633; Watanabe H, Nagai Y (1985) In: Sakurai H (ed) Organosilicon and bioorganosilicon chemistry, Ellis Horwood, Chichester, p 107
65. Diaz A, Miller RD (1985) J Electrochem Soc 132: 834
66. Mochida K, Itani A, Yokoyama M, Tsuchiya T, Worley SD, Kochi JK (1985) Bull Chem Soc Jpn 58: 2149
67. Shono T, Ohmizu H, Kise N (1980) Chem Lett: 1517
68. Evans AG, Jerome B, Rees NH (1973) J Chem Soc Perkin II: 447
69. Curtis MD, Allred AL (1965) J Am Chem Soc 87: 2554; Correa-Duran F, Allred AL, Glover DE, Smith DE (1973) J Organomet Chem 49: 353
70. Watanabe H, Aoki M, Matsumoto H, Nagai Y, Sato T (1977) Bull Chem Soc Jpn 50: 1019
71. Allred AL, Bush LW (1968) J Am Chem Soc 90: 3352
72. Paqeutte LA, Wright III CD, Traynor SG, Taggart DL, Ewing GD (1976) Tetrahedron 32: 1885
73. Shono T, Matsumura Y, Katoh S, Kise N (1985) Chem Lett 463
74. Yoshida J, Muraki K, Funahashi H, Kawabata N (1985) J Organomet Chem 284: C33; Yoshida J, Muraki K, Funahashi H, Kawabata N (1986) J Org Chem 51: 3996
75. Bordeau M, Biran C, Pons P, Léger-Lambert M-P, Dunoguès J (1992) J Org Chem 57: 4705; Pons P, Biran C, Bordeau M, Dunoguès J, Sibille S, Perichon J (1987) J Organomet Chem 321: C27
76. Pons P, Biran C, Bordeau M, Dunoguès J (1988) J Organomet Chem 358: 31
77. Fry AJ, Touster J (1989) J Org Chem 54: 4829
78. Ohno T, Nakahiro H, Sanemitsu K, Hirashima T, Nishiguchi I (1992) Tetrahedron Lett 33: 55/5
79. Corriu RJP, Dabosi G, Martineau M (1980) J Organomet Chem 186: 19, 188: 63; Corriu RFP, Dabosi G, Martineau M (1979) J Chem Soc Chem Commun: 457
80. Dessy RE, Kitching W, Chivers T (1966) J Am Chem Soc 88: 453
81. Hengge E, Litscher G (1976) Angew Chem Int Ed Engl 15: 370; Hengge E, Firgo H (1981) J Organomet Chem 212: 155
82. Allred AL, Bradley C, Newman TH (1978) J Am Chem Soc 100: 5081
83. Corriu RJP, Dabosi G, Martineau M (1981) J Organomet Chem 222: 195
84. Kunai A, Kawakami T, Toyoda E, Ishikawa M (1991) Organometallics 10: 893
85. Kunai A, Kawakami T, Toyoda E, Ishikawa M (1991) Organometallics 10: 2001
86. Shono T, Kashimura S, Ishifune M, Nishida R (1990) J Chem Soc Chem Commun: 1160
87. Shono T, Kashimura S, Murase H (1992) J Chem Soc Chem Commun: 896
88. Umezawa M, Takeda M, Ishikawa H, Ishikawa T, Koizumi T, Fuchigami T, Nonaka T (1990) Electrochim Acta 35: 1867; Biran C Bordeau M, Pons P, Léger M-P, Dunoguès J (1990) J Organomet Chem 382: C17
89. Bordeau M, Biran C, Leger-Lambert M-P, Dunoguès J (1991) J Chem Soc Chem Commun: 1476
90. Boudjouk P, Han B-H, Anderson KR (1982) J Am Chem Soc 104: 4992
91. Mairanovskii SG, Ponomarenko VA, Barashkova NV, Snegova AD (1960) Doklady Akad Nauk S.S.S.R. 134: 387 (CA (1961) 55: 16221g); Mairanovski SG, Ponomarenko VA, Barashkova NV, Kadina MA (1964) Izv Akad Nauk S.S.S.R. Ser Khim: 1951 (CA (1965) 62: 6378c)
92. Bordeau M, Biran C, Pons P, Léger M-P, Dunoguès J (1990) J Organomet Chem 382: C21

Electroenzymatic Synthesis

Eberhard Steckhan

Institut für Organische Chemie und Biochemie der Universität Bonn,
Gerhard-Domagk-Str. 1, D-53121 Bonn, FRG

Table of Contents

The use of redox enzymes in organic synthesis, while having a large potential for broad application in the selective formation of high-value compounds, has been limited by the necessity of cofactor regeneration or enzyme reactivation. Electrochemistry offers an attractive and, in principle, simple way to solve this problem because the mass-free electrons are used as regenerating agents. No

Topics in Current Chemistry, Vol. 170
© Springer-Verlag Berlin Heidelberg 1994

co-substrates are needed and no couple products are formed. Therefore, continous processes can be developed much easier than with other techniques. Applications of this promising and therefore fast-developing technique to the synthetic use of reducing and oxidizing enzymes for regioselective redox reactions, the formation of enantiomerically pure products, and also C-C-bond formations are reported. The review treats electroenzymatic oxidations and reductions organized with respect to the cofactors and prosthetic groups of the different redox enzymes. Because this method is just starting to develop, trends for future research are outlined.

1 Introduction

Biocatalysts like isolated enzymes or whole cells have found widespread application in organic synthesis during the last two decades [1]. The interest in these biocatalysts stems from the fact that complex organic compounds can be generated with high enantioselectivity and often fewer ecological problems [2]. Therefore, numerous applications in asymmetric synthesis [3], C–C-bond formation, synthesis and derivatization of carbohydrates [4], peptide synthesis and protective group strategy [5], and redox reactions [6] have been reported. Until now, preparative synthetic applications of isolated enzymes have been concentrated on hydrolytic reactions. Recently, C–C-bond-forming reactions have attracted increasing attention. The synthetic applicability of redox enzymes, especially in the field of enantioselective reductions, soon found interest [6] but is also connected with intrinsic problems. The difficulty with redox enzymes in synthetic processes is based on the fact that this class of biocatalysts is dependent on freely dissociated (like NADH, NADPH) or enzyme-bound (like FMN, FAD, or PQQ) cofactors (Fig. 1) in stoichiometric amounts to shuttle the redox equivalents from the enzyme to the substrate.

Therefore, for preparative applications of redox enzymes, effective and simple methods for the continuous recycling of the active cofactors have to be available. In addition, such systems must be stable over long time periods and the separation of the product must be simple to render technical processes economically feasible. Until now, this problem has generally been solved by the application of a second enzymatic reaction (enzyme-coupled regeneration, Fig. 2).

Thus, besides a production enzyme for the transformation of the substrate into the product, a regeneration enzyme (sometimes the production enzyme also acts as regeneration enzyme) together with its co-substrate is also necessary to reactivate the cofactor of the production enzyme. The additionally formed co-product then has to be separated from the desired product. In some cases, such enzyme-coupled regeneration systems are highly developed. This is, in particular, the case for NAD(P)H dependent enzymes. However, these systems can be complicated and not stable enough for synthetic purposes. This is especially true for the chemical regeneration using reducing or oxidizing agents. Therefore, it would be extremely interesting to use the reagent-free

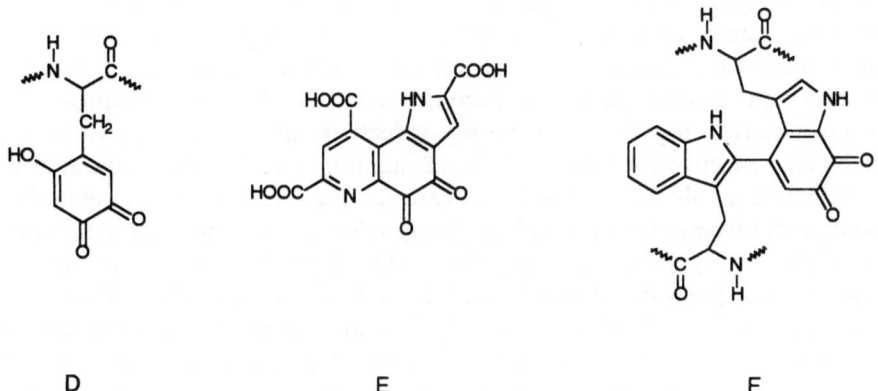

Fig. 1. Prosthetic groups in oxidases (*A*: FAD; *B*: Thio-Tyrosine; *C*: NAD(P)$^+$; *D*: 6-Hydroxy-DOPA; *E*: Methoxanthin (Pyrroloquinoline quinone; PQQ); *F*: Tryptophane-Tryptophan quinone)

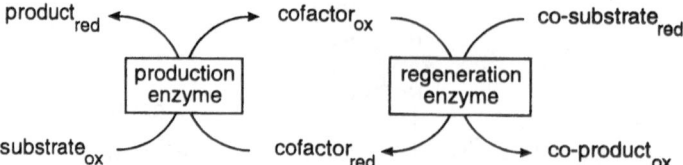

Fig. 2. Principal of an enzyme-coupled cofactor regeneration system for an enzymatic reduction process

electrochemical procedure to regenerate the cofactors. In this case, not only the co-substrate can be replaced by the electrons but also the regeneration enzyme will no longer be necessary. The goal of this review is to present the present status in the field of electroenzymatic reductions and oxidations for the synthesis of complex organic compounds and to point out future chances for their application. The large field of analytical applications of bioelectrochemical systems will not be dealt with. However, it should be pointed out already at this stage that analytical studies and applications present very useful information in the search for synthetic developments.

2 Electroenzymatic Oxidations

The synthetic potential of oxidizing enzymes is extremely large, but has by no means found appropriate applications yet [7]. Especially in the oxidation of alcoholic groups, a large number of applications have been developed or may be foreseen. At first glance, it might look too complicated to use enzymes for the oxidation of alcohols to carbonyl groups on a synthetic scale. Why should it be valuable to apply such complex systems to the transformation of chiral sp^3 carbons to homo- or enantiotopic sp^2 centers while innumerable simple, and easy to use oxidizing agents are available and established in organic synthesis? However, on second glance it becomes obvious that the enantio- and diasteroselectivity together with the regioselectivity of the redox enzymes can also be used effectively for the synthesis of enantiomerically pure compounds in the oxidation of alcohols. This is the case for the enantiomer differentiating oxidation of racemic alcohols and the group selective oxidation of prochiral or *meso*-diols or polyols mainly using the NAD^+-dependent horse-liver alcohol dehydrogenase [8]. On the other hand, the regioselectivity of the oxidation of a particular hydroxy function in a polyol by an enzymatic oxidation can be extremely valuable thus avoiding a complicated protection deprotection sequence. Interesting enzymes for this type of transformation are the NAD^+-dependent glycerol dehydrogenase [9], glycerol-3-phosphate dehydrogenase [10], sorbitol dehydrogenase [11], the hydroxy-steroid dehydrogenases [12], the $NADP^+$-dependent alcohol dehydrogenase from *Thermoanaerobium brockii*

[13], the copper protein dependent galactose oxidase [14], and the FAD-dependent enzymes glycerol-3-phosphate oxidase [15], cholesterol oxidase [16], and pyranose-2-oxidase [17]. Also FAD-dependent alcohol oxidases [18] and PQQ-dependent dehydrogenases [19] may be interesting enzymes for the selective transformation of hydroxy groups into aldehyde functions, carboxylic acids, or ketones. The introduction of hydroxy groups into aliphatic and aromatic compounds can be just as interesting. Nuclear and side-chain oxidations of phenolic compounds by phenol oxidases [20], or p-cresol methyl hydroxylase and 4-ethylphenol methylene hydroxylase [21], respectively, are useful for the synthesis of *ortho*-quinoid systems and chiral p-hydroxy benzylic alcohols. Another interesting class of oxidizing enzymes are amine and amino acid oxidases as well as xanthine oxidase and the similar aldehyde oxidase. These flavoenzymes or copper proteins can be used for synthetically useful desaminations [22], the purification of amino acids from small amounts of the undesired enantiomer [23] or the oxygenation of N-heterocycles [24].

As mentioned before, the main problem for the application of such oxidizing enzymes in organic synthesis is the availability of an effective reactivation method for the prosthetic groups or the freely dissociated cofactors. Especially, the long-term stability of the whole system is the limiting factor [25]. All oxidases mentioned previously can be reactivated by an aerobic

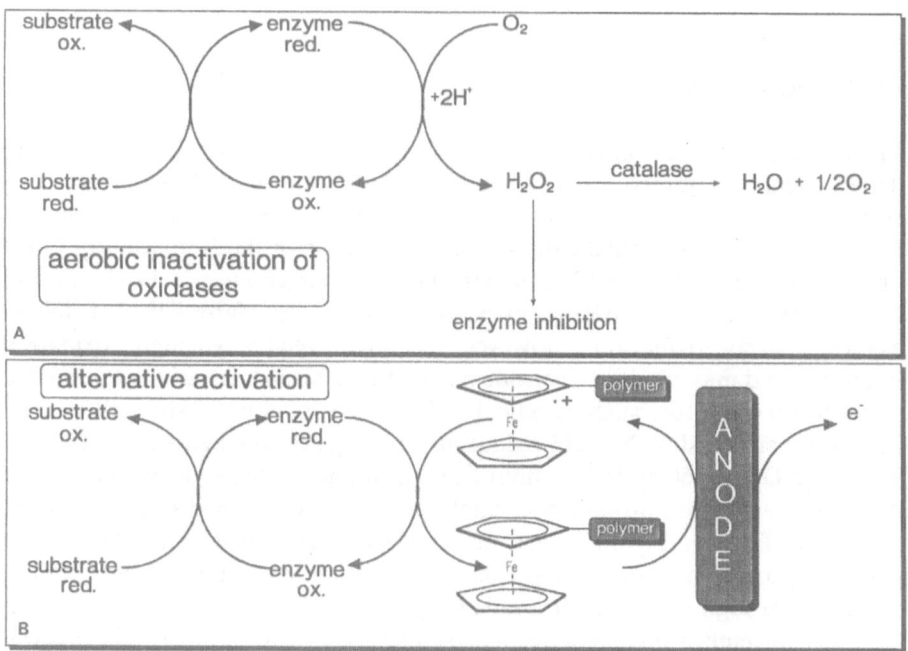

Fig. 3. Aerobic inactivation (**A**) and alternative anaerobic electrochemical activation (**B**)

auto-regeneration sequence with oxygen as final electron acceptor. Usually, hydrogen peroxide is produced via the intermediate superoxide. Hydrogen peroxide and the intermediates, however, are extremely reactive compounds destroying the enzymes after a very short time. The combination of such enzyme systems with catalase to destroy the hydrogen peroxide gives acceptable results in only very few cases [26]. Usually, the life-time of the enzymatic system is not sufficient. In addition to the hydrogen peroxide problem, many enzymes are sensitive to oxygen due to the oxidative degradation of thiol functions, and also the starting materials or the products of the enzymatic reaction may be negatively influenced by oxygen (Fig. 3A). Therefore, the most elegant solution to this problem would be the anaerobic reactivation using the electrode (anode) as final electron acceptor (Fig. 3B).

However, because of the mostly very slow electron transfer rate between the redox active protein and the anode, mediators have to be introduced to shuttle the electrons between the enzyme and the electrode effectively (indirect electrochemical procedure). As published in many papers, the direct electron transfer between the protein and an electrode can be accelerated by the application of promoters which are adsorbed at the electrode surface [27]. However, this type of electrode modification, which is quite useful for analytical studies of the enzymes or for sensor applications is in most cases not stable and effective enough for long-term synthetic application. Therefore, soluble redox mediators such as ferrocene derivatives, quinoid compounds or other transition metal complexes are more appropriate for this purpose.

2.1 NAD(P)$^+$-Dependent Enzymes

In the case of the enzymatic oxidation with NAD$^+$-dependent alcohol dehydrogenases, direct electrochemical oxidation of NADH has been successfully applied. In contrast to the direct electrochemical reduction of NAD(P)$^+$, which is synthetically not useful because of the formation of NAD dimers via intermediate NAD radicals, the direct electrochemical oxidation of NAD(P)H to give NAD(P)$^+$ can be performed successfully [28–32]. Direct electrochemical oxidation, however, requires relatively high oxidation potentials and may result in electrode passivation [28, 34]. By using high surface area carbon electrodes such as reticulated vitreous carbon or especially carbon felt the otherwise slow NADH oxidation can be performed effectively to yield 99.5% NAD$^+$ at 250 cycles per hour [29]. As model systems, the production of D-gluconic acid from β-D-glucose, catalyzed by glucose dehydrogenase, or the formation of pyruvate from lactic acid, catalyzed by lactate dehydrogenase, were performed using the direct electrochemical regeneration of the cofactor NAD$^+$ at a flow-through graphite felt anode (Fig. 4) [30].

The deracemization of lactate, in principle, is also interesting. In this process, pyruvate is reduced at the cathode to racemic lactate which is reoxidized to pyruvate by the cheap L-lactate dehydrogenase combined with the anodic

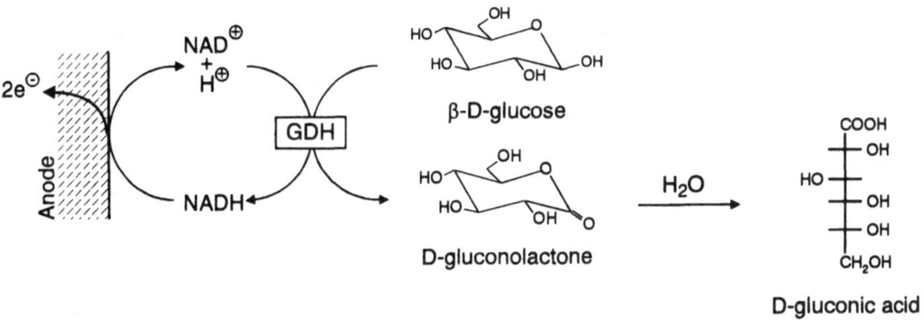

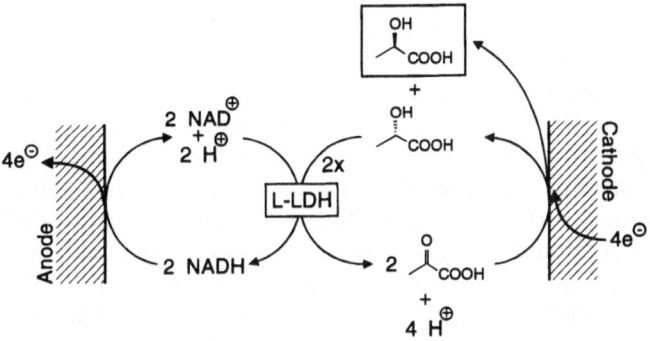

Fig. 4. Electroenzymatic oxidation of β-D-glucose with direct electrochemical NAD⁺ regeneration (GDH = glucose dehydrogenase)

Fig. 5. Coupled electrochemical process for the formation of D-lactate from pyruvate (L-LDH = L-lactate dehydrogenase)

regeneration of the oxidized cofactor NAD^+. As D-lactate is not accepted by the enzyme it is accumulated during the process (Fig. 5) [31].

An immobilized-enzyme continuous-flow reactor incorporating a continuous direct electrochemical regeneration of NAD^+ has been proposed. To retain the low molecular weight cofactor $NADH/NAD^+$ within the reaction system, special hollow fibers (Dow ultrafilter UFb/HFU-1) with a molecular weight cut-off of 200 has been used [32].

Because the direct electrochemical oxidation of NAD(P)H has to take place at an anode potential of $+900\,mV$ vs NHE or more, only rather oxidation-stable substrates can be transformed without loss of selectivity—thus limiting the applicability of this method. The electron transfer between NADH and the anode may be accellerated by the use of a mediator. At the same time, electrode fouling which is often observed in the anodic oxidation of NADH can be prevented. Synthetic applications have been described for the oxidation of 2-hexene-1-ol and 2-butanol to 2-hexenal and 2-butanone catalyzed by yeast alcohol dehydrogenase (YADH) and the alcohol dehydrogenase from *Thermoanaerobium brockii* (TBADH) repectively with indirect electrochemical

89

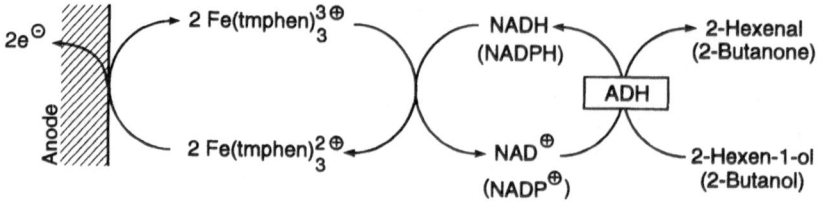

Fig. 6. Schematic representation for the ADH-catalyzed electroenzymatic oxidation of 2-hexene-1-ol and 2-butanol with indirect electrochemical NAD^+ regeneration using (3,4,7,8-tetramethyl-1,10-phenanthroline) iron(II/III) $[Fe(tmphen)_3]$ as redox catalyst

regeneration of NAD^+ and $NADP^+$, respectively, using the tris(3,4,7,8-tetra-methyl-1,10-phenanthroline) iron(II/III) complex as redox catalyst at an anode potential of 850 mV vs NHE (Fig. 6)[33]. Under batch electrolysis conditions using a carbon felt anode the turnover number per hour was 40. The current efficiency was between 90 and 95%.

Mediators acting as one-electron transfer agents towards NADH must possess relatively positive potentials [34]. This is the case for the iron complexes mentioned above with potentials between 800 und 930 mV vs NHE. This limits the application of this system to substrates which do not contain other oxidation labile functions. Therefore, mediator systems with very low oxidation potentials would be interesting for obtaining high chemoselectivities. However, mediators with much lower oxidation potentials cannot react via electron transfer but must act as hydride ion abstracting agents. To reach high reaction rates, the hydride ion abstraction must be fast and very effective. The frequently used chemical regeneration agent for NAD^+ is FMN, a quinoid system. It is mostly applied as a stoichiometric oxidant. In less than stoichiometric amounts, it can be introduced in the presence of oxygen, thus producing hydrogenperoxide, which has to be destroyed by catalase [35].With FMN, however, the problem is a very slow regeneration rate for NAD^+. Therefore, the costly FMN is usually present in very high concentrations and the total turnover number for NAD^+ does not exceed 25 [36]. Many o-quinoid systems are principally quite effective for the NAD^+ regeneration via hydride ion abstraction thus lowering the necessary oxidation potential. A large number of these systems has been studied in the context of the development of an amperometric biosensor for glucose using the NAD^+-dependent glucose-dehydrogenase [37]. Modified electrodes for this analytical purpose have mostly been formed by electrode adsorption of the mediator systems on the electrode surface. Preparative applications have been limited until now due to the limited stability of most of the modified electrodes or the degradation of the quinoid systems under conditions of pH 9 and higher which is the optimum pH value for many dehydrogenases in their oxidative mode. However, some of the systems such as the benzo-phenoxazines or phenoxazines show promising stability at pH 9 and could be interesting for synthetic purposes in the $NAD(P)^+$ regeneration under homogeneous conditions. Another possibility is the use of ferrocene derivatives in combination with a thermostable diaphorase (lipoamide dehydrogenase) which seems to transform the one-electron transfer agent ferrocene into a two-electron oxidant thus accelerating the reaction [38]. A combination of ferrocene derivatives and diaphorase both immobilized at the anode surface has been used for the NAD^+ formation from NADH in the presence of HLADH as production enzyme [39]. The modified electrode was prepared by dip-coating the graphite felt material with poly(acrylic acid) (PAA) of a molecular weight of 1400 kDa in methanol solution followed by a treatment with either aminoferrocene or 2-aminoethyl-ferrocene in DMF in the presence of DCC. After crosslinking the electrodes with hexamethylenediamine, the material contained about 11 μmol/cm^3 of ferrocene derivative and 21–22% free carboxylic acid functions. The oxidation potential of the ferrocene units was $+ 0.24$ and $+ 0.23$ V vs SCE respectively. This material was then treated with diaphorase in the presence of a water soluble carbodiimide. Thus about 8% of the carboxylic functions were modified by diaphorase. This electrode was capable of oxidizing NADH to NAD^+ completely. It is reported to be very stable over several runs. After binding the diaphorase to the PAA network, it was possible to bind HLADH to the remaining free carboxylic acid functions. Thus 3% of the carboxylic functions could be modified by the production enzyme. In the presence of a 1 mM buffer solution of NADH, this ferrocene/diaphorase/HLADH-PAA modified electrode could be applied successfully to the oxidation of cyclohexanol (50 mM) to cyclohexanone or of meso-diols to optically active lactones (Fig. 7).

Other mediators which have been used in combination with diaphorase for the regeneration of NAD^+ are riboflavin and Vitamin K_3, which is 2,3-dimethyl-1,4-naphthoquinone. However, especially riboflavin is not stable enough for synthetic applications [40]. Better stability is exhibited by phenanthrolindiones as mediators. In combination with diaphorase, Ohshiro [41] showed the indirect electrochemical oxidation of cyclohexanol to cyclohexanone using the NAD^+ dependent HLADH with a turnover frequency of two per hour. For an effective enzymatic synthesis, this turnover frequency, however, would be too small. In our own studies, we were able to accelerate the $NAD(P)^+$ regeneration considerably by lowering the electron density within the 1,10-phenanthroline-5,6-dione by complexation with a transition metal ion. In addition, the solubility of the mediators in the aqueous buffer system can be enhanced by this method. The Co^{2+}, Ni^{2+}, and Cu^{2+} complexes with 1,10-phenanthrolin-5,6-dione as a chelating ligand reacted under fast hydride ion transfer from NAD(P)H to the ligand. However, insoluble precipitates of telomers or oligomers of the reduced forms were formed. This behavior was very similar to that of the uncomplexed ligand. Thus, the application of these systems as mediators for the NAD^+ regeneration is not feasible. However, we found two efficient and stable systems for this purpose [42]. The first consists of tris(1,10-phenanthrolin-5,6-dione)Ru(II) perchlorate (1, $E_{p,ox} = - 0.05$ V vs Ag/AgCl). In the second case, we prevented the precipitation of telomers or oligomers of the reduced Co^{2+} complex by using a mixed ligand system consisting of one 1,10-phenanthrolin-5,6-dione (phendi) ligand as catalytically

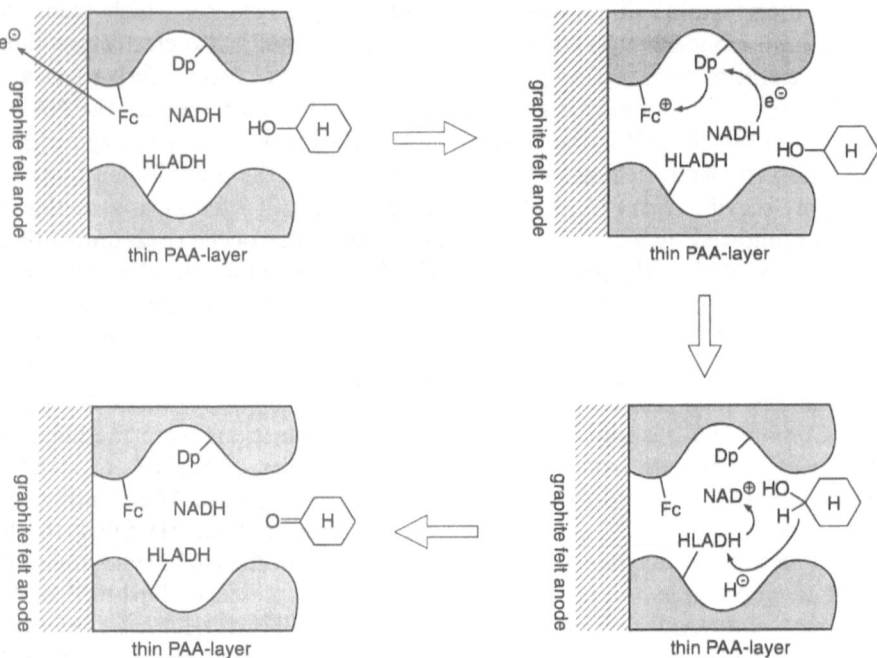

Fig. 7. Horse-liver alcohol dehydrogenase (HLADH) catalyzed alcohol oxidation at a graphite felt anode modified by poly(acrylic acid) (PAA) under coimmobilization of ferrocene derivatives (Fc), diaphorase (Dp), and HLADH [39]

active unit and one *N,N,N*-tris(aminoethyl)amine (tren) ligand to block the other sites of the metal center, [Co(tren)(phendi)(BF$_4$)$_2$ (**2**, E$_{p,ox}$ = + 0.02 V vs Ag/AgCl)]. Cyclic voltammograms showed strong catalytic oxidation peak current enhancements (Fig. 8).

We then coupled the regeneration system **1** to the horse liver alcohol dehydrogenase (HLADH) catalyzed oxidation of cyclohexanol to cyclohexanone as a model system (Fig. 9).

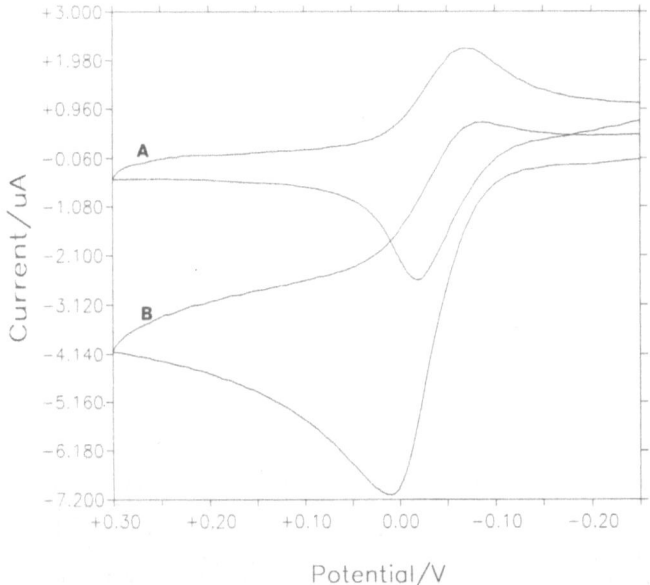

Fig. 8. Cyclic voltammograms of complex **1** (5.7×10^{-5} M in phosphate buffer pH 8.2) in the absence (A) and presence of NADH (B: 7.5 equivalents NADH) (potentials vs Ag/AgCl-reference electrode, scan rate 20 mV/s)

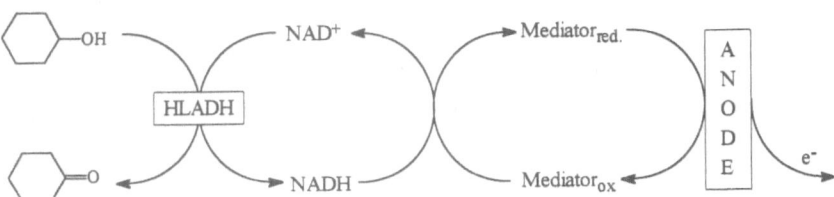

Fig. 9. Schematic representation for the HLADH-catalysed enzymatic oxidation of cyclohexanol with indirect electrochemical NAD^+ regeneration

The time-dependent formation of cyclohexanone according to Fig. 9 is shown in Fig. 10. We obtained a turnover frequency of 28 turnovers per h. This is larger by a factor of 14 than that reported for the indirect electrochemical oxidation using only the free ligand as mediator [41].

In a very special system, the photoelectrochemical regeneration of $NAD(P)^+$ has been performed and applied to the oxidation of the model system cyclohexanol using the enzymes HLADH and TBADH. In this case, tris(2,2'-bipyridyl)ruthenium(II) is photochemically excited by visible light [43]. The excited Ru(II)* complex acts as electron donor for N,N'-dimethyl-4,4'-bipyridinium sulfate (MV^{2+}) forming tris(2,2'-bipyridyl)ruthenium(III) and the MV-cation radical. The Ru(III) complex oxidizes NAD(P)H effectively thus

93

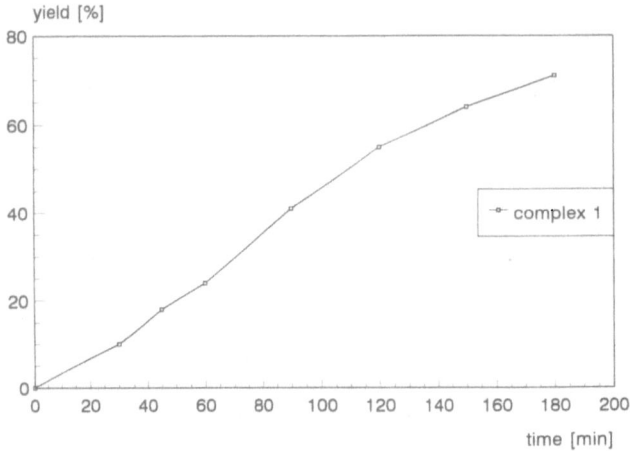

Fig. 10. HLADH-catalyzed conversion of cyclohexanol to cyclohexanone by anaerobic electrochemical regeneration of NAD^+ mediated by complex **1**: 0.1 M phophate buffer of pH 8.0 containing **1** (1×10^{-4} M), NADH (5×10^{-4} M), cyclohexanol (1×10^{-2} M) and HLADH [25 U (EtOH)] at room temperature in an argon atmosphere. The working potential was 100 mV vs Ag/AgCl with a Sigraflex carbon foil anode of 23 cm^2 surface area and a cell volume of 20 ml

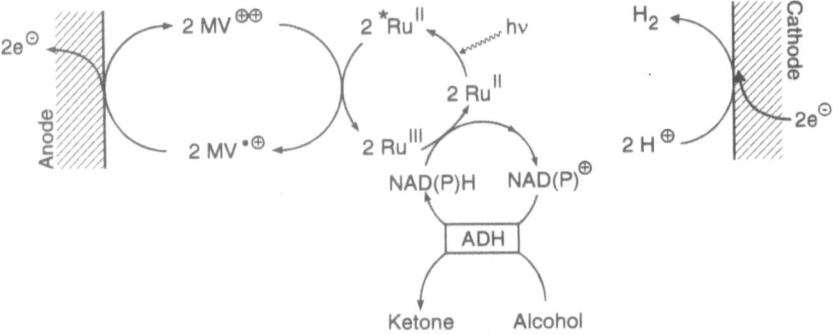

Fig. 11. Application of a double mediator system for the photoelectrochemical regeneration of $NAD(P)^+$ in enzymatic oxidations with dehydrogenases

regenerating the cofactor for the oxidizing enzyme. The MV-cation radical is reoxidized at an anode at a potential around 0 V vs Ag/AgCl (Fig. 11). On the other hand, if the aerobic reoxidation of the MV-cation radical is applied, the system loses its activity at a very fast rate due to the formation of hydrogenperoxide.

2.2 FAD-Dependent and Related Enzymes

In the field of enzymatic oxidations, especially the class of flavo enzymes with bound FAD as cofactor is interesting for synthetic applications. The regeneration of the oxidized FAD within the enzyme can be performed by oxygen. However, as mentioned above, hydrogenperoxide is then formed—this drastically diminishes the enzyme stability and activity, rendering it unsuitable for synthesis. Even the presence of large amounts of catalase to destroy the hydrogenperoxide very often does not lead to a system which is stable enough for the application in synthesis. Therefore, an anaerobic reactivation of flavoenzymes and similar oxidizing enzymes is necessary. The most simple anaerobic oxidation would be the electrochemical procedure. However, as pointed out earlier, large biomolecules do not undergo a fast enough electron exchange with the electrode. For analytical purposes, this limitation can be solved by the application of "promoter" molecules which are adsorbed at the electrode surface [27]. For preparative applications, the most effective way is the indirect electrochemical procedure. As mediators, ferrocenes are most effective with flavo enzymes [44] (Fig. 3).

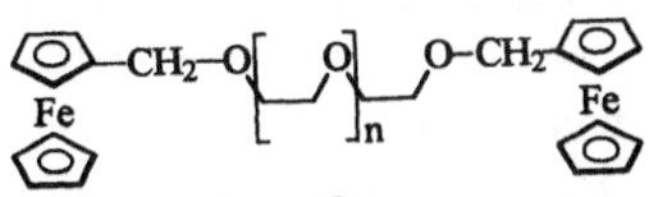

3

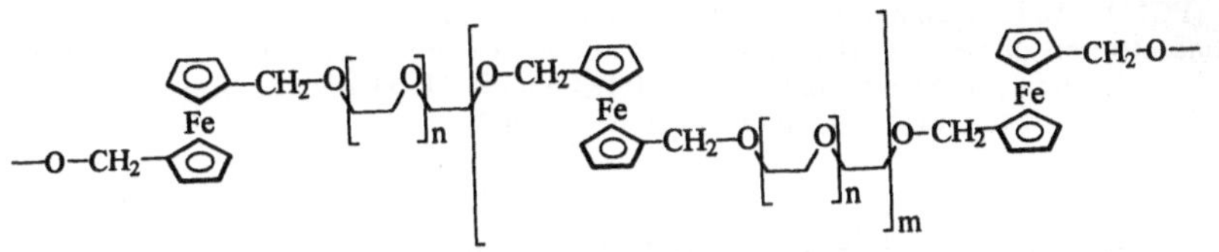

4

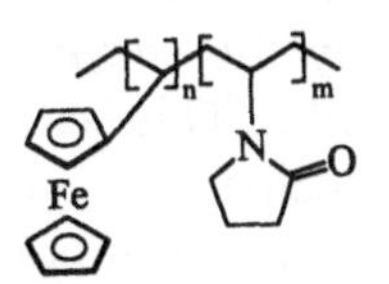

5

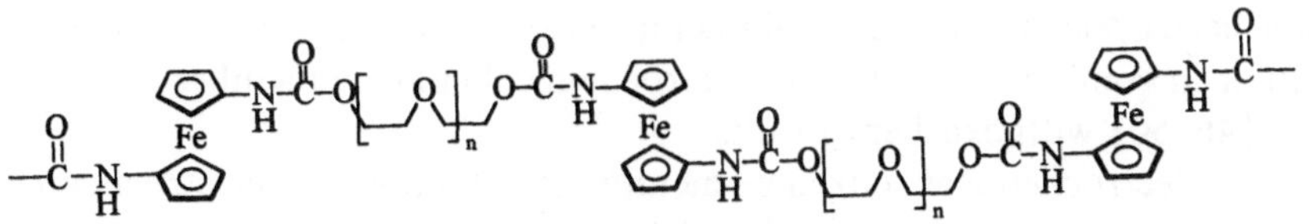

6

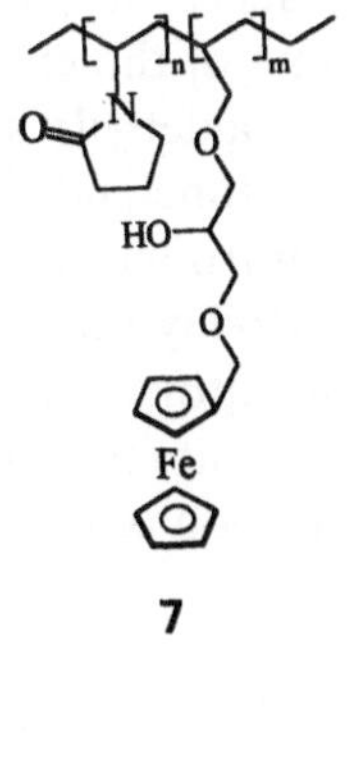

7

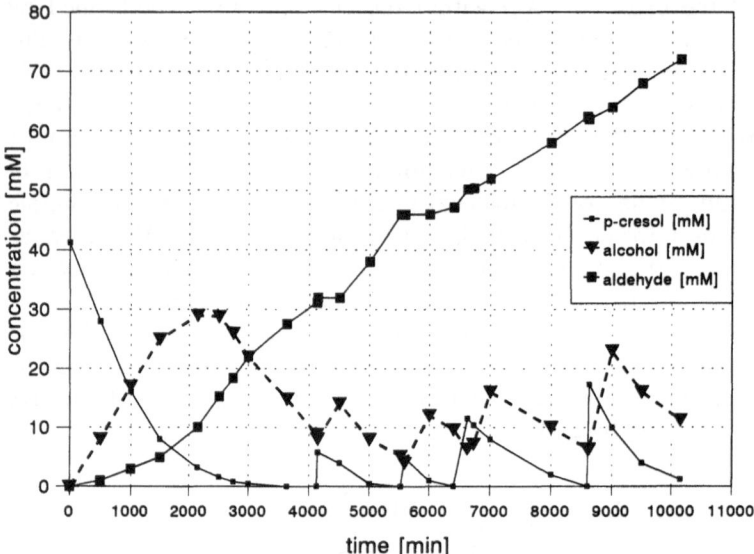

Fig. 12. Electroenzymatic oxidation of *p*-cresol under catalysis by PCMH in a long-time batch electrolysis under formation of *p*-hydroxy benzylalcohol (alcohol) and *p*-hydroxy benzaldehyde (aldehyde) (PCMH: 16 U = 5.6 nmol; PEG-20000 ferrocene **3**: 0.51 mmol = 9.45 μmol ferrocene; starting concentration of *p*-cresol 41.25 mM = 0.66 mmol; additions of substrate after 4140 min (0.0925 mmol), 5590 min (0.0784 mmol), 6630 min (0.184 mmol), 11253 min (0.371 mmol), in 10 ml tris/HCl-buffer of pH 7.6; divided cell: Sigraflex-anode 26 cm^2)

A first application using ferroceneboronic acid as mediator [45] was described for the transformation of *p*-hydroxy toluene to *p*-hydroxy benzaldehyde which is catalyzed by the enzyme *p*-cresolmethyl hydroxylase (PCMH) from *Pseudomonas putida*. This enzyme is a flavocytochrome containing two FAD and two cytochrome c prosthetic groups. To develop a continuous process using ultrafiltration membranes to retain the enzyme and the mediator, water soluble polymer-bound ferrocenes [50] such as compounds **3–7** have been applied as redox catalysts for the application in batch electrolyses (Fig. 12) or in combination with an electrochemical enzyme membrane reactor (Fig. 13) [46, 50] with excellent results.

In an electrochemical enzyme membrane reactor an electrochemical flow-through cell using a carbon-felt anode is combined with an enzyme-membrane reactor. The residence time is adjusted by the flow of the added substrate solution. The off-flow of the enzyme membrane reactor only contains the products *p*-hydroxy benzaldehyde and *p*-hydroxy benzylalcohol. By proper adjustment of the residence time and the potential, total turnover of the *p*-hydroxy toluene, which is introduced into the reactor in 13 mM concentration, can be obtained. In a 10-day run, the enzyme underwent 400 000 cycles and the polymer-bound mediator, which was present in a higher concentration than the enzyme, underwent more than 500 cycles. At the end, the system was still active. By proper selection of the residence time, one can either

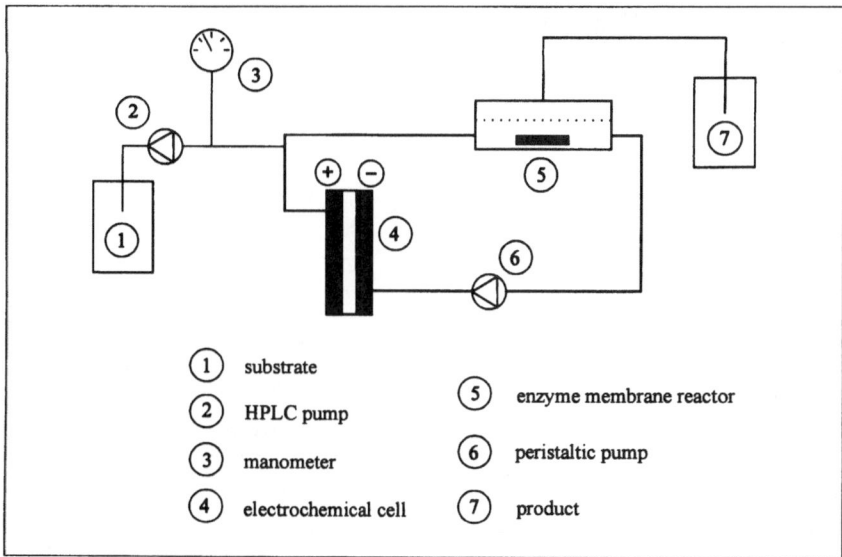

Fig. 13. Electrochemical enzyme membrane reactor (EEMR)

obtain more than 90% of the *p*-hydroxy benzaldehyde at a turnover of more than 95% of the starting material or, at lower residence times and lower turnover of the substrate, the intermediate product *p*-hydroxybenzylic alcohol can be generated as major product [46].

The enzyme *p*-ethylphenol methylene hydroxylase (EPMH), which is very similar to PCMH, can also be obtained from a special *Pseudomonas putida* strain. This enzyme catalyzes the oxidation of *p*-alkylphenols with alkyl chains from C_2 to C_8 to the optically active *p*-hydroxybenzylic alcohols. We used this enzyme in the same way as PCMH for continuous electroenzymatic oxidation of *p*-ethylphenol in the electrochemical enzyme membrane reactor with PEG-ferrocene **3** (MW 20 000) as high molecular weight water soluble mediator. During a five day experiment using a 16 mM concentration of *p*-ethylphenol, we obtained a turnover of the starting material of more than 90% to yield the (*R*)-1-(4'-hydroxyphenyl)ethanol with 93% optical purity and 99% enantiomeric excess (glc at a *β*-CD-phase) (Figure 14). The (*S*)-enantiomer was obtained by electroenzymatic oxidation using PCMH as production enzyme.

Thus, it was shown that flavo enzymes and comparable systems can be used for synthetic applications in a continuous process under anaerobic reactivation of the prosthetic group in the electrochemical enzyme membrane reactor [46, 50].

Another oxidizing enzyme with very interesting synthetic potential is galactose oxidase [14]. This copper protein oxidizes primary hydroxy functions in polyols enantioselectively to the corresponding aldehydes. Thus, sugar alcohols may be transformed into the interesting non-natural L-configurated

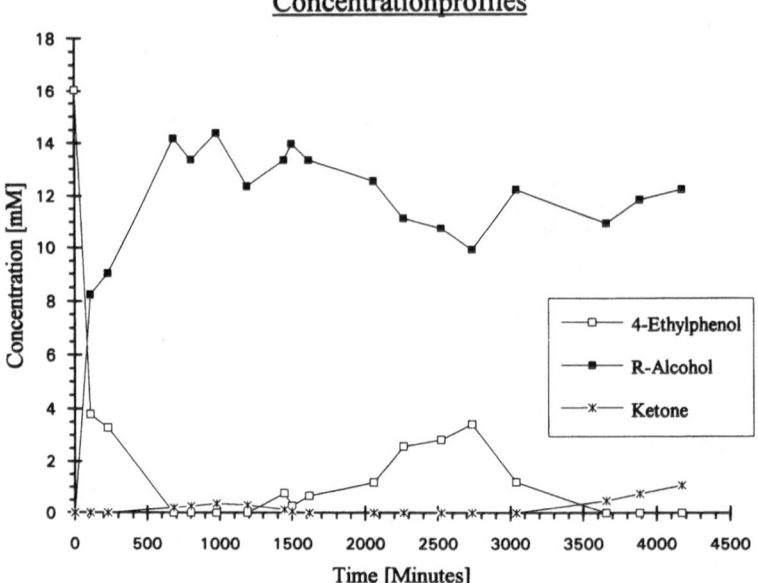

Concentrationprofiles

Fig. 14. Concentration profiles during the continuous indirect electrochemical oxidation of 4-ethylphenol catalyzed by the enzyme EPMH in the electrochemical enzyme membrane reactor

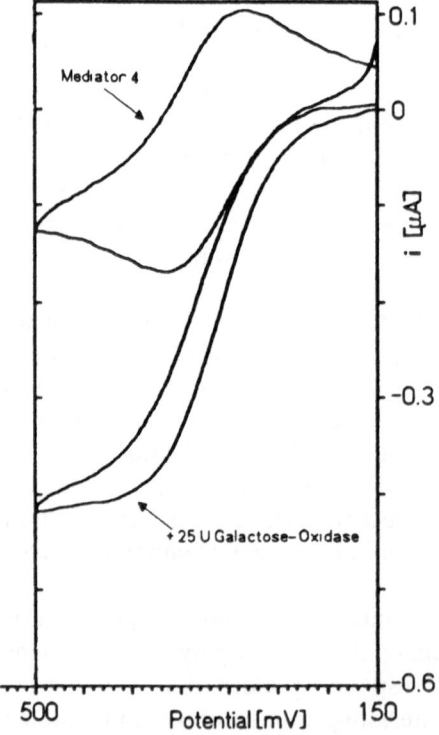

Fig. 15. Cyclic voltammogram of PEG-ferrocene **4** (0.5 mM in ferrocene) in a tris-buffer of pH 9.0 containing xylitol (100 mM) in the absence and presence of 25 U galactose oxidase (potentials vs Ag/AgCl; scan rate 2 mV/s)

carbohydrates. Electrochemical reactivation of this enzyme using mediator systems has been applied in biosensors. As mediators, ferrocenes [47], tetracyano iron 1,10-phenanthroline complexes [48] and the TTF/TCNQ conducting salts [49] have been applied. Recently, we have shown that polymer-bound, water-soluble ferrocenes of high molecular weight like 4 are active mediators for the indirect electrochemical reactivation of galactose oxidase under analytical conditions (Fig. 15) [50]. Strong catalytic effects could be observed with galactose, D,L-threitol, and xylitol in the presence of galactose oxidase. These results are so promising that the application in the above mentioned electrochemical enzyme membrane reactor is under way.

We were also able to electrochemically activate the flavo enzyme L-glycerin-3-phosphate oxidase (GPO) using the polymer-bound ferrocenes 4 and 7 as mediators. Strong catalytic peak current enhancements during cyclovoltammetric measurements could thus be observed with glycerin-3-phosphate in the presence of GPO [50]. GPO is of high synthetic interest because it catalyzes the oxidation of L-glycerin-3-phosphate to dihydroxyacetone phosphate (DHAP) which is the methylene component for the enzymatic C-C-coupling reactions catalyzed by a large number of aldolases. In contrast to other mostly complex or costly methods for the formation of DHAP, this anaerobic indirect electroenzymatic method is a one-step procedure starting from cheap gylcerin-3-phosphate. In a batch electrolysis in a phosphate buffer of pH 7 (18 mL) using polymer-bound ferrocene 4 (1 mM in ferrocene) as redox mediator, a Sigraflex carbon foil anode and a working potential of 500 mV vs. Ag/AgCl, and 180 U GPO (*Pediococcus sp.*, Merck) D,L-glycerin-3-phosphate (390 mM) was effectively transformed to DHAP (90 mM) within 1000 min. DHAP was determined by an enzymatic assay. Finally, the formed DHAP could be used in-situ for the C-C-coupling with propanal in the presence of fructose-1,6-diphosphate aldolase [50]. Thus, continuous aldolase catalyzed reactions in combination with the electroenzymatic generation of DHAP catalyzed by GPO within an electrochemical enzyme membrane reactor can be foreseen.

3 Electroenzymatic Reductions

For enzymatic reductions with NAD(P)H-dependent enzymes, the electrochemical regeneration of NAD(P)H always has to be performed by indirect electrochemical methods. Direct electrochemical reduction, which requires high overpotentials, in all cases leads to varying amounts of enzymatically inactive NAD-dimers generated due to the one-electron transfer reaction. One rather complex attempt to circumvent this problem is the combination of the NAD^+ reduction by electrogenerated and regenerated potassium amalgam with the electrochemical reoxidation of the enzymatically inactive species, mainly NAD dimers, back to NAD^+ [51]. If one-electron

transfer redox catalysts are used as mediators the same problem occurs. The dimerization can only be prevented if an additional regeneration enzyme is applied which must be able to accept two electrons in two steps from the mediator and then to transfer one electron pair to $NAD(P)^+$. This is the case, if ferredoxin-$NADP^+$-reductase (FNR) [52], lipoamide dehydrogenase (diaphorase) [53], enoate reductase [54, 73, 74], or the so-called "VAPOR" enzymes (Viologen accepting pyridine nucleotide oxido-reductases) [55] are used as regeneration enzymes in combination with methyl viologen (N,N-dimethyl-4,4-bipyridinium dichloride) as one-electron mediator.

3.1 Enzyme-Coupled Indirect Electrochemical Regeneration of NAD(P)H

Usually, viologens have been used as one-electron transfer agents in combination with FNR to produce NADPH by an indirect electrochemical way [52]. However, ferredoxin itself (from spinach leaves) can also be used as mediator system for FNR, if special conditions are applied. Ferredoxin is an electron transfer protein with an iron-sulfur cluster as redox center and its surface is negatively charged in neutral solution. Electron transfer can therefore be promoted if the electrode surface is positively charged for example in the presence of bivalent metal ions [56] or by adsorbed or bound viologens [57]. Similarly, ferredoxin does not give an electrochemical response in a buffer solution at an In_2O_3 electrode, however, a well-defined redox reaction was observed in the presence of small amounts of positively charged poly peptides such as poly-L-lysine as promoters. The heterogeneous electron transfer rate constant was $> 5 \times 10^{-3}$ cm s^{-1} and the formal redox potential -0.6 V (vs Ag/AgCl) at pH 7 [58]. Using this system in the presence of FNR, NADPH was effectively produced from $NADP^+$. Ferrodoxin was more efficient than the viologens otherwise used. This indirect electrochemical NADPH production was combined with an NADPH-dependent production enzyme. For example, in the combination with malic enzyme pyruvate was carboxylated by carbondioxide to form malic acid with a current efficiency of 90% on a very small scale (Fig. 15) [59].

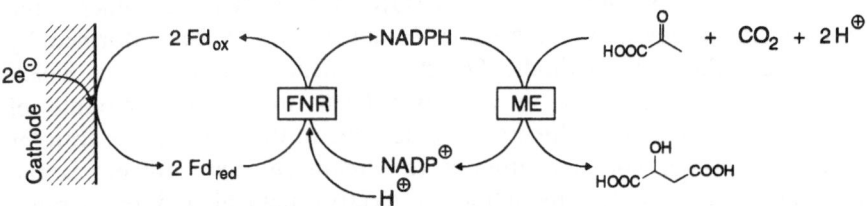

Fig. 16. Ferredoxin (Fd)/ferredoxin-$NADP^+$-reductase (FNR) mediated enzymatic carboxylation of pyruvic acid to form malic acid catalyzed by the NADPH-dependent malic enzyme (ME)

Lipoamide dehydrogenase (diaphorase) can act as electron accepting and transferring enzyme for the formation of NADH using electrogenerated and regenerated viologen radical cations as one-electron donors, if the enzyme is immobilized at the electrode surface. This system works best, if the production enzyme, horse liver alcohol dehydrogenase, is coimmobilized with lipoamide dehydrogenase at the electrode surface [60].

The VAPOR enzymes are flavoenzymes and can be isolated from thermophilic bacilli. They are especially valuable because they allow the regeneration of all four forms of the pyridine nucleotides NADH, NAD^+ NADPH, $NADP^+$ according to the following equations [55, 61]:

$$NAD(P)^+ + 2V^{\cdot +} + H^+ \overset{\text{VAPOR}}{\rightleftharpoons} NAD(P)H + 2V^{++}$$

$$2V^{++} + 2e^- \xrightarrow{\text{cathode}} 2V^{\cdot +} \qquad (V^{++} = \text{viologen})$$

Coimmobilization of the viologen and the VAPOR enzyme at the surface of a carbon cathode can be used for the electrochemical production of NADH from NAD^+ at a rate of $9\ \text{nmol}\,h^{-1}\,cm^{-2}$ [55, 61, 62].

3.2 Indirect Electrochemical NAD(P)H Regeneration Without a Regeneration Enzyme

To be able to regenerate NADP(H) by an indirect electrochemical procedure without the application of a second regeneration enzyme system, the redox catalyst must fulfill four conditions:

1. The active redox catalyst must transfer two electrons in one step or a hydride ion.
2. The electrochemical activation of the catalyst must be possible at potentials less negative than -0.9 V vs SCE since at more negative potentials the direct electrochemical reduction of $NAD(P)^+$ will lead to NAD dimer formation.
3. The active form of the catalyst must transfer the electrons or the hydride ion to $NAD(P)^+$, but not directly to the substrate. Otherwise, this non-enzymatic reduction will lead to low chemoselectivity and/or low enantioselectivity.
4. Only enzyme-active 1,4-NAD(P)H must be formed.

Systems which fulfil these conditions are tris(2,2'-bipyridyl)rhodium complexes [63] and, more effectively, substituted or unsubstituted (2,2'-bipyridyl) (pentamethylcyclopentadienyl)-rhodium complexes [64]. Electrochemical reduction of these complexes at potentials between -680 mV and -840 mV vs SCE leads to the formation of rhodium hydride complexes. Strong catalytic effects observed in cyclic voltammetry and preparative electrolyses are indicating a very fast hydride transfer from the complex to NAD(P)$^+$ under formation of only, 1,4-NAD(P)H and the starting complex as shown in the following reaction scheme [64]:

$$[Cp^*Rh(bpy)L]^{2+} + e^- \rightleftharpoons [Cp^*Rh(bpy)L]^+$$

$$[Cp^*Rh(bpy)L]^+ \rightleftharpoons [Cp^*Rh(bpy)]^+ + L$$

$$[Cp^*Rh(bpy)L]^+ + e^- \rightleftharpoons Cp^*Rh(bpy) + L$$

$$[Cp^*Rh(bpy)]^+ + e^- \rightleftharpoons Cp^*Rh(bpy)$$

$$[Cp^*Rh(bpy)L]^+ + [Cp^*Rh(bpy)]^+ \rightleftharpoons Cp^*Rh(bpy)$$
$$+ [Cp^*Rh(bpy)L]^{2+}$$

$$Cp^*Rh(bpy) + H^+ \rightarrow [Cp^*RhH(bpy)]^+$$

$$[Cp^*RhH(bpy)]^+ + NAD(P)^+ + L \rightarrow [Cp^*Rh(bpy)L]^{2+} + NAD(P)H$$

This system fulfills the four above-mentioned conditions, as the active species is a rhodium hydride which acts as efficient hydride transfer agent towards NAD$^+$ and also NADP$^+$. The regioselectivity of the NAD(P)$^+$ reduction by these rhodium-hydride complexes to form almost exclusively the enzymatically active, 1,4-isomer has been explained in the case of the [Rh(III)H(terpy)$_2$]$^{2+}$ system by a complex formation with the cofactor[65]. The reduction potentials of the complexes mentioned here are less negative than -900 mV vs SCE. The hydride transfer directly to the carbonyl compounds acting as substrates for the enzymes is always much slower than the transfer to the oxidized cofactors. Therefore, by proper selection of the concentrations of the mediator, the cofactor, the substrate, and the enzyme it is usually no problem to transfer the hydride to the cofactor selectively when the substrate is also present [66]. This is especially the case when the work is performed in the electrochemical enzyme membrane reactor.

The successful synthetic application of this electroenzymatic system has first been shown for the in-situ electroenzymatic reduction of pyruvate to D-lactate using the NADH-dependent D-lactate dehydrogenase. Electrolysis at -0.6 V vs a Ag/AgCl-reference electrode of 50 mL of a 0.1 M tris-HCL buffer of pH 7.5 containing pentamethylcyclopentadienyl-2,2'-bipyridinechloro-rhodium(III) (1×10^{-3} M), NAD$^+$ (2×10^{-3} M), pyruvate (2×10^{-2} M), 1300 units D-lactate dehydrogenase (divided cell, carbon foil electrode) after 3 h resulted in the formation of D-lactate (1.4×10^{-2} M) with an enantiomeric excess of 93.5%. This means that the reaction occurred at a rate of 5 turnovers per hour with respect to the mediator with a 70% turnover of the starting material. The current efficiency was 67% [67].

Recently, we adopted the same system for the reduction of 4-phenyl-2-butanone to (S)-4-phenyl-2-butanol using the NADH-dependent horse liver alcohol dehydrogenase (HLADH) and S-ADH from *Rhodococcus sp* [68] with high enantioselectivity (Fig. 17) [69]. As mediator, we applied the low-molecular

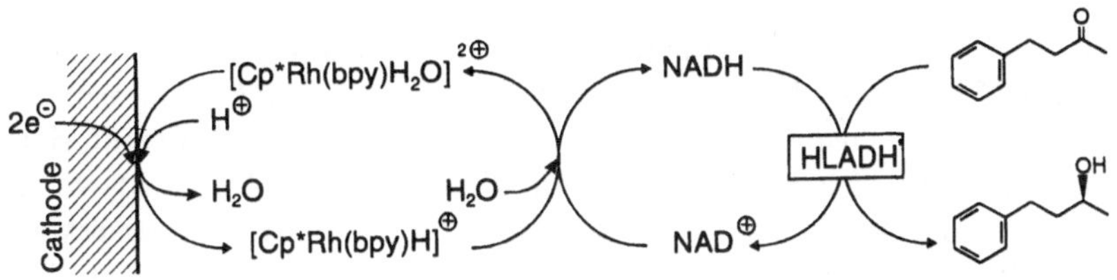

Fig. 17. Electroenzymatic reduction of 4-phenyl-2-butanone catalyzed by HLADH with in-situ indirect electrochemical regeneration of NADH using a Cp*(2,2'-bipyridyl)aquo rhodium(III) complex as mediator

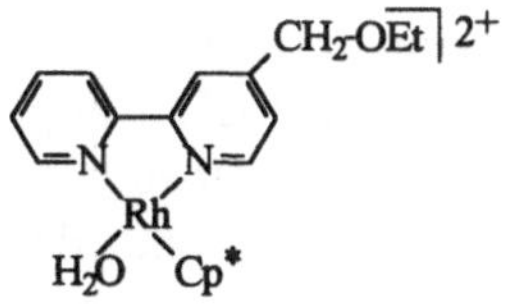

8

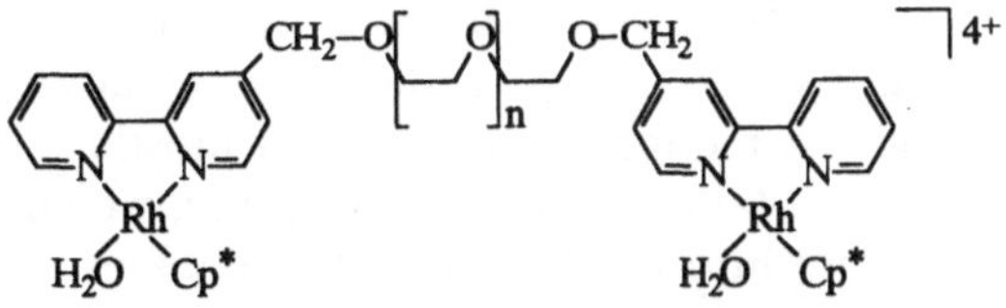

9

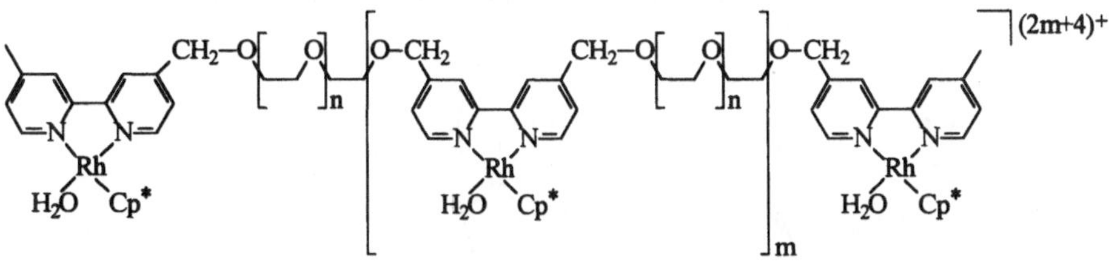

10

weight pentamethylcyclopentadienyl-4-ethoxymethyl-2,2'-bipyridine-chloro-rhodium(III) (**8**) and, to test the possibility for a continuous application in the electrochemical enzyme membrane reactor (see above), the water-soluble polymer-bound rhodium complexes **9** and **10** as mediators. Batch electrolyses of 10 mM solutions of 4-phenyl-2-butanone in 0.1 M tris-HCl buffer of pH 7.5 in the presence of HLADH (50 U, ethanol), mediator **8** (0.5 mM), and NAD$^+$ (2.5 mM) after 5 h resulted in the conversion of 70% of the starting material to the (S)-alcohol with 65% ee. After addition of diaphorase, the turnover increased to 82% with 75% ee. Using S-ADH (30 U with respect to 4-chloroaceto-phenone) and the unmodified Cp*-bipyridine rhodium complex (0.5 mM) under otherwise identical conditions resulted in the conversion of 76% of the starting

103

material to the (S)-alcohol with 77% ee. Continous application of the system in the electrochemical enzyme membrane reactor under application of the polymer bound mediator **10** (0.5 mM), NAD$^+$ (2.5 mM), and S-ADH (100 U with respect to 4-chloroacetophenone) using a 10 mM concentrations of the starting material in tris-HCl buffer of pH 7.5 led to the formation of the (S)-alcohol with 90–95% ee with a conversion between 15 and 30% of the starting material over seven days [69]. Optimization of the system to obtain a higher turnover of the substrate will be necessary.

Several approaches have been undertaken to construct redox active polymermodified electrodes containing such rhodium complexes as mediators. Beley [70] and Cosnier [71] used the electropolymerization of pyrrole-linked rhodium complexes for their fixation at the electrode surface. An effective system for the formation of 1,4-NADH from NAD$^+$ applied a poly-Rh(terpy-py)$_2^{3+}$ (terpy = terpyridine; py = pyrrole) modified reticulated vitreous carbon electrode [70]. In the presence of liver alcohol dehydrogenase as production enzyme, cyclohexanone was transformed to cyclohexanol with a turnover number of 113 in 31 h. However, the current efficiency was rather small. The films which are obtained by electropolymerization of the pyrrole-linked rhodium complexes do not swell. Therefore, the reaction between the substrate, for example NAD$^+$, and the reduced redox catalyst mostly takes place at the film/solution interface. To obtain a water-swellable film, which allows the easy penetration of the substrate into the film and thus renders the reaction layer larger, we used a different approach. Water-soluble copolymers of substituted vinylbipyridine rhodium complexes with N-vinylpyrrolidone, like **11** and **12**, were synthesized chemically and then fixed to the surface of a graphite electrode by γ-irradiation. The polymer films obtained swell very well in aqueous

11　　　　　　　　　**12**

solutions. The penetration of substrates into the film is in some cases almost not hindered at all. It can be influenced by the amount of polymer fixed to the electrode surface and the irradiation dose used for the polymer immobilization. Cyclovoltammetric studies demonstrate the catalytic efficiency of these modified electrodes towards the mediated reduction of NAD^+ to form NADH [72].

3.3 Indirect Electrochemical Activation of Reducing Enzymes Without Reduced Pyridine Nucleotides as Cofactors

There are also some interesting enzyme systems which are able to catalyze stereospecific reductions using electrogenerated viologen cation radicals directly as electron donors even in the absence of reduced pyridine nucleotides. For example, enoate reductase [54, 73, 74] from *Clostridium tyrobutyricum* or from *Clostridium kluyveri* performs stereospecifically a *trans* hydrogen addition to the double bond of a large number of enoates in an electrochemical cell using methyl viologen as mediator. Instead of the isolated enzyme, also resting cells of *C. tyrobutyricum* can be used. Another viologen-accepting enzyme is 2-hydroxy carboxylate viologen oxidoreductase from *Proteus vulgaris*. It is very effective in the reduction of α-keto acids to form chiral (*R*)-α-hydroxy acids using methyl viologen (MV^{++}) as the electron shuttle from the cathode to the enzyme system[54, 73, 74]. This reaction works especially well with resting cells of *Proteus vulgaris* (Fig. 18). Such systems have been named "electromicrobial reductions".

The reduction of carboxylic acids to the aldehydes is catalyzed by reduced viologen accepting tungsten containing enzymes from *Clostridium thermoaceticum* and *Clostridium. formicoaceticum*. This reaction is reversible [73–75]:

$$RCOOH + 2V^{\cdot+} + 2H^+ \xrightarrow{\text{Enzyme}} R–CHO + 2V^{++} + H_2O$$

cathode

2e−

Similarly, the pyruvate dehydrogenase complex (PDC) can be activated directly by electrogenerated methyl viologen radical cations ($MV^{\cdot+}$) as mediator. Thus, the naturally PDC-catalyzed oxidative decarboxylation of pyruvic acid in the

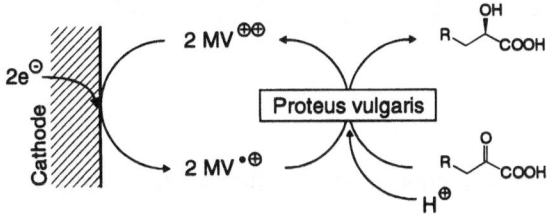

Fig. 18. Electromicrobial reduction of α-keto acids with *Proteus vulgaris* in the presence of methyl viologen as mediator

presence of coenzyme A (HSCoA) to give acetyl-coenzyme A (acetyl-SCoA) can be reversed. In this way, electroenzymatic reductive carboxylation of acetyl-SCoA is made possible:

$$CH_3CO-SCoA + CO_2 + 2MV^{\cdot+} + 2H^+ \xrightarrow{\text{PDC}} CH_3CO-COOH$$

$$+ HSCoA + 2MV^{2+}$$

cathode

$2e^-$

To render this system more effective, it must be coupled with the regeneration of acetyl-SCoA from HSCoA and acetyl phosphate under catalysis by phosphotransacetylase (PTA) (Fig. 19).

A turnover number of 500 for PDC was obtained after 50 h at which point saturation occurred and the reaction stopped. Under the reaction conditions, between 0.55 and 0.98 μmol of pyruvic acid were obtained [76].

The same authors proposed a complex system for the electrochemically driven enzymatic reduction of carbon dioxide to form methanol. In this case, methyl viologen or the cofactor PQQ were used as mediators for the electroenzymatic reduction of carbon dioxide to formic acid catalyzed by formate dehydrogenase followed by the electrochemically driven enzymatic reduction of formate to methanol catalyzed by a PQQ-dependent alcohol dehydrogenase. With methyl viologen as mediator, the reaction goes through the intermediate formation of formaldehyde while with PQQ, methanol is formed directly [77].

Also, the carboxylation of oxoglutaric acid to form isocitric acid catalyzed by isocitric acid dehydrogenase (ICDH) can be driven electrochemically using methyl viologen as mediator without reduced pyridine nucleotides as cofactors. Thus, electrolysis at -0.95 V vs SCE of a 0.2 M tris buffer solution (pH 7.7)

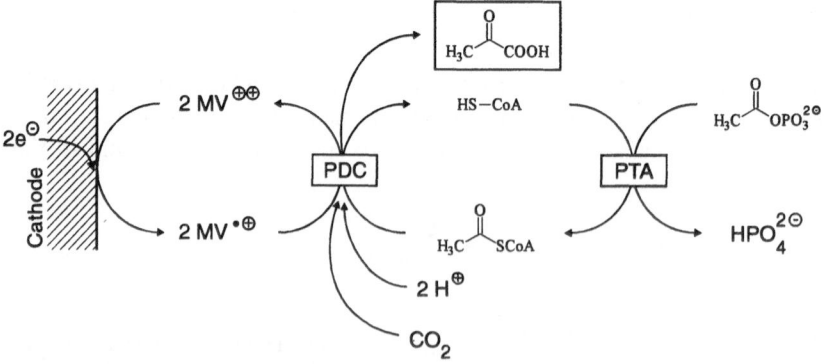

Fig. 19. Reaction scheme for the electrochemically driven enzymatic carboxylation of acetyl-SCoA to form pyruvic acid

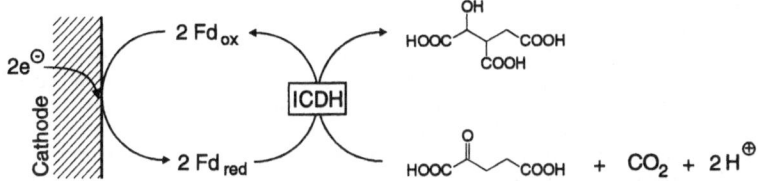

Fig. 20. Electroenzymatic carboxylation of oxoglutaric acid with isocitric acid dehydrogenase (ICDH) using ferredoxin (Fd) as mediator and poly-L-lysine as promoter

containing 0.2 M $NaHCO_3$, 1 unit of ICDH, 1.0×10^{-4} M methyl viologen, and 1×10^{-2} M oxoglutaric acid gave 7.33 μmol of isocitric acid with a current efficiency of almost 100% [78]. Similarly, the ferredoxin system described above (In_2O_3 cathode in the presence of small amounts of poly-L-lysine) can be used as mediator to drive this enzymatic reduction in a more efficient way as indicated by much larger catalytic currents. (Fig. 20) [59].

4 Conclusion

The results presented here demonstrate the first successful applications of the concept of electroenzymatic synthesis. However, it is also obvious that many more systems are potentially very interesting for synthetic applications. This area is wide open for further creative research and only the tip of the iceberg has surfaced. What is especially promising is the fact that electroenzymatic synthesis is an environmentally friendly technique using the electrode as a clean reagent and the enzymes for high selectivity. Closed systems for technical applications are easier to realize than in many other areas. The author is convinced that industrial syntheses of valued small scale products will be seen in the near future.

Acknowledgements. I am greatly indebted to my coworkers, whose names appear in the references of this review article, for their splendid and enthusiastic work in electroenzymatic synthesis. Without their creativity and perseverence our contributions to this field would not have been possible. I am also very thankful for financial support by the Fonds der Chemischen Industrie, the BASF Aktiengesellschft and, in part, the Bundesministerium für Forschung und Technologie. My gratitude also goes to Prof. C. Wandrey, Prof. M.-R. Kula and Priv.-Doz. W. Hummel of the Forschungszentrum Jülich for their kind cooperation.

5 References

1. Suckling CJ, Suckling KE (1974) Chem Soc Rev 3: 387; Jones JB, Sih CJ (1976) In: Perlman D (ed) Techniques of Chemistry, vol 10, part 1 and 2, Wiley, New York; Whitesides GM, Wong

CH (1983) Aldrichimica Acta 16(2): 27; Bürger E (1984) Chem Z 108: 157; Whitesides GM, Wong CH (1985) Angew Chem 97: 617; Angew Chem Int Ed Engl 24: 617; Jones JB (1986) Tetrahedron 42: 3351; Butt S, Roberts SM (1987) Chem Brit 23: 127; Akiyama A, Bednarski MD, Kim MJ, Simon ES, Waldmann H, Whitesides GM (1987) Chem Brit 23: 645; Schneider M, Reimerdes EH (1987) Forum Mikrobiol 65, 302; Yamada H, Shimizu S (1988) Angew Chem 100: 640, Angew Chem Int Ed Engl 27: 622; Davies HG, Green RH, Kelly DR, Roberts RM (eds) (1989) Best Synthetic Methods: Biotransformations in Preparative Organic Chemistry, Academic Press, London; Crout HG, Christen M (1989) In: Scheffold R (ed) Modern Synthetic Methods, vol 5, Springer, Berlin Heidelberg New York, p 1; Wong CH (1989) Science 244: 1145; Poppe L, Novak L (1991) Selective Biocatalysts, VCH, Weinheim

2. Renneberg R (1991) Chem Ind 1991(11): 58; Mechaley GM (1991) Chem Ind 1991(3): 28

3. Pratt AJ (1989) Chem Brit 25: 282; Gais HJ, Hemmerle H (1990) Chem i u Zeit 24: 239; Davies HG, Green RH, Kelly DR, Roberts SM (1990) Crit Rev Biotechnol 10: 129; Faber K, Griengel H (1991) In: Janoscheck R (ed) Chirality, Springer, Berlin Heidelberg New York, chap 6, p 103; Santaniello E, Ferrabochi P, Grisenti P, Manzocchi A (1992) Chem Rev 92: 1071

4. Toone EJ, Simon ES, Bednarski MD, Whitesides GM (1989) Tetrahedron 45: 5365; Waldmann H (1991) Nachr Chem Tech Lab 39: 1408; Bednarksi MD (1991) In: Trost BM (ed) Comprehensive Organic Synthesis, vol 2, Pergamon, Oxford, p 455; Drueckhammer DG, Hennen WJ, Pederson RL, Barbas CF, Gautheron CM, Krach T, Wong CH (1991) Synthesis 1991: 499; Fessner WD (1992) GIT 1992(5): 471; Thiem J, Korf U (1992) Kontakte(Merck) 1992(1): 3; Wong CH, Liu KKC, Kajimoto T, Chen L, Zhong Z, Dumas DP, Liu JLC, Ichikawa Y, Shen GJ (1992) Pure Appl Chem 64: 1197

5. Waldmann H (1991) Kontakte(Merck) 1991(2): 33; Schellenberger V, Jakubke HD (1991) Angew Chem 103: 1440, Angew Chem Int Ed Engl 30: 1437; Jakubke HD (1991) Kontakte(Merck) 1991(3): 61; Jakubke HD (1992) Kontakte(Merck) 1992(1): 46

6. Bowen R, Pugh S (1985) Chem Ind 1985: 323; Simon H, Bader J, Günther H, Neumann S, Thanos J (1985) Angew Chem 97: 541, Angew Chem Int Ed Engl 24: 539; Hummel W, Kula MR (1989) Eur J Biochem 184: 1; Holland HL (1992) Organic Synthesis with Oxidative Enzymes, VCH, Weinheim; Simon H (1992) Pure Appl Chem 64: 1181

7. Frede M, Steckhan E, Hofbauer M (1993) In: Waldmann H, Drauz (eds) Handbook of enzymes in organic synthesis, VCH, Weinheim (in press)

8. Irwin AJ, Jones JB (1977) J Am Chem Soc 99: 556; Jones JB (1985) In: Enzymes in Organic Synthesis, CIBA Foundation Symposium, Pitman, London, p 3; Jones JB (1986) Tetrahedron 42: 3352; Jones JB, Jakovac IJ (1985) Org Synth 63: 10; Jakovak IJ,Ng G, Lok KP, Jones JB (1980) J Chem Soc Chem Commun 1980: 515; Jakovak IJ, Goodbrand HB, Lok KP, Jones JB (1982) J Am Chem Soc 104: 4659; Jones JB, Finch MAW, Jakovac IJ (1982) Can J Chem 60: 2007; Goodbrand HB, Jones JB (1977) J Chem Soc Chem Commun 1977: 469; Jones JB, Jakovac IJ (1982) Can J Chem 60: 19; Bally C, Leuthardt F (1970) Helv Chim Acta 53: 732; Wong CH, Matos JR (1985) J Org Chem 50: 1992; Nakazaki M, Chikamatsu H, Sasaki Y (1983) J. Org. Chem. 48: 2506; Yamazaki Y, Uebayasi M, Hosono K (1989) Eur J Biochem 184: 671; Yamazaki Y, Hosono K (1989) Tetrahedron Lett 30: 5313; Yamazaki Y, Hosono K (1988) Tetrahedron Lett 29: 5769; Santaniello E, Ferraboschi P, Grisenti P, Manzocchi A (1992) Chem Rev 92: 1071; Matos JR, Wong CH (1986) J Org Chem 51: 2388; Grundwald J, (1986) J Am Chem Soc 108: 6732; Gorrebeck C, Spanoghe M, Lanens D, Lemiere GL, Dommisse RA, Lepoivre JA, Alderweireldt, FC (1991) Recl Trav Chim Pays-Bas 110: 231; Snijder-Lambers AM, Vulfson EN, Doddema HJ (1991) Recl Trav Chim Pays-Bas 110: 226

9. Lee LG, Whitesides GM (1986) J Org Chem 51: 25; Lin ECC, Magasanik B (1960) J Biol Chem 235: 1820

10. Wong CH, Whitesides GM (1983) 48: 3199

11. Schmid W, Heidlas J, Mathias JP, Whitesides GM (1992) Liebigs Ann Chem 1992: 95; Borysenko CW (1989) J Am Chem Soc 111: 9275; Eklund H, Horjales E, Jörnvall H, Bränden CI, Jeffery J (1985) Biochemistry 24: 8005

12. Riva S, Bobvara R, Pasta P, Carrea G (1986) J Org Chem 51: 2902; Carrea G, Riva S, Bovara R, Pasta P (1988) Enzyme Microb Technol 10: 333; Green MJ, Abraham NA, Fleisher EB, Case J, Fried J (1970) J Chem Soc Chem Commun 1970: 234; Fried J Green MJ, Nair GV (1970) J Am Chem Soc 92: 4136

13. Lamed R, Zeikus JG (1981) Biochem J 195: 183; Hummel W, Kula MR (1989) Eur J Biochem 184: 1

14. Maradufu A, Perlin AS (1974) Carbohydr Res 32: 93; Schlegel RA, Gerbeck CM, Montgometry

R (1968) Carbohydr. Res 7: 193; Johnson JM, Halsall HB, Heineman WR (1985) Biochemistry 24: 1579; Root RL, (1985) J Am Chem Soc 107: 2997; Drueckhammer DG, Hennen WJ, Pederson RL, Barbas CF, Gautheron CM, Krach T, Wong CH (1991) Synthesis 1991: 499; Klibanov AM, Alberti BN, Marletta MA (1982) Biochem Biophys Res Commun 108: 804

15. Esders TW, Michrina CA (1979) J Biol Chem 254: 2710; Clairborne A (1986) J Biol Chem 261: 14398

16. Khmelnitsky YL, Hilhorst R, Veeger C (1988) Eur J Biochem 176: 265; Yoshimoto T, Ritani A, Ohwada K, Takahashi K, Kodera Y, Matsushima A, Saito Y, Inada Y (1987) Biochem Biophys Res Commun 148: 876; Lee KM, Biellmann JF (1988) Tetradehedron 44: 1135

17. Liu TE, Wolf B, Geigert J, Neidleman SL, Chin JD, Hirnao DS (1983) Carbohydr Res 113: 151; Geigert J (1983) Carbohydr Res 113: 159; Soda K, Yonaha K (1987) In: Kennedy JF (ed) Biotechnology, vol 7a, VCH, Weinheim, p 606; Huwig A, Brans A, Daneel HJ, Köpper S, Giffhorn F (1992) In: Kreysa G, Driesel A (eds) DECHEMA Biotechnol Conferences, vol 5A, VCH, Weinheim, p 49

18. Sahm H, Wagner F (1973) J Biochem 36: 250; Soda K, Yonaha K (1987) In: Kennedy JF (ed) Biotechnology, vol 7a, VCH, Weinheim, p 606; Hamilton GA (1976) In: Jones JB, Sih CJ, Perlman D (eds) Techniques of chemistry - Applications of biological systems in organic chemistry, vol 10, part 2, Wiley, New York, p 875; Farmer VC, Henderson MEK, Russel JD (1960) Biochem J 74: 257; Guillen F, Martinez AT. Martinez MJ (1992) 209: 603

19. Duine JA, Frank J (1980) Biochem J 187: 213; Davis G (1989) In: Turner APF, Karube I, Wilson GS (eds) Biosensors, Oxford Univ Press, Oxford, p 247; Schär HP (1985) FEMS Microbiol Lett 26: 177; Ghisalba O, Schär HP, Tombo GMR (1986) Nachr Chem Tech Lab 34: 973; Groen BW, VanKleef MAG, Duine JA (1986) Biochem J 234: 611; Elferink VHM, Breitgoff D, Kloosterman M (1991) Recl Trav Recl Trav Chim Pays Bas 110: 247; Duine JA (1991) TIBTECH 9: 343

20. Palumbo P, d'Ischia M, Prota G (1987) Tetrahedron 43: 4203; Bhalerao UT, Muralikrishna C, Pandey G (1989) Synth Commun 19: 1303; Kazandjian RZ, Klibanov AM (1985) J Am Chem Soc 107: 5448

21. McIntire WS, Hopper DJ, Singer TP (1985) Biochem J 228: 325; McIntire WS, Bohmont C (1987) In: Edmundson DE, McCormick DB (eds) Flavin and Flavoproteins, de Gruyter, Berlin p 677 McIntire WS, Hopper DJ, Craig JC, Everhart ET, Webster RV, Causer MJ, Singer TP (1984) Biochem J 224: 617; Hill HAO, Oliver BN, Page DJ, Hopper DJ (1985) J Chem Soc Chem Commun 1985: 1469; Reeve CD, Carver MA, Hopper DJ (1989) Biochem J 263: 431; Reeve CD, Carver MA, Hooper DJ (1990) Biochem J 269: 815

22. Battersby AR (1985) In: Ciba Foundation Symposium 111, Pitman, London, p 22; Oda O, Manabe T, Okuyama T (1981) J Biochem 89: 1317; Youdin MBH, Harshak N, Yoshioka M, Araki H, Mukai Y, Gotto G (1991) Biochem Soc Trans 19: 224; Cragg JE, Herbert RB, Kgaphola MM (1990) Tetrahedron Lett 31: 6907; Equi AM, Brown AM, Cooper A, Ner SK, Watson AB, Robins DJ (1991) Tetrahedron 47: 507; Santaniello E, Manzocchi A, Biondi PA, Secchi C, Simonic T (1984) J Chem Soc Chem Commun 1984: 803

23. Parkin K, Hultin HO (1979) Biotechnol Bioeng 21: 939; Nakajima N (1990) J Ferment and Bioeng 70: 322; Nakajima N, Esaki N, Soda K (1990) J Chem Soc Chem Commun 1990: 947; Hecht SM, Rupprecht KM, Jacobs PM (1979) J Am Chem Soc 101: 3982; Demynck C, Bolte J, Hecquet L, Sanaki H (1990) Carbohydr Res 206: 79; Bjurling P (1989) J Chem Soc Perkin 1 1989: 1331; Asada Y, Tanizawa K, Sawada S, Suzuki T, Misono H, Soda K (1981) Biochemistry 20: 6881

24. Tramper J (1987) In: K Mosbacer (ed) Methods in Enzymology, vol 136, Academic, London p 254; Tramper J, Nagel A, van der Plas HC, Müller F (1979) Recl Trav Chim Pays-Bas 98: 224; Van der Plas HC, Tramper J, Angelino SAGF, de Meester JWG, Naeff HSD, Müller F, Middelhoven WF (1984) Innov Biotechnol 20: 93; Kusmierek JT, Czochralska B, Johannson NG, Shugar D (1987) Acta Chem Scand B41: 701; Lee CH, Gilchrist JH, Skibo EB (1986) J Org Chem 51: 4784; Angelino SAGF, Buurman DJ, van der Plas HC, Müller F (1982) Recl Trav Chim Pays-Bas 101: 342; Angelino SAGF, Buurman DF, van der Plas HC, Müller F (1984) J Heterocyl Chem 21: 749; Dastoli F, Price S (1967) Arch Biochem Biophys 118: 163; Pelsy G, Klibanov AM (1983) Biochem Biophys Acta 742: 352

25. Chenault HK, Simon ES, Whitesides GM (1988) Biotechnol Genet Engineering Rev 6: 221

26. Huwig A, Brans A, Daneel HJ, Köpper S, Giffhorn F (1992) In: Kreysa G, Driesel A (eds) DECHEMA Biotechnol Conferences. vol 5A, VCH, Weinheim, p 49, Nakajima N, Conrad D, Sumi H, Suzuki K, Esaki N, Wandrey C, Soda K (1990) J Ferment and Bioeng 21: 939

27. Allen PM, Hill HAO, Walton NJ (1984) J Electroanal Chem 178: 69; Armstrong FA, Hill HAO, Walton NJ (1988) Acc Chem Res 21: 407; Guo LH, Hill HAO, Lawrance GA, Sanghera GS, Hopper DJ (1989) J Electroanal Chem 226: 379
28. Aizawa M, Coughlin RW, Charles M (1975) Biochim Biophys Acta 385: 362
29. Fassouane A, Laval JM, Moiroux J, Bourdillon C (1990) Biotechnol Bioeng 35: 935
30. Bonnefoy J, Moiroux J, Laval JM, Bourdillon C (1988) J Chem Soc, Faraday Trans 1 84:941; Laval JM, Bourdillon C, Moiroux J (1984) J Am Chem Soc 106: 4701; Laval JM, Bourdillon C, Moiroux J (1987) J Biotechnol Bioeng 30: 157; Laval JM, Moiroux J, Bourdillon C (1991) Biotechnol Bioeng 38: 788
31. Biade AE, Bourdillon C, Laval JM, Mairesse G, Moiroux J (1992) J Am Chem Soc 114: 893
32. Coughlin RW, Aizawa M, Alexander BF, Charles M (1975) Biotechnol Bioeng 17: 515
33. Komoschinski J, Steckhan E (1988) Tetrahedron Lett 29: 3299
34. Carlson BW, Miller LL, Neta P, Grodkowski J (1984) J Am Chem Soc 106: 7233; Kitani A, Miller LL (1981) 103: 3595
35. Jones JB, Taylor KE (1973) J Chem Soc Chem Commun 1973: 205; (1976) Can J Chem 54: 2069, 2974; Jones JB, Jakovac IJ (1985) Org Synth 63: 10
36. Lee LG, Whitesides GM (1985) J Am Chem Soc 107: 6999; Chenault HK, Simon ES, Whitesides GM (1988) Biotechnol Genet Eng Rev 6: 221
37. Degrand C, Miller LL (1980) J Am Chem Soc 102: 5728; Fukui M, Kitani A, Degrand C, Miller LL (1982) J Am Chem Soc 104: 28; Lau NK, Miller LL (1983) J Am Chem Soc 105: 5271; Cenas NK, Kanapieniene JJ, Kulys JJ (1984) Biochem Biophys Acta 767: 108; Cenas NK, Pocius AK, Kulys JJ (1984) Bioelectrochem Bioenerg 12: 583; Huck H, Schmidt HL (1981) Angew Chem 93: 421, Angew Chem Int Ed Engl 20: 402; Ueda C, Tse DCT, Kuwana T (1982) Anal Chem 54: 850; Kitani A, Miller LL (1981) J Am Chem Soc 103: 3595; Jaegfeldt H, Torstennson A, Gorton L, Johansson G (1981) Anal Chem 53: 1979; Carlson BW, Miller LL, Neta P, Grodkowski J (1984) J Am Chem Soc 106: 7233; Günther H, Simon H (1987) Appl Microbiol Biotechnol 26: 9; Matsue T, Masayuki S, Uchida I, Kato T, Akiba U, Osa T (1987) J Electronal Chem 234: 163; Gorton L, Johansson G, Torstensson A (1985) J Electroanal Chem 196: 81; Gorton L (1986) J Chem Soc Faraday Trans 1 82: 1245; Gorton L, Bremle G, Csoeregi E, Joensson-Pettersson G, Persson B (1991) Anal Chim Acta 249: 43; Gorton L, Torstensson A, Jaegfeldt H, Johansson G (1984) J Electroanal Chem 161: 103; Bremle G, Persson B, Gorton L (1991) Electroanalysis 3: 77; Persson B, Gorton L (1990) J Electroanal Chem 292: 115; Goss CA, Abruna HD (1985) Inorg Chem 24: 4263; Huck H (1982) Fres Z Anal Chem 313: 548; Atta NF, Galal A, Karagözler E, Zimmer H, Rubinson JF, Mark HB (1990) J Chem Soc Chem Commun 1990: 1347; Gorton L, Csoeregi E, Dominguez E, Emneus J, Petterson GJ, Mark-Varga G, Persson B (1991) Anal Chim Acta 250: 203; Alberry WJ, Bartlett PN (1984) J Chem Soc Chem Commun 1984: 234; Bennetto HP, Dew ME, Stirling JL (1981) Chem & Ind 1981: 776; Bennetto HP, Stirlin JL, Tanaka K (1985) Chem & Ind 1985: 695
38. Matsue T, Yamada H, Chang HC, Uchida I, Nagata K, Tomita K (1990) Biochim Biophys Acta 1038: 29
39. Kashiwagi Y, Osa T (1993) Chem Lett 1993: 677; Kashiwagi Y, Osa T, 183rd Meeting of The Electrochemical Society, Honolulu, Hawaii, U.S.A. May 16–21, Extended Abstracts, vol 93-1
40. Ozawa S, Ikeda T, Senda M (1991) Anal Sci 7 (Suppl Proc Int Congr Anal Sci 1991, Pt 2): 1689; Chem Abstr 117: 22482v; Ikeda T, Kobayashi D, Ozawa S, Yanai H (1993) 183rd Meeting of The Electrochemical Society, Honolulu, Hawaii, U.S.A. May 16–21, Extended Abstracts, vol 93-1
41. Itoh S, Fukushima H, Komatsu M, Oshiro Y (1992) Chem Lett 1992: 1583
42. Hilt G, Steckhan E (1993) J Chem Soc Chem Commun 1993: 1706
43. Ruppert R, Steckhan E (1989) J Chem Soc Perkin Trans 2 1989: 811; Ruppert R, Franke M, Herrmann S, Komoschinski J, Steckhan E (1988) DECHEMA Monogr. 112: 13
44. Cass AEG, Davis G, Green MJ, Hill HAO (1985) J Electroanal Chem 190: 117
45. Hill HAO, Oliver BN, Page DJ, Hopper DJ (1985) J Chem Soc Chem Commun 1985: 1469
46. Steckhan E, Frede M, Herrmann S, Ruppert R, Spika E, Dietz E (1992), DECHEMA Monographien, Vol. 125, pp 712–752, VCH, Weinheim; Frede M, Steckhan E (1991) Tetrahedron Lett 32: 5063; Brielbeck B, Frede M, Steckhan E, Biocatalysis (in print); Brielbeck B, Spika E, Frede M, Steckhan E (1993) Bioforun, GIT Verlag (in print)
47. Dicks JM, Aston WJ, Davis G, Turner APF (1986) Anal Chim Acta 182: 103
48. Johnson JM, Halsall HB, Heineman WR (1985) Biochemistry 24: 1579
49. Hale PD, Skotheim TA (1989) Synth Met 28: C853
50. Frede M (1993) PhD thesis, Bonn
51. Drakesmith FG, Gibson B (1988) J Chem Soc Chem Commun 1988: 1493

52. Ito M, Kuwana T (1971) J Electroanal Chem 32: 415
53. DiCosimo R, Wong C-H, Daniels L, Whitesides GM (1981) J Org Chem 46: 4622
54. Simon H, Bader J, Günther H, Neumann S, Thanos I (1985) Angew Chem 97: 541; Angew Chem Int Ed Engl 24: 539; Thanos I, Bader J, Günther H, Neumann S, Krauss F, Simon (1987) In: Mosbach K (ed) Methods Enzymol 136: 302
55. Günther H, Paxinos AS, Schulz M, vand Dijk C, Simon H (1990) Angew Chem 102: 1075; Angew Chem Int Ed Engl 29: 1053; Nagata S, Feicht R, Bette W, Günther H, Simon H (1987) Eur J Appl Microbiol Biotechnol 26: 263
56. Armstrong FA, Hill HAO, Walton NJ (1982) FEBS Lett 145: 241
57. Landrum HL, Salmon RT, Hawkridge FM (1977) J Am Chem Soc 99: 3154; Crawley CD, Hawkridge FM (1981) Biochem Biophys Res Commun 99: 516; van Dijk C, van Eijs T, van Leeuwen W, Veeger C (1984) FEBS Lett 166: 76
58. Taniguch I, Hayashi K, Tominaga M, Muraguchi R, Hirose A (1993) J Electrochem Soc Jpn 61: 774
59. Taniguchi I (1993) 183rd Meeting of The Electrochemical Society, Honolulu, Hawaii, May 16–21, 1993, Extended Abstracts vol 93-1, p 2272; Taniguch I (1991) Proceedings of the International Symposium on Chemical Fixation of Carbon Dioxide, Nagoya, Japan, December 2–4, 1991, Abstract A 12, p 81; Taniguch I (1993) Proceedings of the 5th International Symposium on Redox Mechanisms and Interfacial Properties of Molecules of Biological Importance, Schultz E, Taniguchi I (eds), The Electrochemical Society, Pennington, NJ, USA, p 9
60. Matsue T, Chang C-H, Uchida I, Osa T (1988) Tetrahedron Lett 29: 1551; Chang C-H, Matsue T, Uchida T, Osa T (1989) Chem Lett 1989: 1119; Chang H-M, Matsue T, Uchida I (1991) In: Little RD, Weinberg NL (eds) Electroorganic Synthesis: Festschrift for Manuel M, Baizer M, Dekker, New York, p 281
61 Nagata S, Günther H, Bader J, Simon H (1987) FEBS Lett 210: 66; Bette H, Günther H, Frank C, Bückmann AF, Simon H (1990) DECHEMA Biotechnol. Conferences vol 4, VCH, Weinheim, p 285; Simon H (1988) GIT Fachz Lab 1988(5): 458
62. Günther H, Paxinos AS, Schulz M, Simon H, DECHEMA Biotechnol Conferences vol 4, VCH, Weinheim, p 281
63. Wienkamp R, Steckhan E (1982) Angew Chem 94: 786; Angew Chem Int Ed Engl 21: 782; Angew Chem Suppl 1982: 1739; Franke M, Steckhan E (1988) Angew Chem 100: 295; Angew Chem Int Ed Engl 27: 265; Ruppert R, Franke M, Herrmann S, Komoschinski J, Steckhan E (1988) DECHEMA Monogr 112: 13
64. Steckhan E, Herrmann S, Ruppert R, Dietz E, Frede M, Spika E (1991) Organometallics 10: 1586; Steckhan E, Frede M, Herrmann S, Ruppert R, Spika E, Dietz E (1992) DECHEMA-Monographien, Vol. 125, pp 723–752, VCH, Weinheim
65. Umeda K, Ikeda H, Nakamura A, Toda F (1992) Chem Lett 1992: 353
66. Westerhausen D, Herrmann S, Hummel W, Stechan E (1992) Angew Chem 104: 1496; Angew Chem Int Ed Engl 31: 1529
67. Ruppert R, Herrmann S, Steckhan E (1987) Tetrahedron Lett 28: 6583
68. Hummel W (unpublished results); Westerhausen D, Herrmann S, Hummel W, Stechan E (1992) Angew Chem 104: 1496; Angew Chem Int Ed Engl 31: 1529
69. Brielbeck B, Frede M, Steckhan E, Biocatalysis, in print; Brielbeck B, Spika E, Frede M, Steckhan E (1993) Bioforum, GIT Verlag, in print; Spika E (1994) PhD thesis, Bonn, in preparation; Spika E, Steckhan (unpublished results)
70. Beley M, Collin JP (1993) J Mol Cat 79: 133
71. Cosnier S, Gunther H (1991) J Electroanal Chem 315: 307
72. Steckhan E, Höfer E, Heineman WR, Ramos B (1993) 44th Meeting of the International Society of Electrochemistry, Berlin, Germany, Sept 5–10, 1993, Book of Abstracts, Abstract P: I.3.3, p 66
73. Simon H (1992) Pure Appl Chem 64: 1181
74. Thanos I, Simon H (1987) J Biotechnol 6: 13; Simon H (1988) GIT Fachz Lab 1988(5): 458
75. White H, Strobl G, Feicht R, Simon H (1989) Eur J Biochem 184: 89; White H, Feicht R, Huber C, Lottspeich F, Simon H (1991) Biol Chem Hoppe Seyler 372: 999
76. Kuwabata S, Morishita N, Yoneyama H (1990) Chem Lett 1990: 1151
77. Yoneyama H, Tsuda R, Nishida K, Kuwabata S (1993) Proceedings of the 5th International Symposium on Redox Mechanisms and Interfacial Properties of Molecules of Biological Importance, Schultz E, Taniguchi I (eds), The Electrochemical Society, Pennington NJ, USA
78. Sugimura K, Kuwabata S, Yoneyama H (1989) J Am Chem Soc 111: 2361; (1990) Bioelectrochem Bioenergy 24: 241

Electrochemistry for a Better Environment

Pierre M. Bersier[1] (formerly Ciba-Geigy), **Lars Carlsson**[2] and **J. Bersier**[1]

[1] Gstaltenrainweg 61, 4125 Riehen, Switzerland
[2] ElectroCell Systems AB, S-18366 Täby, Sweden

Table of Contents

Topics in Current Chemistry, Vol. 170
© Springer-Verlag Berlin Heidelberg 1994

To many, pollution is how the environment pays for industrial inefficiency, motorized transportation, waste consumption, and some agricultural practices. For the (chemical) industry "the solution
to pollution is dilution" is not acceptable any more. The musts for the chemical industry and other
industries in the 1990s and beyond are: (i) avoidance, minimization of wastes and (ii) waste
elimination. Electrochemistry – which uses a mass-free reagent and often does not need additional
chemicals and, on scale-up, produces more acceptable wastes – is inherently an environmentally
friendly technology. Thus preparative electrochemistry in the preventive mode (zero effluent technology = electrosynthesis) and curative mode (waste treatment: elimination, recycling, recovery of
pollutants) offers unique solutions to the problems of industry. Examples from the laboratory, pilot
plant and full-size commercial operation highlight the state-of-the-art, the scope and limitations and
the pros and cons of advanced preparative eectrochemistry in the synthesis of inorganic and organic
species (preventive mode) and the elimination, recovery and recycling of inorganic pollutants and
the destruction of organic pollutants (curative mode). Electrochemical recovery and recycling (e.g. of
metals, salt splitting) become a raw material credit.

1 Introduction

Growing pressure for the preservation of resources and the environment are
enhancing the competitive prospects of electricity driven processes [1]. In this
scenario, electrochemistry by its nature can offer several solutions

– analytical applications to environmental issues,
– nonpolluting fuel synthesis
– cleaner technology for energy production and conversion [2].

The public is so alert that attention to such problems is becoming a common
parameter to be included in any further technological design. The "solution to
pollution is dilution" philosophy is not acceptable any more. Ideal pollution

control technology is recover, recycle, reuse of substances before they become pollutants. Even the more recent philosophy of waste minimization is being transformed now to the much more acceptable approach of pollution prevention and closed loop recycling [3]. Today, companies are looking for new processes that are environmentally friendly and energy efficient. At its best, electrochemistry which deals with a clean reagent – the electron – can fulfil both of these targets.

Landmarks of electrosynthesis are summarized in Fig. 1.

Preparative electrochemistry in both its modes –preventive (electrosynthesis) and curative (elimination, recovery, and recycling of wastes) is finally catching up.

This review article summarizes the broad area of electroorganic synthesis, (selected electroorganic synthetic reactions, with a special emphasis on those that have been commercialized or investigated in pilot plants) and selected applications of electrochemical techniques for waste-water and effluent treatment. There are a number of modern textbooks and updated reviews [4–53] of electroorganic chemistry that include much more detail on organic reactions and their mechanisms than it is appropriate to discuss here.

1801	*Achim von Armin*: Production of a sour smelling product (acetic acid) after dipping the ends of a voltaic pile into beer or wine [3b], cf. also [3a]
1834	Faraday: reduction of acetic acid
1891	*Brown and Walker*: Electrosynthesis of hemiesters of dicarboxylic acids to obtain diesters having twice as many CH_2 groups according to

$$2ROOC - (CH_2)_n - COO^- \rightarrow ROOC-(CH_2)_{2n}-COOR + 2CO_2 + 2e^-$$

1900	First industrial attempts: (a) Reduction: nitro aromatics, unsat. aldehydes/ketones, heterocycles; (b) Oxidation of aromatics
1922	Schall, Melzer at Pt in H_2O, H_2SO_4: transformations of acetic acid to hydroxyacetic acid; of glyoxalic acid to oxalic acid; of formaldehyde to formic acid
1933	Yokahama at PbO_2 in H_2O, H_2SO_4: transformation of cyxclyhexane to cyclohexanol; cyclohexanone, adipic acid, maleic acid
1960	Pioneering industrial works in US companies
1960–1990	Development of more than 100 processes
1800	beginning of organic synthesis
1890	12000 compounds known
1990	6–7000000 compounds synthesized/described
about 1980	Development of *commercialy available cells* for laboratory, bench, pilot plant and production

Fig. 1. Historical landmarks of electrosynthesis

2 Electrolysis

2.1 Fundamentals/Features

In the first edition of "Industrial Electrochemistry", (1985), Pletcher [8a] stated: ⋯ the future of electrochemical technology is bright and there is a general expectation that new applications of electrochemistry will become economic as the world responds to the challenge of more expensive energy, of the need to develop new materials and to exploit different chemical feedstocks and of the necessity of protecting the environment. In the second edition the same author notices seven years later that ⋯ " to our great pleasure ⋯ despite the fact that energy has not become more expensive, the progress in terms of both improved technology and complete new processes and devices is very substantial ⋯ new processes for manufacture of low-tonnage organic and inorganic chemicals, ⋯ many electrolytic processes for effluent treatment, the commercial availability of several families of electrochemical cells, etc ⋯ are symptons of a healthy technology" [10].

Electrolysis is an environmentally friendly technology. Properly designed electrochemical processes which use a mass-free reagent (the electron without the need of additional chemicals at ambient temperature) do not produce unwanted effluents, they often do not need toxic compounds (such as CN^-) and do not introduce hazards of their own.

Advantages of electrochemical processes over chemical methods are the following:

– closer control of each reaction step
– lower processing temperature
– lower costs, cf. Table 1
– fewer waste byproducts
– more often than not, safer operating conditions.

Safer reactions can be substituted for reactions that are currently performed with hazardous reagents, such as lithium aluminium hydride and hydrogen peroxide. In situ generation of strong bases to substitute for reagents like n-butyl lithium, which are dangerous to use on a large scale. Rainer Engels of Royal Dutch/Shell has described the electrosynthesis of isocyanates and carbamates without the use of phosgene [35]. Unlike most redox reagents, electrode reactions do not present toxicity, fire or explosion hazards and do not lead to large volumes of toxic effluents which must be treated prior to discharge [10]. For example in the in situ generation of the potentially hazardous Fremy's salt, as depicted in Scheme 1 for the chlorogenic acid oxidation [47].

Use of stoichiometric quantities of manganese (III) or chromium (VI) would not only be widely uneconomic for most organic oxidations, for example, but the dumping of spent manganese (II) or chromium (III) would be unthinkable.

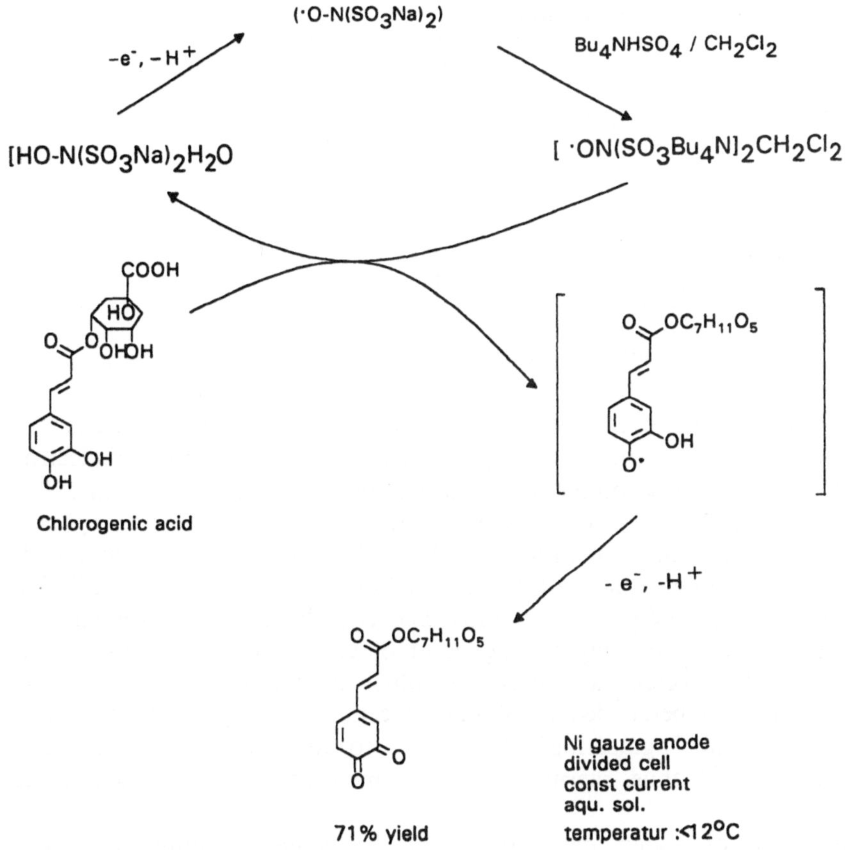

$(\cdot O\text{-}N(SO_3Na)_2)$

$-e^-, -H^+$

$Bu_4NHSO_4 \, / \, CH_2Cl_2$

$[HO\text{-}N(SO_3Na)_2H_2O$

$[\; \cdot ON(SO_3Bu_4N]_2CH_2Cl_2$

COOH

Chlorogenic acid

$O \;\; OC_7H_{11}O_5$

OH

O$\cdot$

$- e^-, -H^+$

$O \;\; OC_7H_{11}O_5$

O

71% yield

Ni gauze anode
divided cell
const current
aqu. sol.
temperatur :≤12°C

Scheme 1. [47]

The use of such reagents in an electrochemical closed loop, however, is entirely practical [35]. In this context see Sect. 3.1.2.1.

Compared on the mole basis, the electron is the cheapest, but also the purest, most versatile redox reagent. Prices for various reagents are summarized in Table 1, (cf. also Table 20).

The *wide applicability* of the electrochemical processes in the chemical processing industry (CPI) derives from the fact that the electron is a versatile reagent. Thus the electron can – unlike standard chemical reagents – be readily removed (oxidation) or added (reduction). Depending on its potential, the electrode can either oxidize or reduce various chemical species to convert them into profitably salable products without the undesirable byproducts.

The *great versatility* of electrochemical processes, on the other hand, comes from the range of electrode sizes available today. Electrodes are routinely used as small as $10^{-12} \, m^2$, e.g. as sensors used for monitoring electrochemical processes [53], or, as large as 16 and more m^2 in production cells for synthesis or

Table 1. Prices of various redox reagents [52]

Reagent	£ per tonne equivalent
Electron	8
Iron powder	27
Zinc dust	29
Sodium borohydride	59
Potassium permanganate	95
Sodium dichromate	164
Lithium aluminium hydride	417

effluent treatment. In a full scale commercial plant, the total electrode area may be greater than 10^4 m^2.

The practical application of electrosynthesis in the industrial scale is still questioned. In 1976 Krumpelt et al. [54a] remarked: " ⋯ Success of electro-organic processes on a commercial scale has been sporadic, except in a few well-known cases." In 1993, electrosynthesis still too often remains unexplored territory for many chemical producing industries. The reason are manifold. Electrochemical engineering is still not a well-defined and accepted discipline outside the circle of the practioners [13]. The available literature is invariably written from a chemical engineering point of view, and thus chemists and industrial workers often experience difficulties in reading the texts. These considerations have persuaded Walsh in his recent monograph "A First Course in Electrochemical Engineering" [13] " ⋯ " to adopt a simple mathematical approach ⋯ ", thus, sacrifying rigour somewhat in order to widen the readership.

The number of modern monographs and the increasing number of papers and of case studies published in the last few years is evidence of the strong, primarily academic interest. Additionally, there is a growing awareness of electrochemical techniques among organic chemists. Normally, there are several possible ways to carry out any transformation. The economic assessment will need to consider the relative merits of all possible syntheses. Generally, electrochemistry will be selected only if it is the cheapest route, and generally, as emphasized by Pletcher and Walsh" ⋯ sometimes only when electrolysis is a technology familiar to the company" [10]. Often, the competition between electrochemical and other routes is not well established. On the other hand – except electricity producers – the chemical industry in general is very reluctant to provide information on their present activity in this field, thus there is a lack of documented case histories of processes on an industrial scale.

For comparison, the processes are best divided into two classes [31]:
The first class: There are no technical convenient alternative processes or the conventional route causes severe environmental problems.

Examples:
(1) Selective reduction at a Pb cathode in H_2SO_4 of phthalic anhydride (**1**) to yield dihydrophthalic acid (**2**). According to the literature the only other

process is the reduction with sodium amalgam [31]. The estimated production amounted to about 100 tonnes per year, cf. Table 8. see also Table 5. The electrochemical process is probably discontinued due to loss of product market,

(1)   **(2)**

cf. Ref. [10].

(2) Electrofluorination of aliphatic carboxylic and sulfonic acid chlorides or fluorides to perfluorinated products [31].

$$C_7H_{15}COCl + 16\ HF \xrightarrow{\ 30e^-\ } C_7F_{15}COF + 15H_2 + HCl$$

The conventional procedure involves 5 steps via telomerization of perfluoro-alkyl iodide with perfluoroethylene and ethylene followed by oxidation. The estimated scale of perfluorination of $C_8H_{17}COOH$ or CH_3SO_3H at a Ni anode in liquid HF in the US, Japan and UK amounts to about 100 tonnes per year cf. Table 8.

(3) Environmental problems are often caused by the use of auxiliary materials, e.g. processes using Clemmensen reduction with amalgamated Zn as reducing agent [31, 35].

	Clemmensen	Electrochemical
Yield	ca. 95%	95%
Reduction	1.7 kg Zn/Hg/kg conc. HCl	5.5 kWh/kg
Pollution	1 kg Zn^{2+}/kg; 0.01 kg Hg/kg Zn^{2+} = Grey List of EC must be removed as far as possible Hg^{2+} = Black List of the EC	

The second class: One or more conventional routes compete with the electrochemical route. Typical advantages and disadvantages are decisive for the process.

Examples:

(1) Synthesis of 2,5-dimethoxy-2,5-dihydrofuran (desinfectant) [4] [31, 33].
Conventional

(3)   **(4)**

119

Electrochemical

(3)   (4)

	Conventional	Electrochemical
Yield	70–75%	85%
		Current yield: 80–85%
		Energy consumption 2.5 kWh/kg
Auxiliary material:	1.6 kg Br$_2$/kg	0.005 kg Br$^-$/kg

The classical synthesis needs more than equimolar amounts of bromine, whereas the electrochemical synthesis needs only traces of bromide and electrical energy [31]. Estimated scale of the electrochemically produced (4) is about 100 tonnes per year in Europe. (4) is commercially produced by BASF [54].

(2) Reduction of anthranilic acid (5) to o-aminobenzylalcohol (6). The expensive LiAlH$_4$ can be used as reducing agent for valuable fine chemicals, e.g. pharmaceuticals [31].

(5)   (6)

Conventional	Electrochemical
Yield 95%	about 50%
Reduction 0.8 kg LiAlH$_4$/kg	about 25 kWh/kg
	Price: 0.1DM/kWh

2.1.1 Catalytic vs Electrochemical Processes [48]

Competion between catalytic and electrochemical processes, for instance, is heating up.

Choosing between the two technologies can mean trade-offs. When working with complex molecules, the high temperatures needed for conventional heterogeneous catalytic reactions can affect functional side groups. Proponents see electrodes as "tuneable catalysts" for redox reactions. Reactions can be driven by adjusting the "overvoltage" (overvoltage = difference between applied voltage and the desired reaction equilibrium potential). Similarly, although electrosynthesis is run at low temperatures, the high energy intermediates it forms can also affect the side group. Thus, opponents view the electrode as a "sledge hammer" capable only of hammering out high energy intermediates, such as cation radicals, during the electrontransfer reactions (Blum. cf. Ref. [48]).

Many remarks advanced against heterogeneous catalytic processes do not apply to homogenous catalysis. According to opponents only few electrochemical processes are so far really competive with corresponding catalytic processes.

Experts also disagree on how easily electrochemical processes can be scaled up from laboratory to plant scale. Some see electrochemical processes limited in that the scale is only the two-dimensions based on the electrode surface area, whereas conventional catalytic processes are scaled in three dimensions based on the volume of the reaction unit [48].

Scale up need, however, as pointed out by Samdani and Gilges [48], not be an issue, as commodity products such as adiponitrile (200 000 t/year), chloralkali, or chlorate are already produced on a very large scale electrochemically.

In 1990 about 6–7 million organic compounds were known or had been synthesized. About 60 000 are in practical use. A fair estimate indicates that about the same number of compounds or metabolites reach the environment via the atmosphere or wastewaters or effluents. Most organic compounds are manufactured on a small scale. In 1975, only about 200 organic components were made in the USA on a scale exceeding 10 000 t/year, thousands were made in smaller quantities [48, 55]. Product numbers in 1982 given by the EC are as follows [56]

1000 products		< 100 t/year
190	about	100 t/year
44		> 1000 t/year
25		> 10000 t/year

2.1.2 Paired Reactions [32, 57]

Electrochemical processes can be made more competitive when products are made at both electrodes simultaneously. Paired electro-organic reactions are divided into three categories.

1) Both the anodic and cathodic reactions are employed to give the final product (Fig. 2A).
2) The anodic and cathodic reactions lead to two different desired inorganic or organic products (Fig. 2B, C). Problems arise, when there is a major imbalance in demands of the two products, e.g. chlorine and caustic soda.

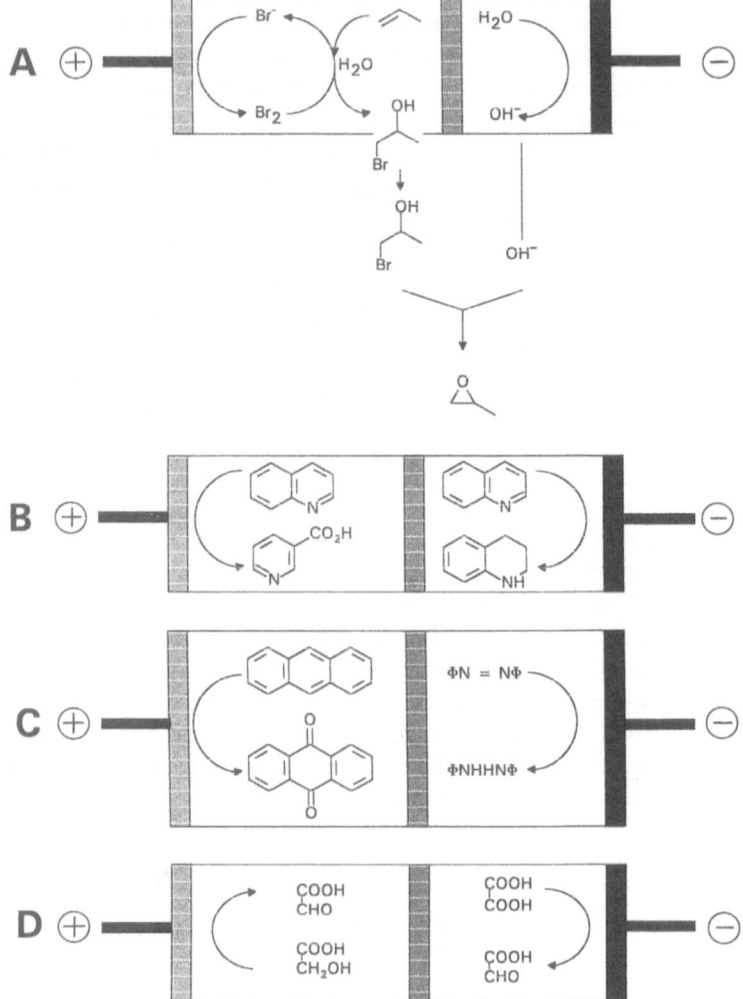

Fig. 2. Types of "paired synthesis" [57]

3) Both the anodic and cathodic reactions lead to the same final product, (Fig. 2D), e.g. simultaneous anodic and cathodic production of glyoxylic acid [57].

Scheme 2 summarizes the electrosynthesis of M-hydrol, using both the anodic and cathodic cell compartments for the production of the same compound.

Paired synthesis can be academically very elegant and industrially interesting because the cost of the process is shared between two products and the production per cell is increased [57].

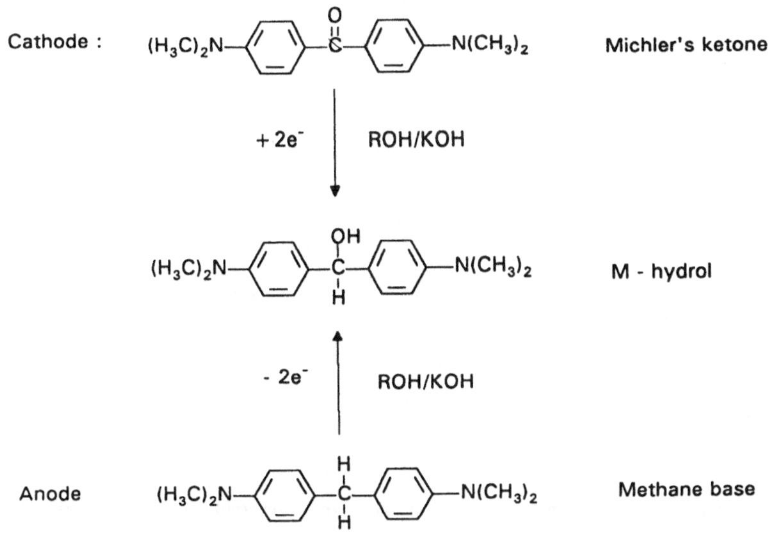

Cathode :   (H$_3$C)$_2$N—⟨benzene⟩—C(=O)—⟨benzene⟩—N(CH$_3$)$_2$ **Michler's ketone**

$+ 2e^-$ | ROH/KOH

(H$_3$C)$_2$N—⟨benzene⟩—C(OH)(H)—⟨benzene⟩—N(CH$_3$)$_2$ **M - hydrol**

$- 2e^-$ | ROH/KOH

Anode   (H$_3$C)$_2$N—⟨benzene⟩—C(H)(H)—⟨benzene⟩—N(CH$_3$)$_2$ **Methane base**

Scheme 2. Paired reaction: Electrochemical alternative for the M-hydrol synthesis (Chemical reaction: reduction with Zn/ROH; yield 90%)

Reaction of the three types have been evaluated in the laboratory, on the pilot plant scale and for commercialization, as illustrated by the following examples.

Preparation of propylene oxide. The mechanism suggested involves a hypobromination:

at the anode:

$$Br^- - e^- \rightarrow 1/2\ Br_2$$

$$Br_2 + H_2O \rightarrow HOBr + HBr$$

$$HOBr + CH_3–CH = CH_2 \rightarrow H_3C - \underset{\underset{OH}{|}}{CH} - CH_2Br$$

and the cathode

$$H_2O + e^- \rightarrow OH^- + 1/2\ H_2$$

This is followed by the reaction of the anodic and cathodic products.

$$OH^- + H_3C - \underset{\underset{OH}{|}}{CH} - CH_2Br \rightarrow H_3C - \underset{\diagdown O \diagup}{CH - CH_2} + H_2O + Br^-$$

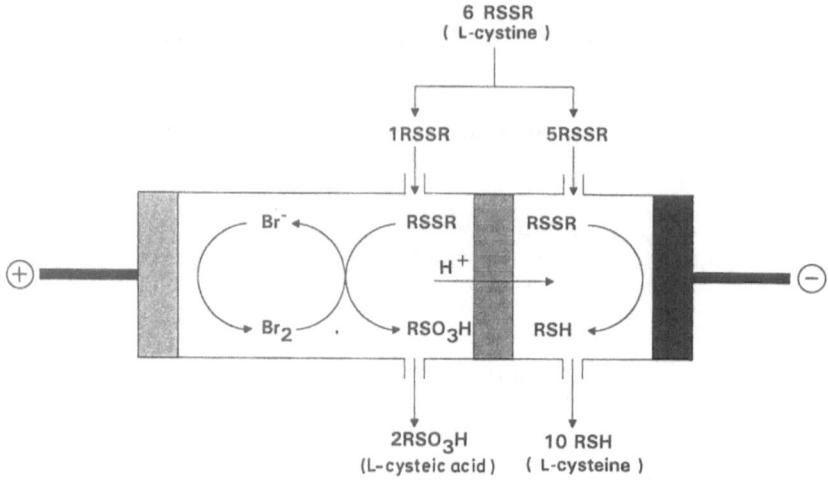

Fig. 3. Process flow for the paired synthesis of cysteic acid and L-cysteine. Partial and overall reactions [57]

Table 2. Comparison of "paired" (B) and individual syntheses (A, C) [57]

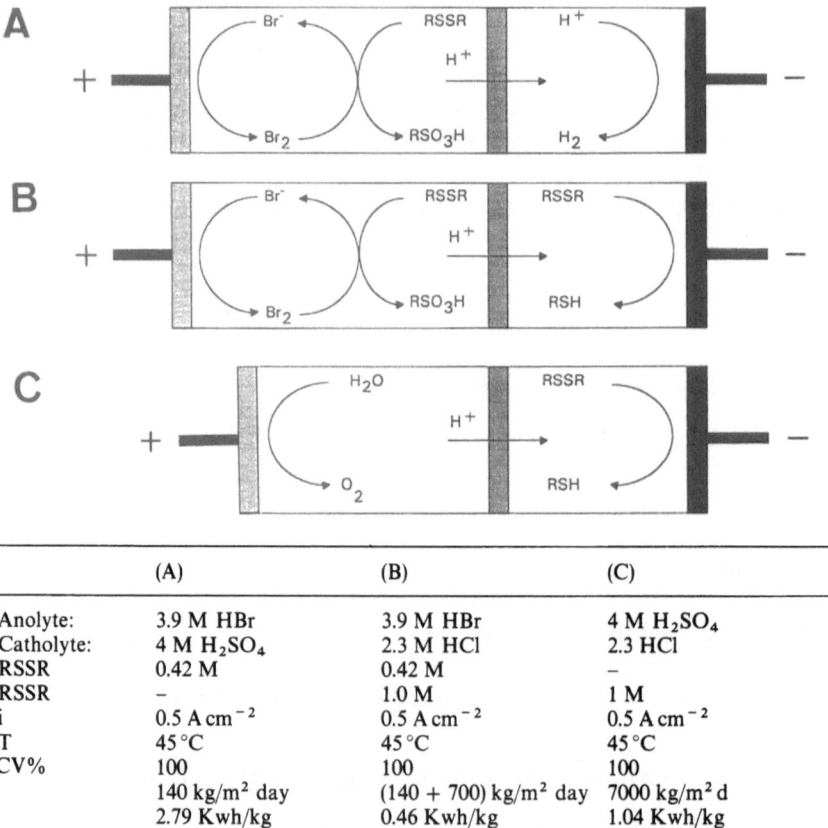

	(A)	(B)	(C)
Anolyte:	3.9 M HBr	3.9 M HBr	4 M H_2SO_4
Catholyte:	4 M H_2SO_4	2.3 M HCl	2.3 HCl
RSSR	0.42 M	0.42 M	–
RSSR	–	1.0 M	1 M
i	0.5 A cm^{-2}	0.5 A cm^{-2}	0.5 A cm^{-2}
T	45 °C	45 °C	45 °C
CV%	100	100	100
	140 kg/m² day	(140 + 700) kg/m² day	7000 kg/m² d
	2.79 Kwh/kg	0.46 Kwh/kg	1.04 Kwh/kg

New "paired reaction" processes of type 2 being examined currently are:

cathode: ethylene glycol; anode: terephthalic acid [60]
cathode: H_2O_2; anode: O_3
cathode: H_2O_2; anode: chlorate

Sanchez–Cano et al. have proposed paired synthesis for obtaining L-cysteic acid and L-cysteine from L-cystine which greatly improves the economical parameters [57]. The global process-flow for the paired synthesis, with L-cystine and water as starting materials is shown in Fig. 3. Table 2 compares the results for the paired (B) and the individual syntheses (A, C).

An example of industrial interest is the benzanthrone (9) synthesis. Benzanthrone derivatives are manufactured by cathodically reducing anthraquinone derivatives that may contain electronegative substituents [61]. In the *cathode* compartment the reduction of anthraquinone (7) in 85% H_2SO_4 to oxanthrone (8) occurs which in presence of glycerol reacts to form benzanthrone (9), which is an important dye intermediate [40, 61].

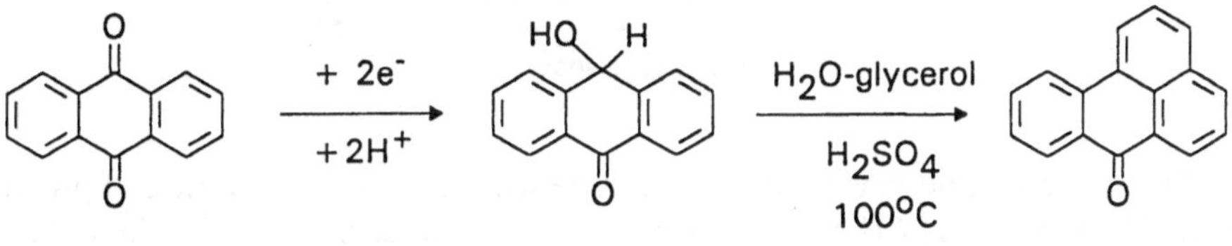

In the *anode* compartment the oxidation of Mn(II) to Mn(III) takes place which can be used for the synthesis of dioxoviolanthrone (10), which is reduced with $NaHSO_3$ to dihydroxyviolanthrone (11), an intermediate for vat dyes [40, 62].

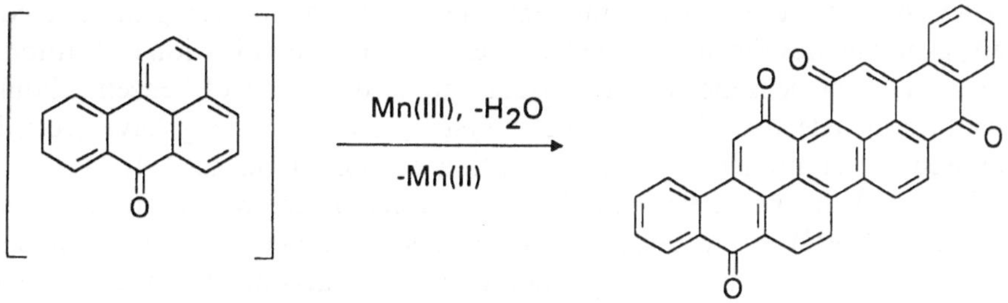

Direct production of benzoquinone (BQ) from benzene is one of the targets in industrial chemistry. Considerable efforts have been made to develop the electrochemical oxidation of benzene to p-benzoquinone to the industrial scale thus forming a basis for a new hydroquinone process [40]. Benzene in aqueous emulsions containing sulfuric acid (1:1 mixture of benzene and 10% aqueous H_2SO_4) forms, at the anode, p-benzoquinone which can be reduced cathodically to yield hydroquinone in a paired synthesis. A divided cell with PbO_2 anodes is used.

Anode : benzene $\xrightarrow[\text{PbO}_2 \text{ anode}]{\text{H}_2\text{O - H}_2\text{SO}_4}$ benzoquinone

Cathode : benzoquinone $\xrightarrow[\text{Pb cathode}]{\text{H}_2\text{O - H}_2\text{SO}_4}$ hydroquinone

Conversion of benzene :	< 10%
Selectivity for hydroquinone :	79%
Current efficiency :	38%

The process is, however, not competitive with the H_2O_2 oxidation of phenol or the Hock process [40]. Selectivities up to 90% and current efficiencies of 80% were reported by Dow using special porous electrodes of PbO_2 and polytetrafluoroethylene. Danly has published a project study for a 10000 tons/year hydroquinone plant [63]. Pistorius and co-workers chose a route via tetramethyl ketal intermediates. Benzene treated at 25 °C with methanol in an electrolytic cell containing $Me_4P^+F^-$ gave up to 80% benezene conversion to p-benzoquinone bis(dimethyl acetal), with a current efficiency of 48%, [65a, 65b]. Similar treatment of 2-ClC$_6$H$_4$OMe [65a] and (3-MeC$_6$H$_4$)$_2$O [65a, 65b] gave similar results. Better results (ca. 85% conversion, with a currunt efficiency of 55%) were obtained with anisole as starting material [65d]. A great deal of effort to optimize the electrolytic conditions and to improve the Faradaic yield to near the theoretical value has been made, but in vain [66–68]. Degner [64] has reviewed the different direct and indirect (using metal mediators) processes described in the patent literature. The combination of "cathodic oxygenation" of benzene with the usual anodic oxidation of benzene should constitute an effective electrosynthetic system. Ito et al. [66] have recently proposed the combination of the oxidation of benzene using Ag(I)/Ag(II) as mediator with the cathodic oxidation of benzene via the reduction of oxygen using Cu(I)/Cu(II) as mediator in a single electrolytic cell to produce p-benzoquinone selectively in both the anodic and the cathodic chamber, Scheme 3 (duet electrosynthesis). cf. also Ref. [66a]

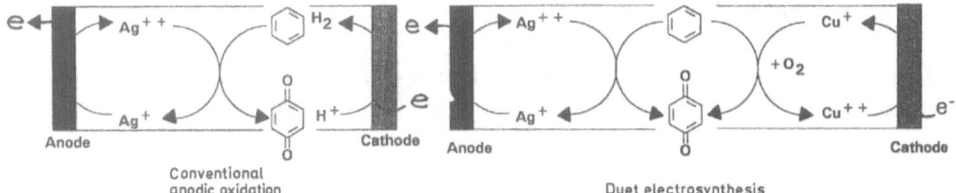

Scheme 3 Anodic oxidation   Paired (duet) electrosynthesis

The actual oxidants in the *cathode* chamber are hydroxyl radicals, which are formed according to the following reactions [66, 67]

$$2Cu(I) + O_2 + 2H^+ \rightarrow 2Cu(II) + H_2O_2$$

$$Cu(I) + H_2O_2 + H^+ \rightarrow Cu(II) + OH\cdot + H_2O$$

The anodic reaction used is an indirect oxidation of benzene by Ag(I)/Ag(II) as redox mediator, because of its high faradaic yield. The high yield of BQ of 84% (of the theoretical yield) compared to the yield of the direct oxidation on the PbO_2 anode of 62% indicates that some mechanism to minimize side reactions such as formation of *o*-BQ is operative. The highest yields are achieved with $AgClO_4$, (cf. Table 2, Ref. [66]). The use of $AgClO_4$ excludes its application in large scale synthesis.

The selective formation of BQ suggests that the active species introducing the second oxygen atom into the benzene nucleus is the bulky AgO^+ species, which is generated by the disproportionation of Ag(II) to Ag(I) and Ag(III) [66, 68].

$$2Ag(II) \rightarrow Ag(I) + Ag(III)$$

$$Ag(III) + H_2O \rightarrow AgO^+ + 2H^+$$

An elegant example of a "paired" mediated reaction has been reported by Chaussard and Lahitte [69] EDF (Electricité de France), who use Cr(VI) generated at the anode of a divided cell to oxidize the methyl side chain of a nitro-aromatic and Ti(III) generated at the cathode to reduce the nitro group. The reduction step, due to the faster homogeneous rate can be performed within the cell, whereas the oxidation has to be performed in a stirred tank reactor.

Anode :   Cr(III) $\longrightarrow$ Cr(VI) + 3e-

Tank : benzene ring—CH_3, NO_2 $\xrightarrow{Cr(VI)}$ benzene ring—COOH, NO_2

Cathode. benzene ring—COOH, NO_2 $\xrightarrow{Ti(III)}$ benzene ring—COOH, NH_2

127

2.2 Assessment of Economic Factors

Too often, researchers turn to electrochemistry only as last resort when other methods will not work. However, electrochemistry should be considered from the time a project is being designed, not when problems occur, because at that point the company has already invested in an alternative method [10].

Danly [70, 71] has presented three criteria on which commercialization of electroorganic process depends:

– product selectivity
– electrical power usage
– electrolysis system capital

Other important considerations are current efficiency, the number of electrons involved in the electrode process, the molecular weight of the reactant and the cell voltage.

In order to define an optimal commercial process considerations including mass transfer, heat transfer, power usage, electrode life, electrolyte quality, etc. require attention [71].

A detailed consideration of process economics is only possible for a specific process where operating conditions are completely specified. Table 3 outlines some of the factors that are likely to determine the success of an electrochemical process.

The salient costs as well as its technical feasibility (competing technology, patents and company experience pertinent to electrolysis) must be considered. Attention must also be paid to:

(1) investment costs, which include the cell, the cell auxiliary equipment such as storage tanks, pumps, unit processes, rectifier, computer to monitor and oversee the process,
(2) life-time of the process components, particularly of the cell,
(3) costs of replacements,

Table 3. Factors determining the economic viability of an electrosynthetic/electrochemical or an electrolytic effluent treatment process [49]

1) Achievable process objectives:
   Low/zero effluent synthesis
   Decrease of metal-ion or organic pollutant concentrations below the regulatory limit or discharge
2) Reproducible process: Performance with all effluents to be treated, alowing variations of the oxidants pollutant concentration, pH, complexing agents presents, etc.
3) Value of products manufactured, of metals or chemicals recovered or recycled
4) Cost and lifetime of the (electrochemical) hardware (cells, components, etc.)
5) Competing technology
6) Patents and company experience pertinent to electrochemical technologies

(4) energy costs, and
(5) value of chemicals vis-a-vis the cost of treatment and disposal.

For the industrial realization of a process, following items must also be considered [31]:

(1) existing plant which has depreciated
(2) availability of raw materials in the company
(3) captive use of the product
(4) electrical energy situation, legislation
(5) personal and subjective opinions

2.3 Electrochemical Hardware

Traditionally, scale-up comes after the chemists have finished and passed the process on to engineers. In electrochemistry, scale-up has to come early in the laboratory. The rule is to look at the entire process and then decide what use can be made of both electrodes before scale-up occurs. There is a long way from the polarogram or voltammogram run in the laboratory, to the batch type glass vessel in the laboratory and to the final full scale technical production cell [72]. This long way often results in much disillusionment. To ascertain whether any environmental and/or economic advantages can be gained, it is necessary for each user to be able to evaluate the proposed electrochemical method on the laboratory, pilot and production scales. This necessitates the availability of efficient and versatile cell equipment suitable for a great spectrum of demands. A prerequisite for evaluation in the technical laboratory is thus off-the-shelf equipment for the laboratory, bench, pilot and production scale [73, 73a]. Electrosynthetic processes have long suffered from a failure to develop the technology essential to successful operation. It is only recently that academic development of electrochemical reaction engineering has brought rationality to cell design (Refs. [13, 46] and references therein). The target for an electrosynthetic process is now most likely to be a product required on any scale from 1 to 10 000 t/year [10]. Since the 1980s, membrane cells have provided electrosynthesis with the same energy efficiency and environmental control that have revolutionized chlor-alkali production.

Several companies now market readily assembled parallel-plate electrolytic devices. The sizes range from 16 to 21 m^2 of flat-plate electrodes. An excellent modern overview on:

(i) electrochemical reactors and their performance;
(ii) electrochemical reactor design

can be found in a recent monograph "A First Course in Electrochemical Engineering" [13] and references therein. Cf. also Refs. [10], [26], [75].

Commercially available technical scale cells:

Cell type	Producer
FM-21	ICI Chemicals and Polymers, Ltd. Cheshire, England [74, 75]
DEM (Dished Electrode Membrane)	Electrolytic Inc., Union, N.Y., USA [76]
X	Reilly Tar and Chemical Industries, Indianapolis, USA., [77, 77a,77b]
Electro Prod	ElectroCell Systems AB, Täby, Sweden [72, 78–81, 99]

Development at ElectroCell Systems AB has resulted in a multipurpose cell design that meets the specific requirements of electrochemical synthesis and environmental protection. To facilitate electrochemical scale-up of potentially promising processes four different cells have been developed and are commercially available off-the-shelf from ElectroCell Systems AB.

Micro Cell	laboratory cell
Electro MP-Cell	multipurpose cell for the laboratory
Electro Syn Cell	laboratory, pilot, production cell
Electro Prod Cell	pilot, full-scale production unit

The four sized modular cells of stacked plate and frame filterpress design are shown in Fig. 4.

These cells have been designed for a great number of alternative configurations, including:

(1) electrolysis and electrodialysis
(2) use of solid, expanded and three-dimensional electrodes (porous, felt and percolated metals, carbons, graphite, etc.)
(3) variable electrode gaps, and
(4) uniform electrolyte distribution at different rates in flow-by, flow-through or flow-across mode without risk of stagnant zones.

The cell construction provides (i) a uniform internal distribution of up to four separate electrolytes, (ii) cooling and heating facilities (useful temperature range: ca. $-40\,°C$ up to $+250\,°C$), (iii) gas supply, and (iv) different turbulent promotors to improve transport performances. The versatility of off-the-shelf cells, paired with increasing experience of integrating electrolytic cells into industrial processes thus reduces the obstacles and risks for the scale-up. Furthermore, electrochemical units lend themselves well to modular construction, thus CPI plant expansion is a chance for this new technique.

Specifications for the FM-21 ICI-Cell and the ElectroProd Cell, two typical commercial-scale production cells, are summarized in Table 4.

Fig. 4. Types of ElectroCell Systems AB modular filter press cells commerically available

Table 4. Specification for the FM21 (from ICI) and ElectroProd (from ElectroCell Systems AB) off-the-shelf cells [49]

	FM21	Electro Prod
Electrode area, m^2	0.21–21	0.4–16
Number of electrode pairs/module	1–100	1–40
Interelectrode gap, mm	2	0.5–4
Maximum current density, mA/cm^2	500–1000	500–1000
Possible number of flows	2 or 3	2–4
Frame materials	inert metal	polypropylene, polyethylene, polyvinylidene-difluoride, ethylenechlorotrifluoroethylene (ECTFE), Halar, Ryton PPS
Method of gasketing	flat gaskets	O-rings
Gasket materials	various	various
Turblance promotors	various	various

Both cells either undivided or as membrane-separated cell.
Construction material is selected for corrosion stability.

It is beyond the scope of this chapter to discuss all devices in detail. The discussion is thus focused on the electrode type and material and separators.

Electrode materials and types:
Properties of electrode materials, influence of electrode materials on process performance and on the mechanism of electrode reactions, selection of electrode materials for electrosynthetic processes and new development of electrode

materials have been reviewed by Cooper, Pletcher et al. [75] (417 references), cf. also Ref. [75a, b].

Conditions and the choice of electrodes for selected organic and inorganic syntheses are summarized in Table 5.

The range of electrode materials available to process design has expanded greatly in the last years. In the following the various materials are discussed:

Table 5a. Conditions and the Choice of Electrodes for Some Organic Electrosyntheses [75]

manufacture of	cell	cathode electrode	cathode reaction	cathode electrolyte	anode electrolyte	anode reaction	anode electrode
adiponitrile	undivided bipolar narrow-gap stack	Cd	$2CH_2CHCN + 2H_2O + 2e^- \rightarrow (CH_2CH_2CN)_2 + 2OH^-$	neutral phosphate buffer + additives		$2H_2O - 4e^- \rightarrow O_2 + 4H^+$	carbon steel
1,2-dihydro-phthalic acid	membrane cell	Pb	(structure) $+ 2H^+ + H_2O + 2e^- \rightarrow$ (structure with COOH)	5% H_2SO_4 in dioxane/H_2O	5% H_2SO_4 in H_2O	$2H_2O - 4e^- \rightarrow O_2 + 4H^+$	PbO_2/Pb
tetrafluoro-xylene	FM21 cell with membrane	Pb or Zn	(structure) $+ 2H_2O + 4e^- \rightarrow$ $2OH^- + N(CH_3)_3 +$ (structure)	none added	satd NaCl	$2Cl^- - 2e^- \rightarrow Cl_2$	RuO_2/Ti
4-pyridyl-methyl-amine	membrane cell with lead shot cathode	Pb shot	(structure with CN) $+ 4H^+ + 4e^- \rightarrow$ (structure with CH_2NH_2)	H_2SO_4 in MeOH/H_2O	H_2SO_4 in H_2O	$2H_2O - 4e^- \rightarrow O_2 + 4H^+$	PbO_2/Pb
3-amino-4-hydroxy-benzoic acid	membrane cell	Cu	(structure with NO_2) $+ 4H_2O + 6e^- \rightarrow$ $6OH^- +$ (structure with NH_2)	NaOH	NaOH	$4OH^- - 4e^- \rightarrow O_2 + 2H_2O$	Ni
glyoxylic acid	membrane cell	Pb	(COOH-COOH) $+ H_2O + 2e^- \rightarrow$ (COOH-COOH) $+ 2OH^-$	none added	H_2SO_4	$2H_2O - 4e^- \rightarrow O_2 + 4H^+$	PbO_2/Pb
intermediate to p-hydroxy-phenyl-acetic acid	membrane cell	Pb	(structure with CCl_3) $+ 2e^- \rightarrow$ $OH^- + Cl^- +$ (structure)	H_2SO_4 in CH_3CN/H_2O	H_2SO_4 in H_2O	$2H_2O - 4e^- \rightarrow O_2 + 4H^+$	PbO_2/Pb
intermediate for 4-hydroxy-benzal-dehyde	undivided bipolar carbon stack	C	$2CH_3OH + 2e^- \rightarrow 2CH_3O^- + H_2$	CH_3OH/KF	(structure) $+ 2CH_3OH - 4e^- \rightarrow$ (structure with $CH(OCH_3)_2$) $4H^+ +$ (structure)			C
sebacic acid	undivided narrow-gap cell	graphite or steel	$2CH_3OH + 2e^- \rightarrow 2CH_3O^- + H_2$	partial neutralized carboxylic acid	$(CH_2)_4$(COOMe)(COO$^-$) $+ 2e^- \rightarrow$ $(CH_2)_8$(COOMe)$_2$ $+ 2CO_2$		Pt/Ti or Pt
substituted dimethoxy-furan	undivided narrow-gap cell	steel	$2CH_3OH + 2e^- \rightarrow 2CH_3O^- + H^+$	$CH_3OH/NaBr$	(furan structure) $+ 2CH_3OH - 2e^- \rightarrow$ (structure) $+ 2H^+$		C
picolinic acid	membrane cell with lead shot anode	Pb	$2H^+ + 2e^- \rightarrow H_2$	H_2SO_4	H_2SO_4/NaHSO$_4$/H_2O	(structure) $+ 2H_2O - 6e^- \rightarrow$ $6H^+ +$ (structure with COOH)	PbO_2

Electrochemistry for a Better Environment

Table 5b. Conditions and Choice of Electrodes for the Electrosynthesis of Some Inorganic Compounds [75]

manufacture of	cell	cathode electrode	reaction	electrolyte	anode electrolyte	reaction	electrode
bromate	undivided parallel plate with external holding tank	steel or Cu	$2H_2O + 2e^- \rightarrow H_2 + 2OH^-$	NaBr, pH 11		$Br^- + 3H_2O - 6e^- \rightarrow BrO_3^- + 6H^+$	C, Pt/Ti or PbO_2/C
perchlorate	undivided parallel plate	stainless steel	$2H^+ + 2e^- \rightarrow H_2$	$NaClO_3$, pH 0–1		$ClO_3^- + H_2O - 2e^- \rightarrow ClO_4^- + 2H^+$	Pt/Ti or PbO_2/C
persulfate	undivided tubular cell	stainless steel	$2H^+ + 2e^- \rightarrow H_2$	$NaSO_4$, pH 1		$2SO_4^{2-} - 2e^- \rightarrow S_2O_8^{2-}$	Pt or Pt/Ti
ceric sulfate	bipolar membrane cell stack	Pb alloy	$2H^+ + 2e^- \rightarrow H_2$	H_2SO_4	H_2SO_4	$Ce^{3+} - e^- \rightarrow Ce^{4+}$	Pb alloy
manganese dioxide	Undivided open-tank cell	stainless steel	$2H^+ + 2e^- \rightarrow H_2$	H_2SO_4		$Mn^{2+} + 2H_2O - 2e^- \rightarrow MnO_2 + 4H^+$	graphite, Ti PbO_2/Pb
hydrogen peroxide	diaphragm cell with trickle bed cathode	carbon black on graphite	$O_2 + 2H_2O + 2e^- \rightarrow H_2O_2 + 2OH^-$	NaOH		$4OH^- - 4e^- \rightarrow 2H_2O + O_2$	Ni
silver nitrate	membrane cell	stainless steel	$2H^+ + 2e^- \rightarrow H_2$	HNO_3	pure HNO_3	$Ag - e^- \rightarrow Ag^+$	Ag
stannate	membrane cell	steel	$2H_2O - 2e^- \rightarrow H_2 + 2OH^-$	KOH	KOH	$Sn + 4OH^- - 2e^- \rightarrow SnO_2^{2-} + 2H_2O + O_2$ oxidation	Sn
dichromate	membrane cell	stainless steel	$2H^+ + 2e^- \rightarrow H_2$	H_2SO_4	$H_2SO_4^{2-}$	$2Cr^{3+} + 14H_2O - 6e^- \rightarrow Cr_2O_7^{2-} + 14H^+$	PbO_2/Pb alloy
ozone	undivided tubular cell	dispersed Pt on C (porous)	$O_2 + 4H^+ + 4e^- \rightarrow 2H_2O$	48% HBF_4		$3H_2O - 6e^- \rightarrow O_3 + 6H^+$	vitreous carbon
permanganate	undivided parallel plate; anode/cathode areas = 100/1	Fe or steel	$2H_2O + 2e^- \rightarrow H_2 + 2OH^-$	K_2MnO_4 + KOH		$MnO_4^{2-} - e^- \rightarrow MnO_4^-$	Ni or monel

Metals: Pb, Cu, and Ni
   Ti electrodes have appeared on the market, offering better durability and performance than conventional graphite electrodes
Alloys: Steels, Monels
Coated electrodes: Pt on Ti; PbO_2 on Ti or C
   RuO_2-based coatings on Ti show excellent properties as dimensionally stable anodes (DSA).
Carbon electrodes are commercially available in many forms. These include plates, foams, felts, cloths, fibers, spherical and other particles suitable for beds or powders. Graphite or amorphous carbons exhibit quite different performances. Porosity, surface area and pretreatment are important variables to be considered in designing carbon electrodes.

Metal oxide coatings confirm the importance of electrocatalysis and show the way to innovative design of anodes. IrO_2-based coatings exhibit superior properties for oxygen evolution. A critical requirement is an anode material with high oxygen overpotential and stability to corrosion at high positive potentials. Sb doped SnO_2 coatings deposited on Ti hold great promise for organic oxidations, with highly enhanced chlorine overpotentials [82, 83], cf. also Refs. [83a, 83b]. It is suggested that such electrodes may be competitive with wet oxidation or combustion for high COD streams. Figure 5, compares anodic Tafel lines during oxygen evolution in H_2SO_4 on Pt, PbO_2 and Sb (4 mol%)-doped SnO_2 electrodes.

Conducting ceramics [75, 85–89]
New electrodes based on conducting ceramics, such as Ti_4O_7/Ti_5O_9 (Magneli phase suboxides) are black, more conductive than graphite, and highly cor-

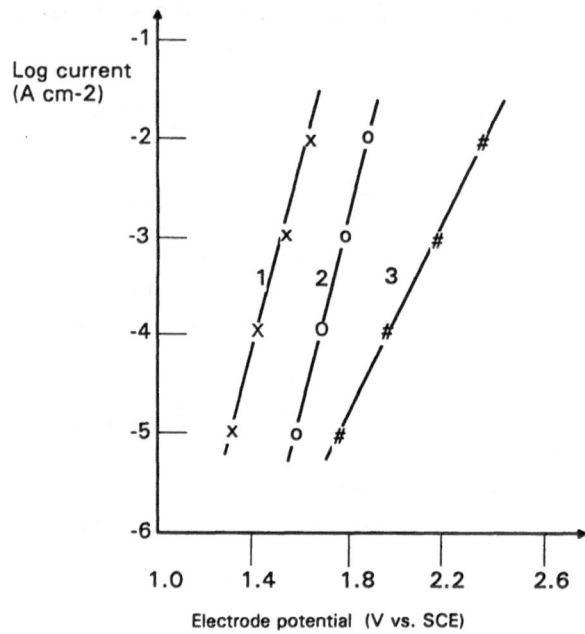

Fig. 5. Comparison of anodic Tafel lines during oxygen evolution in 1 N H_2SO_4 on (1) Pt, (2) PbO_2 and (3) Sb (4 mol%) doped SnO_2 [82, 84]

rosion resistant. These electrodes (marketed under the trade mark Ebonex) are showing promise in the harsh environments of many electrochemical processes. Magneli phase titanium oxides are the first materials to really challenge precious metal coated electrodes in the last 20 years. Ebonex in the form of Ti_4O_7 is the most useful ceramic to be discovered to date for use in aqueous electrochemistry. Ceramic electrodes are likely to become much more important in electrosynthesis and in electrochemial wastewater and effluent treatment, especially in presence of fluoride ions. F-ions in an electrolyte represent a major problem to the anodes of an electrochemical reactor, as it is extremely difficult to find anode substrate materials with appropriate corrosion resistance and the required electrical and mechanical properties, cf. Ref. [90].

A comparison of weight loss during unpolarized immerision tests in various electrolytes [85, 89] for Ebonex and titanium metal electrodes is given below.

Electrode	Weight loss in:		
	nitric + 0.1% HF acid	nitric + 0.4 HF acid	40% HF
Ti 25% porous	22%/150 h	52%/150 h	100%/150 h
Ebonex	0.3%/3500 h	2.4%/3500 h	12.7%/3500 h

The lifetime of PbO_2 (a ceramic) on a Ti substrate is limited by a more or less complete crumbling off of the coating, as the physical properties such as thermal expansion of PbO_2 are different compared to metallic Ti. Ebonex – itself also a ceramic – has comparable physcial properties to the PbO_2 [90]. As it is resistant to acids or strongly oxidizing electrolytes undermining processes cannot take place. For a detailed discussion of Ebonex materials, cf. Refs. [85, 89].

Three-dimensional electrodes are configured as static or solid electrodes (porous or packed bed), or as dynamic or fluid bed electrodes (fluidized bed and moving bed), cf. Table 6.

They all improve mass tansport within the cell via the mechanism of efficient turbulence promotion and also greatly increase the electrode area.

Three-dimensional electrode materials that fit well into parallel-plate [75, 91, 92, 93] reactors are: (i) reticulated metals [75, 91–93], (ii) metalized plastics (metalization of polyurethane foams) [94] and (iii) carbon [95].

For divided cells, appropriate **separators** are important [96, 97]:
In selecting the separator inside a cell, it is critical to use the right membrane [48]. Copolymers of styrene with acrylic acids or of styrene and divinylbenzene which are substituted by sulphonic acid or aminomethyl groups are still widely used in electrolysis cells where they are not exposed to harsh conditions. In electrolysis cells exposed to oxidizing agents and to high or low pH their use leads to problems of long-term stability. Modern perfluorated membranes – highly resistant to chemical attack – are mostly based on similar structures with a polytetrafluoroethylene (PTFE) backbone and side chains ending in an ion exchange center.

Table 6. Configuration of three-dimensional electrodes

Static or solid electrodes	
(1) Porous:	Perforated plates; stacked meshes, cloths, felts, foams
(2) Packed bed:	Granules and flakes, microspheres, spheroids, fibers, rods, Raschig rings

Dynamic of fluid electrodes	
(1) Fluidized bed:	Microspheres, spheroids
(2) Moving beds:	Slurries, inclined bed, turnbled beds, vibrated beds, beds with pulsed flow, Raschig rings

Table 7. Membranes for the cell divider

Supplier	Type of membrane
Du Pont	Nafion: strong acid, and strong or weak acid bilayer cation membranes, with or without reinforcing polymer net
Asahi Glass	Flemion: weak acid and bilayer, cation membranes
	Selemion: Acid, (mono-anion selective), nonperfluorinated polymer, reinforced
Tokuyama	Neosepta-F weak and strong acid, cation membranes
Tosoh	Tosflex anion membranes
Asahi Chemicals	Weak and strong cation membranes
Sybron	Ionac: cation and anion membranes, nonperfluorinated strong acid groups, reinforced
Morgane	Novelect: Cation and anion membranes

Fig. 6. Mobile pilot plant electrolyzer (ElectroCell Systems AB pilot rig with rectifier)

Commercially available perfluorinated membranes are summarized in Table 7.

Their availability has greatly expanded the potential for electrolytic processes in synthesis and fuel cells as well as in environmental control. Perfluorinated cation exchange membranes such as Nafion outlast the material that preceeded them by up to four and a half years [48]. Unfortunately very little has been published on their behaviour outside their use in chlor-alkali electrolysis.

2.4 Technical Assistance from Suppliers

Equipment suppliers such as Electrosynthesis Co., Inc. [98], ElectroCell Systems AB [99], EA Technology [100] et al. do provide technical support.

Various companies licence technology, lease laboratory-scale equipment for companies to try out, conduct short- or long-term feasibility or pilot studies. EDF [101], e.g., lends free of charge movable pilot plants, cf. Fig. 6.

3 Synthesis

3.1 Organic Synthesis: Pilot Plant and Commercial Products

Although still queried, the past years have seen a steady increase in the number of organic electrosynthetic processes within all parts of fine-chemical industry.

There are successful examples relevant to most branches of the chemical industry, due to a growing awareness of electrochemical techniques among synthetic organic chemists and chemical engineers [10]. The interest is illustrated by different papers discussing industrial applications. Furthermore, electrosynthesis should become more attractive as developed economies shift their production focus away from commodity chemicals. Economics of electrosynthesis favor high value products. It is for the synthesis of low tonnage organic products that electrolysis has had, and will continue to have the greatest impact.

Europe and Japan are far ahead of the US in scaling up electrochemical processes. Companies active in the area are: BASF AG (Ludwigshafen, Germany), Hoechst AG (Frankfurt, Germany), SNPE (Paris, France), ICI PLC (Cheschire, UK), Asahi Chemical Co (Nobeoka, Japan), LTEE d'Hydro Quebec, et al.

To date, electrosynthesis is being used in more than 60 commerical processes to make about 100 different compounds. Table 8 [48] lists electrosynthetic reactions that have yielded a product of sufficient practical interest to be commercialized. See also Table 6.2 in Ref. [10]. Table 9a lists commercial intermediates and products evaluated in a pilot plant by BASF, Table 9b lists products produced electrochemically by SNPE.

The electrochemical processes currently make use of direct and indirect electrode reactions, such as (1) hydrodimerization, (2) hydrogenation of heterocycles, (3) hydrogenation of nitriles, (4) reduction of carboxylic acids, (5) pinacolization, (6) cathodic cleavage, (7) nitro group reduction, (8) base generation, (9) selective hydrogenation, (10) sugar chemistry, (11) fluorination, (12) methoxylation, (13) oxidation of polynuclear aromatic hydrocarbons, (14) benzene oxidation, (15) oxidation of methyl aromatics, (16) epoxydation of olefins, (17) Kolbe coupling of half esters, etc.

Details of many synthetic processes are never reported and, hence, as noticed by Pletcher and Walsh [10], any contribution of electrosynthesis remains speculative. Crucial factors are generally the availability and costs of the starting materials, the material yield, a simple product isolation, the stability of the electrolysis medium and acceptable current densities.

There is a preponderance of reactions in aqueous solutions, although organic solvents are more fitted to organic reactions, giving higher selectivity and allowing the use of higher substrate concentrations. This is in sharp contrast to academic/journal literature [10]. Several oxidations are run in methanol whereas only few reactions in aprotic solvents appear. The major reason for the use of aqueous solutions – which is seldom the preferred medium for organic reactions – is the stability of the electrolysis medium. The oxidation/reduction of water yields O_2/H^+ and H_2/OH^- the reduction of methanol H_2 and CH_3O^-. Hence these reactions can conveniently be used to maintain the pH of the electrolyte at a constant level while not contaminating the system and the environment. The oxidation/reduction of aprotic solvents, or, the oxidation of methanol commonly lead to a complex mixture of products which will build up in the electrolyte and also lead to the loss of expensive solvent (Ref. [10]). Electrosynthesis in aprotic systems also requires special cells, cf. Sect. 3.1.2.2.

Table 8. Electrosynthetic processes running on or near a commercial scale [48]

Reaction	Location	Estimated scale (m.t./yr)	Conditions
Hydrodimerization:			
Acrylonitrile → Adiponitrile	U.S., Europe, Japan	3×10^5	Cd cathode, steel anode
Hydrogenation of heterocycles:			
Pyridine → Piperidine	Europe	100	Pb cathode in H_2SO_4
2-methylindole → 2-methylindoline	Europe	20	Pb cathode in H_2SO_4
Hydrogenation of nitriles:			
4-cyanopyridine → 4-(aminomethyl) pyridine	U.S.	30	Pb cathode in acid sulfate
2-cyanopyridine → 2-(aminomethyl) pyridine	U.S.	70	(both)
Reduction of carboxylic acids:			
3-hydroxybenzoic acid → 3-hydroxybenzyl alcohol	Japan	100	Pb cathode in aqueous acid
2-hydroxybenzoic acid → 2-hydroxybenzaldehyde	India	20	Pb cathode in acid
Phthalic anhydride → 1,2-dihydrophthalic acid	Europe	100	Pb cathode in H_2SO_4
Oxalic acid → Glyoxylic acid*	Europe, Japan	–	Pb cathode in aqueous solution
Maleic acid → Succinic acid	India	30	Pb cathode in H_2SO_4
Pinacolization:			
Formaldehyde → Ethylene glycol*	U.S., Canada	–	Neutral solution with R_4N^+; modified graphite cathode
Acetone → Pinacol*	U.S., Japan, Eurpoe	–	Pb cathode in acid
Cathodic cleavage:			
2,2,2-trichloro-1-(4-hydroxyphenyl)ethanol → 2,2-dichloro-1-(4-hydroxyphenyl)ethylene	Japan	120	Pb cathode
1,4-bis(trimethylammoniomethyl)-2,3,5,6-tetrafluorobenzene dichloride → 1,4-dimethyl-2,3,5,6-tetrafluorobenzene	Europe	200	Zn cathode
Cystine → Cysteine	Europe, Japan	30	Pb cathode in acid
3,4,5,6-tetrachloropicolinic acid → 3,6-dichloropicolinic acid	U.S.	n/a	Ag cathode
Cephalosporin sulfone → Azetidinone sulfinic acid*	U.S.	–	–

Process	Region	Tonnage	Conditions
Nitrogroup reduction:			
4-nitrobenzoic acid → 4-aminobenzoic acid	India	3	Pb cathode in acid
Nitrobenzene → 4,4′-diaminobiphenyl	India	30	—
Nitrobenzene → 4-aminophenol*	U.S, Europe, Japan, India	—	Cu or Monel cathode
Base generation:			
Tetramethylammonium chloride → Tetramethylammonium hydroxide	Japan, U.S.	24	—
Selective hydrogenation:			
Steroids*	Europe	—	Ni cathode
Sugar chemistry:			
Protected sorbose → Corresponding acid*	Europe	1,000	Ni anode in aqueous base
Glucose → Gluconic acid	Europe, India		Indirect oxidation via Br_2
Glucose → Sorbitol	U.S., India	n/a	Pb cathode in aqueous solution
Fluorination:			
Perfluorination of $C_8H_{17}COOH$, or CH_3SO_3H	U.S., Japan, U.K.	100	Ni anode in liquid HF
Methoxylation:			
Furan → 2,5-dimethoxydihydrofuran	Europe	100	Indirect via Br_2
Furfuryl-1-ethanol → 2,5-dimethoxydihydrofurfuryl-1-ethanol	Japan	100	Indirect via Br_2
1-formylpiperidine → 2-methoxy-1-formylpiperidine	Europe	25	Direct oxidation
Oxidation of polynuclear aromatic hydrocarbons:			
Anthracene → Anthraquinone	Europe, U.S.	100	Pb anode; Indirect via $Cr_2O_2^{2-}$ indirect via Ce^{4+}
Naphthalene → 1,4-naphthoquinone*	Canada, Europe	—	
Naphthalene → 1-naphthalene acetate	Europe	n/a	Modified graphite anode
Benzene oxidation:			
Benzene → 1,4-benzoquinone*	Europe	—	PbO_2 anode in H_2SO_4
Oxidation of methyl aromatics:			
4-t-butoxytoluene → 4-t-butoxybenzaldehyde dimethyl acetal	Europe	1,000	CH_3OH in 1% KF
3,4,5-trimethoxytoluene → 3,4,5-trimethoxybenzyl methyl ether	Japan	120	NaOH in CH_3OH
2-methylpyridine → Picolinic acid	U.S., Europe	10	PbO_2 in H_2SO_4
4-nitrotoluene → 4-nitrobenzoic acid	India	30	Indirect via $Cr_2O_2^{2-}$

Table 8. (continued)

Reaction	Location	Estimated scale (m.t./yr)	Conditions
Epoxidation of olefins: Propylene → Propylene oxide*	U.S., Europe, Japan	–	Indirect oxidation via BrO⁻, or ClO⁻
Hexafluoropropylene → Hexafluoropropylene oxide 1-(4-fluorophenyl)-1-(2-chlorophenyl)ethylene → 1-(4-fluorophenyl)-1-(2-chlorophenyl)ethylene oxide	Europe	5	PbO₂ anode
Kolbe coupling of half esters: Monomethyl adipate → Dimethyl sebacate	Europe, India, Japan	n/a	Pt anode in methanol

*Pilot-scale process (all others listed are commercial)
(Source: Electrosynthesis Co.)

Table 9a. BASF range of electrochemical intermediates [54]

Commercial products

4-Methoxybenzaldehyde
CAS No.: 123-11-5
EINECS
TSCA
MITI

4-tert.-Butylbenzaldehyde
CAS No.: 939-97-9
EINECS
TSCA
MITI

**2,5-Dimethoxy-2,5-
dihydrofuran**
CAS No.: 332-77-4
EINECS
TSCA

Pilot scale products

p-Tolylaldehyde
CAS No.: 104-87-0
EINECS
TSCA
MITI

o-Tolylaldehyde
CAS No.: 529-20-4
EINECS
MITI

Phthalaldehyde
CAS No.: 643-79-8
EINECS
TSCA
MITI

Terephthalaldehyde
CAS No.: 623-27-8
EINECS
TSCA

4-Ethoxybenzaldehyde
CAS No.: 10031-82-0
EINECS
TSCA
MITI

**p-Benzoquinonetetra-
methylketal**
CAS No.: 15791-03-4
EINECS
TSCA
MITI

Lab scale products

**4-(1,1-Dimethylpropyl)-
benzaldehyde**
CAS No.: 67468-54-6

4-Phenoxybenzaldehyde
CAS No.: 67-36-7
EINECS

**2-tert.-Butyl-p-benzoqui-
none tetramethylketal**
CAS No.: 134962-83-7

2,2-Dimethoxycylohexanol
CAS No.: 63703-34-4

**1,1-Dimethoxy-1-phenyl-2-
butanol**
CAS No.: 882-53-1

**2-(Dimethoxymethyl)-2,5-
dihydro-2,5-dimethoxyfuran**
CAS No.: 59906-91-1
EINECS

**2,5-Dimethoxy-2,5-dihydro-
2-furanmethanol**
CAS No.: 19969-71-2

**2,5-Dihydro-2,5-dimethoxy-
furfurylamine**
CAS No.: 14496-27-6

1,1,6,6-Tetramethoxyhexane
CAS No.: 54286-89-4
EINECS

1,1,8,8-Tetramethoxyoctane
CAS No.: 7142-84-9

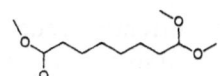

5-Methoxy-2-pyrrolidinone
CAS No.: 63853-74-7

**2-Methoxy-1-pyrrolidine-
carboxaldehyde**
CAS No.: 61020-06-2

Table 9b. Products manufactured by indirect electrochemical technologies using consumable electrodes [115]

Aromatic carboxylic acids

Arylacetic acids

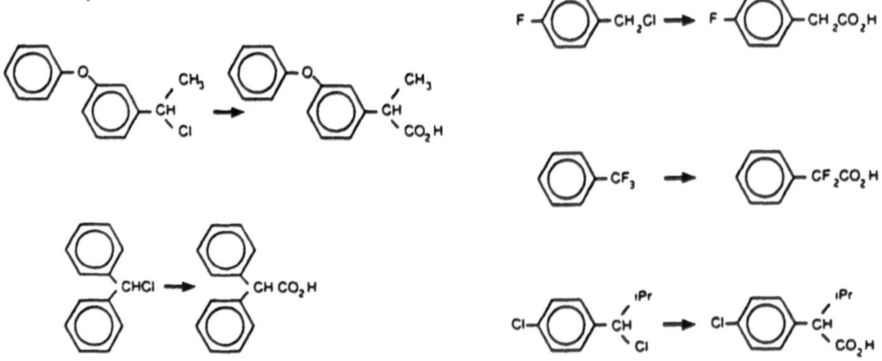

Arylacetones

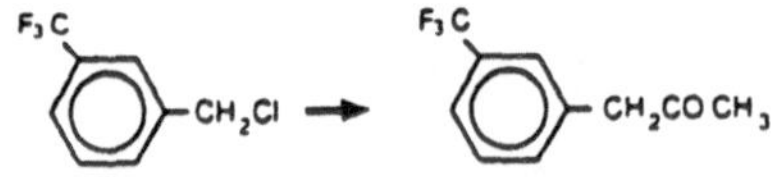

Aromatic aldehydes

Biaryles

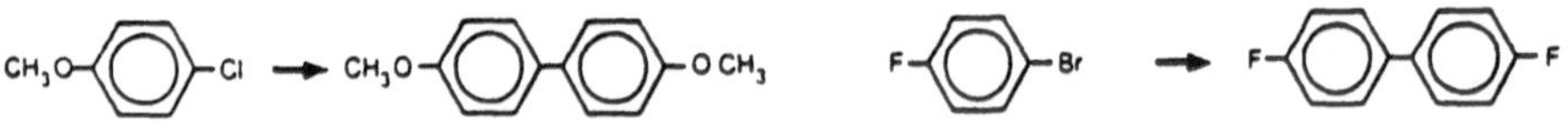

Cysteine derivatives : cysteine base, salts, carbocysteine, acetylcysteine; thiazolidin carboxylic acid.

3.1.1 Direct Electrochemical Processes for Organic Compounds on Production and Pilot Plant Scale

Adiponitrile is still the only large-tonnage organic product currently prodcued electrochemically. The world output of adiponitrile by electrolysis now exceeds 200 000 t year^{-1}. The electrochemical route for conversion of acrylonitrile to adiponitrile – adiponitrile is a convenient intermediate for the production of both hexamethylenediamine and adipic acid – is a cathodic dimerization.

$$2CH_2 = CHCN + 2H_2O + 2e^- \rightarrow \quad \begin{matrix} CH_2CH_2CN \\ | \\ CH_2CH_2CN \end{matrix} \quad + \; 2OH^-$$

The basic chemistry, the early Monsanto process brought on stream in 1965 by Baizer [102–104], and the new Monsanto process based on an undivided cell, combining simpler cell design, less complex product extraction and reduced energy consumption have all been summarized by Pletcher and Walsh [10] (see Scheme 4).

There are many possible reaction pathways between acrylonitrile and adiponitrile and, in each, there are several possible rate-determining steps. None of the reaction intermediates has yet been detected electrochemically or spectroscopically thus indicating very fast chemical processes with intermediates of half-lives of $< 10^{-5}$ s. Bard and Feiming Zhou [104a] have recently detected the $CH_2 = CHCN^{\overline{\cdot}}$ radical by Scanning Electrochemical Microscopy (SCEM) using a 2.5 μm radius Au electrode (1.5 mol $CH_2 = CHCN$ in $MeCN/TBAPF_6$). The dimerization rate has been determined to $\sim 6.10^7 \ M^{-1} S^{-1}$.

Ethylene glycol (the potential market is said to be of the order of 10 billion kilos per annum [104b]) is made by electrodimerisation of formaldehyde from

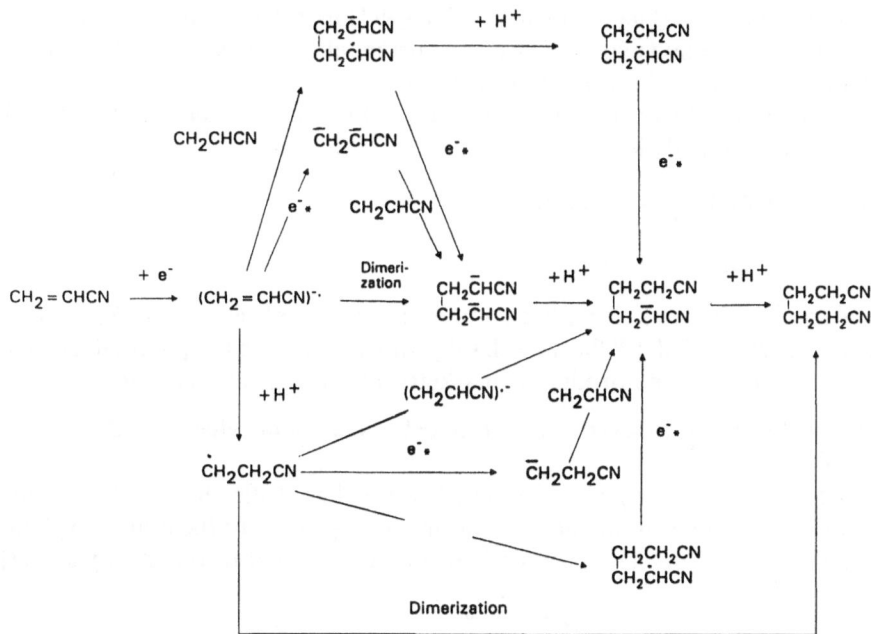

Scheme 4. Possible reaction pathways for the hydrodimerization of acrylonitrile to adiponitrile. The asterisk* indicates that electron transfer can be from the cathode or from $[CH_2CHCN]^{-\cdot}$ in homogeneous solution

synthesis gas [60, 105–109]. The electrodimerization

$$2CH_2O + 2H^+ + 2e^- \rightarrow HOCH_2\text{–}CH_2OH$$

proceeds in close to quantitative yields and current efficiency. Prerequisite for optimum process conditions are: (i) aqueous formaldehyde solutions of high concentration, (ii) a pH range of about 5 to 8, (iii) the presence of a quarternary ammonium salt, and (iv) reaction temperatures in excess of about 80 °C [107]. The process works well only at graphite, at which the enzyme-like specificity is believed to be related to the carbon surface oxide structure [106, 107]. Weinberg of Electrosynthesis Co. thinks that a 20% cost advantage over current technology can be expected. Weinberg and Mazur [108] have recently reviewed the electrohydrodimerization of acrylonitrile to adiponitrile and electrohydrodimerization of formaldehyde to ethylene glycol.

In the following, *selected examples* taken from the more recent literature of pilot plant and production electroorganic syntheses will be presented. For a modern review the reader is also referred to recent monographs, such as "Industrial Electrochemistry" by Pletcher and Walsh, 2nd edn. [10] and modern reviews.

(1) Processes improved by a simpler product isolation [10]

Aliphatic carboxylic acids are difficult to reduce electrochemically. Reduction of a 10% oxalic acid in 10% H_2SO_4 at 15 °C at a mercury cathode (Refs. [494, 532] in Ref. [29]), a lead or amalgamated lead cathode (Ref. [495] in Ref. [29]) or at a sodium amalgam (Na(Hg)) cathode (Ref. [497] in Ref. [29]) produces glyoxylic acid with a material yield of 88% and a current efficiency of 70%. The glyoxylic acid formed is stabilized by hydration [29].

A patent by BASF describes the direct reduction of oxalic acid to glyoxylic acid with no added electrolyte

in a special divided cell with an ion-exchange membrane at 30 A dm^2/6.5 V using a mercury (Ref. [497a] in Ref. [29]) or lead cathode [10]. No salt is added to the electrolyte. The product is marketed as an aqueous solution.

(2) The direct and indirect electrochemical oxidation of toluenes and alkoxytoluenes

This process was examined in detail by BASF [110], cf. also Ref. [10]. The direct oxidation of toluene in methanol leads in a 4e$^-$ process to the dimethoxylation of the methyl group. The product is readily hydrolyzed to the aldehyde [29].

Typical process conditions are:
Undivided cell: bipolar stack of carbon electrodes with a narrow electrode gap
10–15% substrate; methanol/2% HF; temperature $= 40\,^\circ$C;
Material yield $= 95\%$; current efficiency $= 70\%$.
The cathode reaction is used to maintain the pH of the electrolyte constant.

The *tert*-butoxy group was chosen as the protective group for the formation of *p*-hydroxy benzaldehyde [29].

Anode:

Cathode : $2CH_3OH + 2e^- \text{---------->} 2CH_3O^- + H_2$

tert-Butanol is introduced readily into *p*-cresol in the presence of an acidic ion-exchange resin as catalyst. The *tert*-Bu group is easily hydrolysed after electrolysis and can be recycled [10]. Otsuka Chemical in Japan [10, 35, 111] has commercialized a methoxylation process for an intermediate (11) for the manufacture of the food additive maltol.
The electrode reactions are:
Anode:

(11)

Cathode :
$$2CH_3OH + 2e^- \text{-------------->} 2CH_3O^- + H_2O$$

The electrolyte is a 20% solution of the furan in methanol containing sodium bromide. The bromide ion plays an essential role since bromine and brominated furans are probably important intermediates in the anodic processes [10]. Organic yields as high as 97% with current efficiencies of 80 to 95% are reported for electrolytes run below $10\,^\circ$C.

(3) Cathodic processes
Reilly and Tar Chemicals [77] developed different processes in sulfuric acid, using Pb electrodes.

(12)   (13)

The electrolysis in aqueous sulfuric acid with methanol as a cosolvent was perfomed in a filterpress membrane cell stack developed at Reilly and Tar Chemicals. Because of the low current density of the process, a cathode based on a bed of lead shot was used. A planar PbO_2 anode was used. The organic yield was 93% with approximately 1% of a dimer. The costs of the electrochemical conversion were estimated as one-half of the catalytic hydrogenation on a similar scale.

L-Cysteine is a high value α-amino acid used world-wide in a scale of 1200–1500 t year^{-1} as additive in foodstuffs, cosmetics or as intermediate or active agent (as antidote to several snake venoms) in the pharmaceutical industry. Chemical routes generally lack the efficiency of electrochemical techniques, or they produce mixtures of L- and D- forms rather than the L-isomer. The most common electrochemical route is the cathodic reduction of L-Cystine in acid (usually HCl) solution to produce the stable hydrochloride. In Table 10, the charateristic data for a laboratory bench, laboratory pilot and a product pilot reaction using a DEM filter press are compared [13]. A production scale study was carried out in a filterpress reactor divided by a cation exchange membrane with a total area of 10.5 m^2. The typical product inventory was 450 kg/24-hour batch time. For more details see Ref. [13].

Electrosynthesis Co. Inc. [109, 112, 113] has piloted a process that reduces cystine in aqueous ammonia solution, using an ElectroSyn cell.

Cystine   Cysteine

Table 10. Comparison of filterpress reactor sizes for the electrosynthesis of L-cysteine hydrochloride. Catholyte = 2 M HCl at 25 °C; $j = 2$ kA m^{-2}; $v = 0.35$ m s^{-1}; Pb cathode in presence of turbulence promotor; DEM cell [13, p. 327]

	Laboratory bench reactor	Labratory pilot reaction	Product pilot reactor
Number of cathodes	1	1	6
Cathode size (m^2)	0.01	0.05	0.175
Total cathode area (m^2)	0.01	0.05	1.05
Nominal current (kA)	0.02	0.1	2.1
Cell voltage (V)	5.4	4.0	4.1
% Current efficiency at $X_a = 0.99$	66	68	68
Production rate per unit area (kg m^{-2} h^{-1})	5.3	6.1	6.1
Specific energy consumption (kWh/kg)	1.75	1.3	1.35

Figure 7 shows a production plant for cysteine equiped with a Prod Cell. Cf. also Ref. [114].

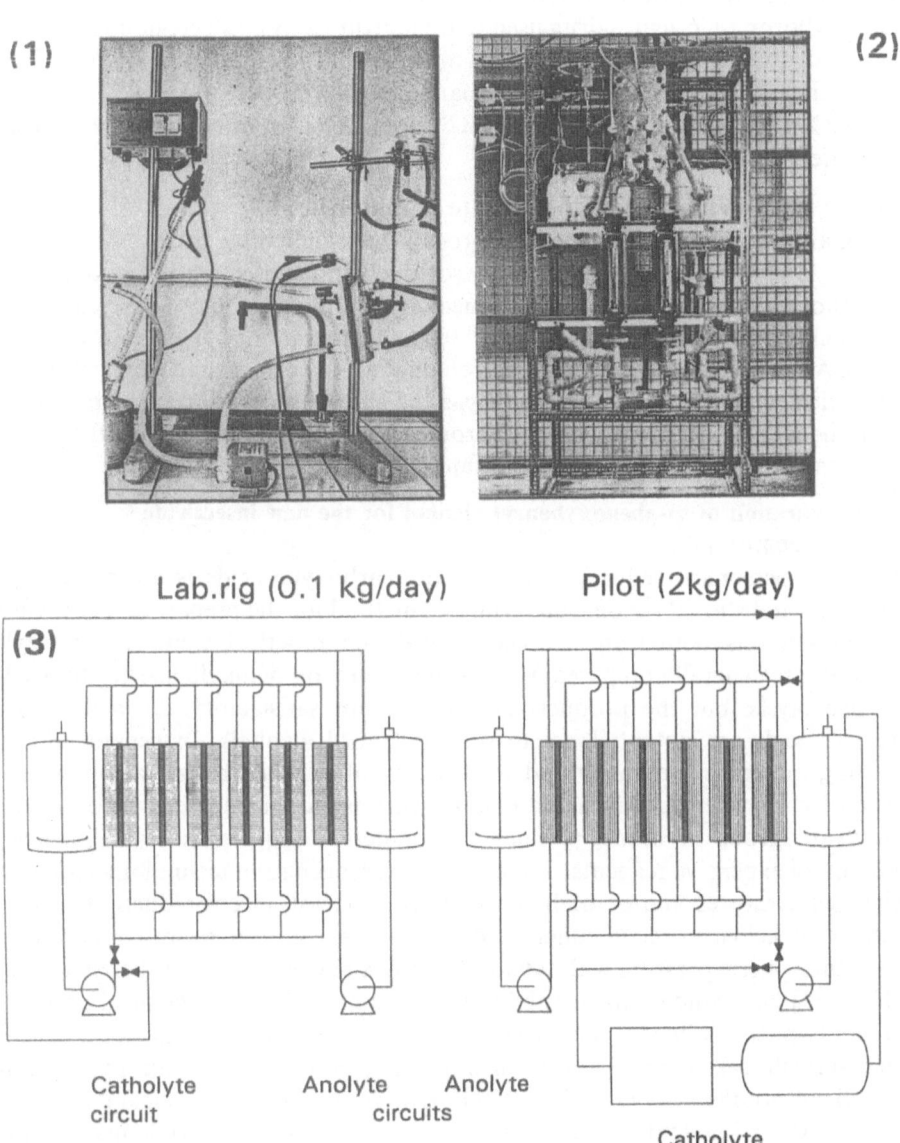

Fig. 7. Electrosynthesis of cysteine: From the laboratory (1) to the pilot plant scale (2), to the technical production (3), using a laboratory cell (1), and ElectroSyn (2) and Electro prod cells (3) Electro cell Systems AB, Sweden

Isochem has set up a multipurpose electrochemical unit devoted to organic electrosynthesis. The unit located in Pithiviers (France) [115] is dimensioned for the production of 80 tons a year of cystein derivatives, such as cysteine base salts, carbocysteine, acetylcysteine, and thiazolidin carboxylic acids.

Monsanto [117] has developed a way to electrosynthesize by reductive carboxylation the optically active precursor to Naproxen, (S)-2-(6'-methoxy-2'-naphthyl)propionic acid, a drug used to treat arthritis. A more economical route was needed since the US patent expires in 1993 while the market is growing. The electrochemical process is said to cut manufacturing costs by over 50%. Since it uses CO_2 instead of the hazardous HCN used in conventional synthesis, it is also safer.

(4) The oxidation of sodium D-gluconate to D-arabinose

Using a fluidized bed electrode, this process was studied by Jircny 1985 [118]. Jircny [119] worked with a laboratory scale cell and subsequently a pilot plant. The pilot plant was designed to produce one ton of D-arabinose per year. The electrochemical reactor was $0.3 \times 0.6 \times 0.6$ m and contained five 225 A cells in series. A major advantage of the electrooxidation over the usual chemical route (oxidation with sodium perchlorate) was the ease of separation of D-arabinose from the reactor outflow. In chemical routes, the separation is made difficult by the presence of large amounts of sodium chloride.

(5) Preparation of m-phenoxybenzyl alcohol for the new insecticide Ethofenprox (17)

The electrochemical reduction of aromatic carboxylic acids is an interesting challenge and should become an attractive method for the preparation of useful benzyl alcohols both from the economical and practical viewpoint. Benzyl alcohols are generally prepared from the halogenation of alkylbenzene followed by hydrolysis, but the product-selectivity is not satisfactory. Aromatic carboxylic acids are potential precursors for benzyl alcohols. In general, these starting materials are cheap and readily available, while benzyl alcohols are expensive. However, chemical reduction of carboxylic acids to benzyl alcohols is disadvantageous in industry, since the reduction requires a stoichiometric amount of expensive reagents. The catalytic hydrogenation technology requires high temperatures and/or high pressure. In contrast, electroreduction of the carboxylic acids proceeds under mild conditions, e.g. near room temperature and atmospheric pressure without any expensive reagents [120]. Takenaka et al. [120–122] have studied the feasibility of a commerical electrochemical procedure for m-phenoxy benzyl alcohol (16) which is a useful intermediate for medical and agricultural chemicals, and an important component of the pyrethroid insecticide ethofenprox (17) (TREBON). Due to its excellent pesticidal activity along with its low toxicity to fish and to mammals, there is a wide market for ethofenprox, especially in agricultural pest control.

The electrochemical process developed using Pb electrodes in aqueous sulphuric acid yields a practical procedure for m-hydroxy-benzyl alcohol (14). Yields up to 88% of (16) were achieved using 1,3-dimethyl-2-imidazolidinone

(DMI) for the condensation of the m-hydroxy-benzyl alcohol (15) with chloro-benzene.

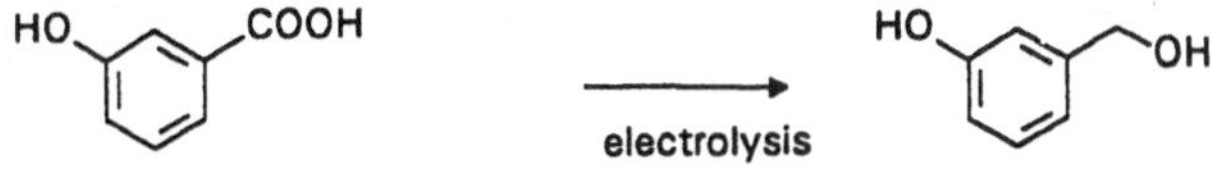

(14)   electrolysis   (15)

condensation

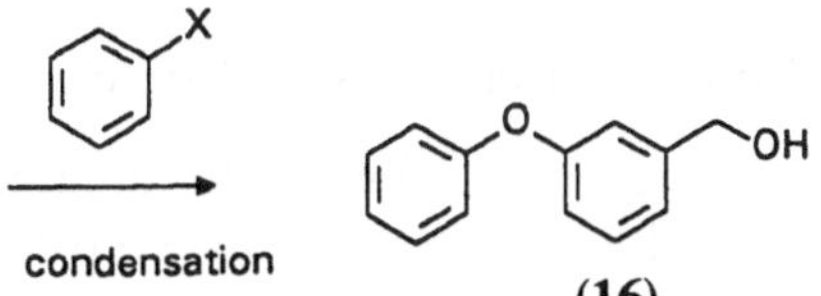

(16)

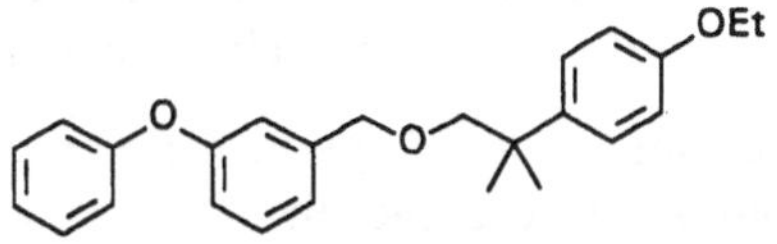

(17)
Ethofenprox

Electrosynthesis of *m*-phenoxybenzyl alcohol (16) successfully went through the pilot plant test in 1987 and started commercial production at the beginning of 1988.

(6) Fluorination
The 3M company uses electrochemical fluorination (ECF) technology for the production of over 100 t/year perfluorocarboxylic acid, sulphonic acids, alkanes, trialkylamines, alkyl ethers [48].
Hoechst has used electrosynthesis since 1985 to produce roughly 100 t/year hexafluoropropylene oxide [116].

3.1.2 Indirect Electrosynthesis

As different research groups have shown, electrochemical reactions can be mediated by metal powders, metal ions, metal oxides, semi-metals, halogens, inter halogens, halogen oxides, organics, or organometallics [123].

3.1.2.1 Indirect Processes Using (Metal) Mediators

Most suitable redox couples for indirect processes are inorganic, including:

for oxidation: Ce(III)/Ce(IV), Cr(III)/Cr(VI), Mn(II)/Mn(III), Mn(II)/Mn(IV), Br$^-$/Br$_2$, Cl$^-$/ClO$^-$, Ni(OH)$_2$/NiOOH
for reduction: Sn(IV)/Sn(II), Cr(III)/Cr(II), Ti(IV)/Ti(III), Zn(II)/Zn, Na(I)/NaHg

A few organic mediators are used such as Ar_3N/Ar_3N^+ for oxidations and viologens for reductions. One special example is the electromicrobial reduction of an α-keto acid to an α-hydroxy acid using the enzyme system of *Proteus vulgaris* (E_F) and methyl viologen (MV) as the electron shuttle (Scheme 5, Fig. 8) [124].

During the indirect process, a redox couple is used as catalyst or "electron carrier" for the oxidation or reduction of another species in the system. The redox reagent is continuously reconverted electrochemically.

Two modes of indirect electrosynthesis are currently used:

(i) **"In-cell mode"**: the reaction between the organic substrate and the redox reagent together with the conversion of the mediator to the initial valency state is performed within the cell.

(ii) **"Ex-cell mode"**: the chemical reaction is carried out in a reactor separate to the electrochemical cell.

Economics encourage the simpler "in-cell" mode. The "ex-cell" mode is advantageous because the electrode reaction and chemical step can be optimized separately. The electrolyte can be purified/conditioned between the reactor and the cell. Sn(II) is used as mediator for the reduction of nitro tegretol [125, 126] in 6 M HCl/ethanol. The Sn(IV) formed is reduced electrochemically after stripping off the alcohol, either to the Sn(II) state using a percolated Sn electrode or to the tin metal on a rotated carbon electrode. The reduction to the metal has the advantage that the mediator can be purified/washed before being recycled to the process [126].

Pletcher has reviewed papers on indirect electrosynthesis when phase transfer catalysis is used to allow the chemical step to occur homogeneously in an organic solvent immiscible with the aqueous electrolyte [10, 127].

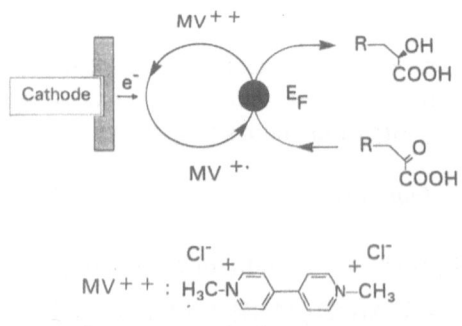

MV^{++} : H_3C-N ⟨ ⟩ ⟨ ⟩ $N-CH_3$ with Cl^-

E_F: proteus vulgaris enzyme systen

Scheme 5

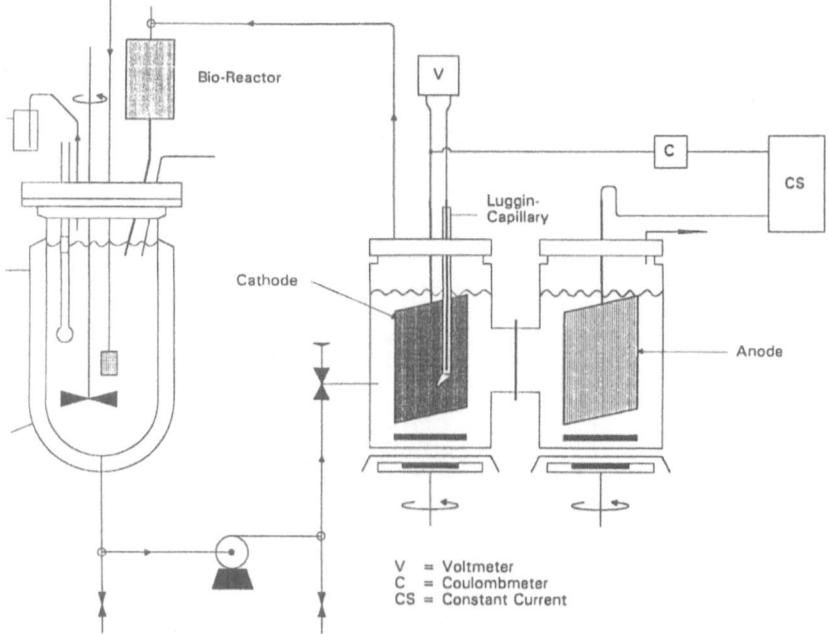

Fig. 8. Reduction of oxo acids with Proteus vulgaris, with methyl viologen (MV) as organic mediator. Display of the bench-scale rig

Selected examples of indirect electrosynthesis which have found technical or pilot plant scale applications are discussed in the following:

1) Synthesis of anthraquinones
The potential market size of anthraquinone was given in 1991 as follows [104b, 132]:

Paper pulping	50 000 t/annum
Dyestuffs	30 000
Others	20 000

Anthraquinone has been produced commercially from anthracene by indirect electrooxidation based on the Cr^{3+}/Cr^{6+} couple, cf. Fig. 9.

In the most recent plants, the electrolysis is performed in a membrane cell while the chemical step is carried out by allowing the chromic acid to trickle through a column of solid anthracene. The product – anthraquinone – is also insoluble in the aqueous acid so that the organic conversion is effectively completed in the solid state. The reaction goes to completion provided the particle size of the anthracene falls within a suitable range. The spent redox reagent is then passed through an activated carbon bed to remove traces of

P.M. Bersier et al.

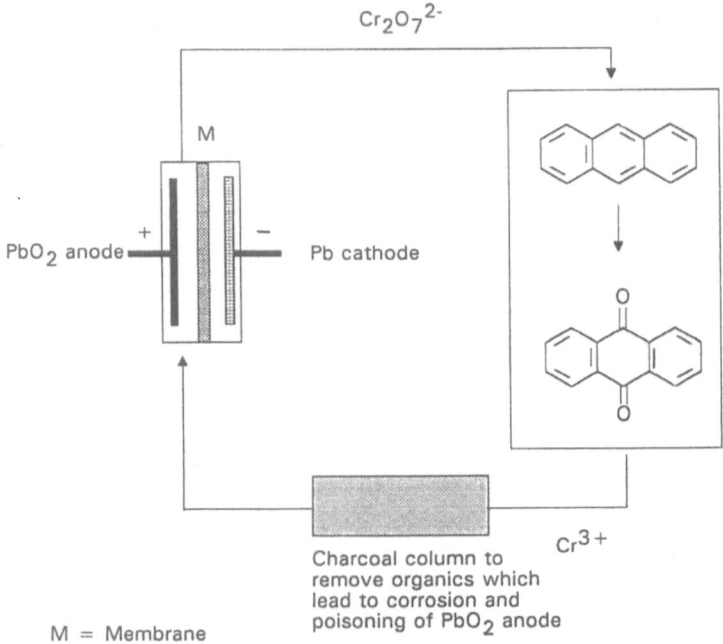

$Cr_2O_7^{2-}$

M

PbO$_2$ anode

+ −

Pb cathode

Charcoal column to
remove organics which
lead to corrosion and
poisoning of PbO$_2$ anode

Cr^{3+}

M = Membrane

Fig. 9. Indirect electrosynthetic process for the oxidation of anthracene to anthraquinone [10]. The charcoal column serves to remove organics which lead to corrosion and poisoning of the lead dioxide (PbO$_2$) anode

organic material which otherwise would lead to a loss in current efficiency and the chromium (III) solution is recyled to the cell [10].

2a) Indirect oxidation of naphthalene with Ce(IV)

Pulping additives such as quinoid compounds increase the yield of the pulp mass up to 4% [128]. For commercial application the most promising additives are anthraquinone (AQ) or the more convenient soluble salt of tetrahydroanthraquinone (THAQ). If AQ or THAQ could be obtained at a price below $ 2/kg it would find a substantial market as a pulping additive [129]. Commercial production of THAQ is now based on the partial thermochemical oxidation of naphthalene. In recent years, however, the lure of the pulp market has promoted several attempts to develop a process for the electrosynthesis of THAQ based on the indirect electrooxidation of naphthalene to naphthaquinone (NQ) with Ce^{4+}, according to the stoichiometry of the reactions 1, 2 and 3.

$$6Ce^{3+} \longrightarrow 6Ce^{4+} + 6e^- \tag{1}$$

$$6Ce^{4+} + C_{10}H_8 + 2H_2O \longrightarrow NQ + 6Ce^{3+} + 6H^+ \tag{2}$$

$$NQ + butadiene \longrightarrow THAQ \tag{3}$$

152

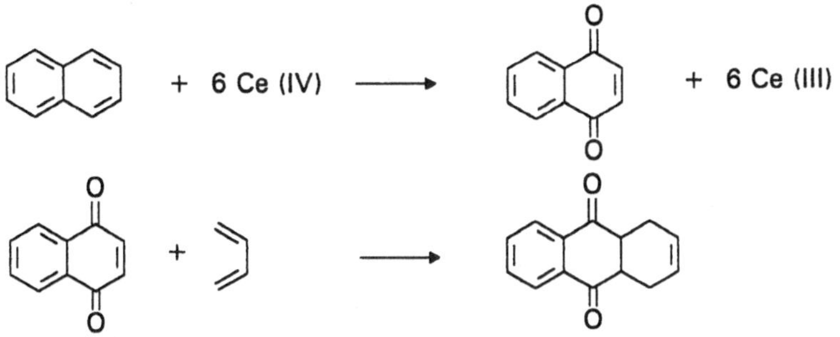

$+ 6\ Ce\ (IV) \longrightarrow$ ... $+ 6\ Ce\ (III)$

Naphtoquinone   butadiene   1,4,9a,10a tetrahydro-
(NQ)   9,10-anthraquinone
                                             (THAQ)

The main contributions to this development stem from by B.C. Research (BCR) [129, 130] for Pacific Northern Gas Ltd. and W.R.-Grace/Hydro-Quebec [129–131] in the period 1980 till 1992. The economic viability of this process depends largely on the performance of the electrochemical reactors for the generation of Ce^{4+}, which should produce concentrated solutions of Ce^{4+} (> 0.3 M) at high current density ($> 1\ kA/m^2$), high current efficiency ($> 70\%$) and long life (> 2 years).

The work on the electrochemical generation of a solution of ceric sulphate from slurry of cerous sulphate in 1–2 M sulphuric acid was abandoned by BCR due to difficulties encountered in handling slurried reactants. A 6 kW pilot reactor operated at 50 °C using a Ti plate anode and a tungsten wire cathode (electrolyte velocity: about 2 m s^{-1}) produced 0.5 M $Ce(SO_4)_2$ on the anode with a current efficiency of 60%. The usefulness of Ce(IV) has been limited by the counter anions [131, 132]. Problems include: instability to oxidation, reactivity with organic substrates and low solubility. Grace found that use of cerium salts of methane sulfonate avoids the above problems. Walsh has summarized the process history, Scheme 6 [133].

The use of methane sulfonic acid as electrolyte eliminates the slurry electrolyte, allows higher Ce^{4+} concentrations up to 1 M and simplifies process-design

1977	ECRC, Capenhurst/L B Holliday, Huddersfield: pilot and early commercial trials on naphthoquinone and anthraquinone
1980	British Columbia Research: experiment with ceric sulfate in suspension
1984	WR&Co: develop ceric methane sulfonate (rather than ceric ions in H_2SO_4)
1989	LTEE (Laboratoire des Technologies Electrochimiques et des Electrotechnologies) at Hydro-Quebec: pilots higher efficiency process using Ce(IV) in methanesulfonic acid and develops a commercial process using modern filter press call technology
1992	LTEE: offers process for licence to chemical manufacturers

Scheme 6. Process history of indirect naphthoquinone synthesis using electrogenerated Ce(IV) [133]

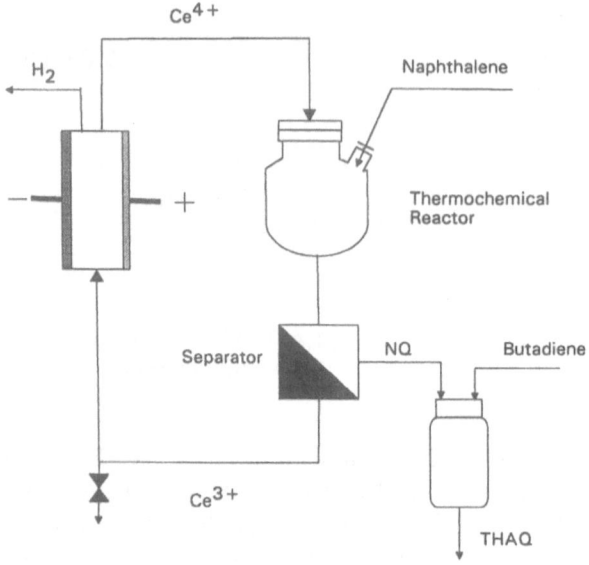

Scheme 7. Process schematic for indirect naphthoquinone synthesis using electrogenerated Ce(IV) [133]

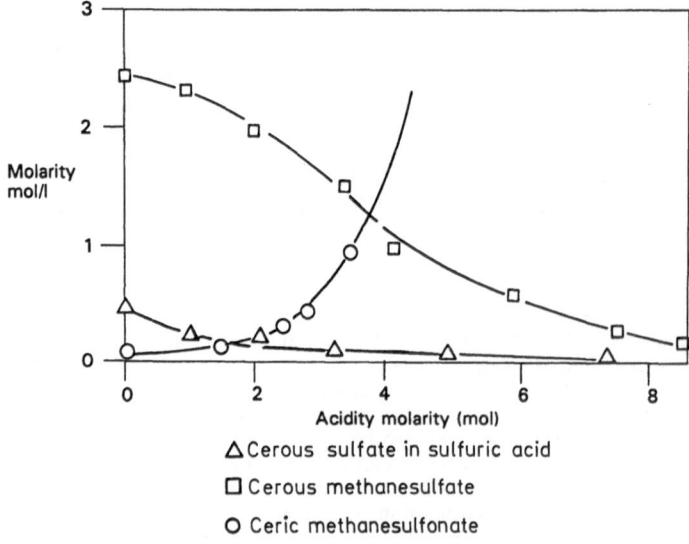

△ Cerous sulfate in sulfuric acid

□ Cerous methanesulfate

○ Ceric methanesulfonate

Fig. 10. Solubility of Ce(III/IV) in aqueous methane sulfonic acid and sulfuric acid. [132], cf also Ref. [134, 135] (see Scheme 7 [133]). Problematic is, however, the cost of the methane sulfonic acid. The solubility of Ce(III) in methane sulfonic acid is compared with sulfuric acid in Fig. 10.

The benefits of the cerium methane sulphonic process are: [132] (1) low cost, (2) methane sulphonic acid is unreactive with reactants and products, (3) stable to anodic and Ce(IV) oxidations, (4) high solubility of both Ce(III) and Ce(IV), (5) high current efficiency (> 90%) at high current density (> 4000 A m^2), and (6) fast and allows selective oxidations.

The process is nearing the end of a 3000 hour pilot trial at the Shawinigan laboratory of Hydro-Quebec in a commercial scale Electro Prod Cell, or an ICI FM21 SP cell, divided by a membrane. The pilot plant based on commercially available electrochemical cells has a design capacity of 100 t/year [132].

Compounds examined on the laboratory scale using the Ce(IV) methane sulfonic acid process are summarized in Table 11.

2b) Cleavage of the p-methoxybenzyl protecting group by Ce(IV)

This process in the synthesis of penem azetidin-2-one-intermediates is performed effectively with Ce(NH$_4$)$_2$(NO$_3$)$_6$(CAN) [46, 126, 136, 137].

Fig. 11. Hydro Quebec electrochemical plant for the synthesis of hydroquinone, using an Electro Prod Cell (photograph supplied by Hydro-Quebec)

Table 11. Compounds oxidized by electrogenerated Ce(IV) in methane sulfonic acid on the laboratory scale (Results obtained under non-optimized conditions) [132]

Reagents	Products	Conversion %	Yields %
Anthracene	Anthraquinone	98	95
Naphthalene	1,4-napthoquinone	100	98
2-methylnaphthalene	2-methyl-1,4-naphthoquinone and	94	67
	6-methyl-1,4-naphthoquinone		17
2-t-butylnaphthalene	6-*tert*-butyl-1,4-naphthoquinone and	100	65
	2-*tert*-butyl-1,4-naphthoquinone		26
1-nitronapthalene	2-nitronaphthoquinone		
Toluene	Benzaldehyde	98	93
	Benzaldehyde and	10	57
	Benzylalcohol		37
p-chlorotoluene	*p*-chlorobenzaldehyde	99	87
p-xylene	*p*-tolualdehyde	98	80
m-xylene	*m*-tolualdehyde	91	77
1,2,3.5-tetra-methyl benzene	2,4,6-trimethylbenzaldehyde	26	63
p-methylanisole	*p*-anisaldehyde	93	84

A serious drawback is the large amount of CAN (up to 2.5 molar amounts) needed. Cerium salts are highly toxic pollutants and must be removed from industrial effluents and wastewaters. Cerium (III) solutions from penem pilot plant solutions containing up to 1.2 M Ce(III) were recycled in a two compartment Electro Syn Cell. Typical recycling conditions: Nafion diaphragm with coated Ti-anode, applied current densities = 50–150 A/cm^2; yield > 90%; processed amount: about 475 kg CAN [46, 126, 136, 137]. The simultaneous determination of Ce(III) and Ce(IV) in the pilot plant solution and in solid CAN can be performed polarographically. As little as 0.3% Ce(NH$_4$)$_2$(NO$_3$)$_5$ can be determined in Ce(NH$_4$)$_2$(NO$_3$)$_6$ [136].

2c) Oxidation of toluene and substituted toluenes to aldehydes by ceric ions
[13, 35, 72, 132, 138]
The oxidation of Ce(III) sulphate in sulfuric acid has been economically evaluated in the Electro Syn Cell at various flow rates and cerous ion concentrations by Carlsson [72]. Walsh [139] has presented the calculation for the Ce(III) reoxidation process.

3) Oxidation of toluenes and its derivatives to corresponding benzaldehydes with electrogenerated manganic sulfate in H_2SO_4, cf. Sect. 3.1.1.

Vaudano [140, 140b] et al and Comninellis et al. [141–144] have described the electrooxidation of manganous to manganic sulfate on a pilot plant scale.

The scale up to an industrial plant is presented.

The conventional chemical procedure for the oxidation of toluenes to the corresponding aldehydes according to the following equation produces 302 kg $MnSO_4$ and uses up to 196 kg of sulfuric acid plus about 20% waste originating from the technical manganese dioxide consumed which in fact is a mixture of Mn(IV) and Mn(III) [145].

$$\text{CH}_3\text{-}\underset{R}{\bigcirc} + 2\,MnO_2 + 2\,H_2SO_4 \longrightarrow \text{CHO-}\underset{R}{\bigcirc} + 2\,MnSO_4 + 3\,H_2O$$

$$MnO_2 + Mn^{2+} + 4H^+ \longrightarrow 2Mn^{3+} + 2H_2O$$

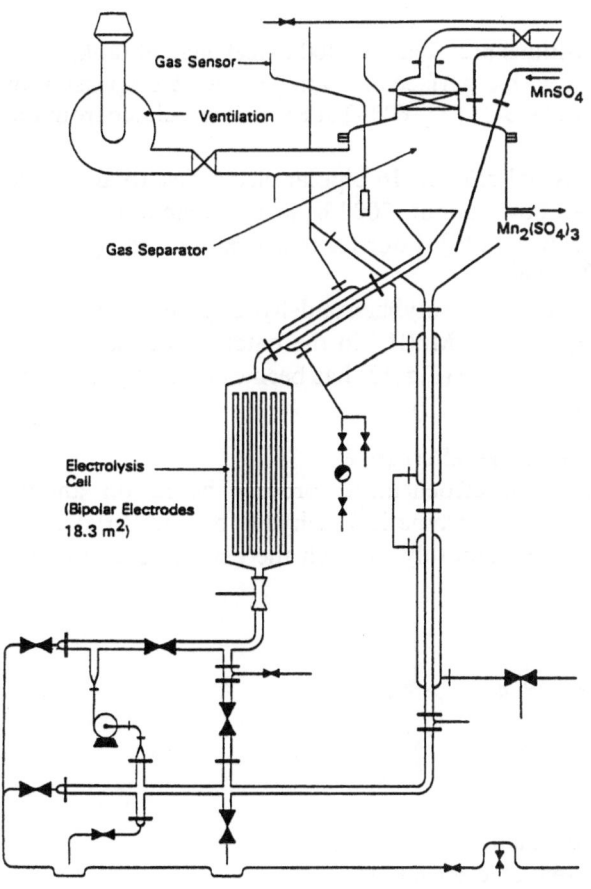

Fig. 12. Indirect electrooxidation of toluene and toluene derivatives using Mn(III) as mediator [140] Industrial plant

The electrooxidation is based on the following reaction:
(i) reaction at the anode:

$$4MnSO_4 + 2H_2SO_4 \longrightarrow 2Mn_2(SO_4)_3 + 4H^+ + 4e^-$$

(ii) reaction at the cathode:

$$4H^+ + 4e^- \longrightarrow 2H_2$$

(iii) side reaction:

$$2H_2O \longrightarrow 4H^+ + O_2 + 4e^-$$

The target homogenous chemical reaction is the oxidation of the toluene derivative to the benzaldehyde:

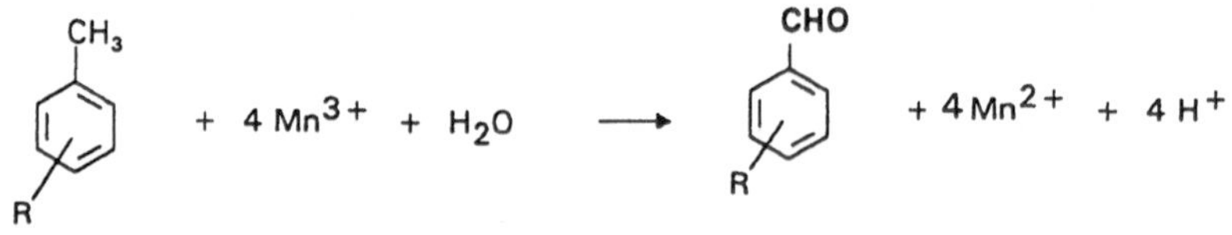

Typical pilot process conditions are: Reactor: 400 l steel enamel tank. Electrolyzer: ·10 bipolar electrodes with a surface of 0.9 m^2/300 A. A suspension of 1 kmol MnSO$_4$: H$_2$O in 535 kg H$_2$SO$_4$ [55–60%] at 85 °C is circulated from the tank to the electrolyzer.

Characteristics of the industrial reactor: 16 bipolar electrodes, total surface of 18.3 m^2. Current: 4000 A; flow: about 700 l/h Mn^{2+} reaction mass at 80–85 °C. Yield: about 90% Mn^{3+} [140]. Four technical cells are in operation [140b] with an output of 100 t/year.

The indirect preparation of p (tert-butyl)benzaldehyde (TBB) from p(tert-butyl)toluene and the electrogenerated Mn(III) in the batch mode (630 l tank) and continuous process using a cascade (Fig. 13) has been examined [140]. The data are summarized in Table 12.

Direct electrochemical oxidation as an alternative

BASF has developed a direct electrochemical process based on anodic acetoxylation for the production of aromatic aldehydes on industrial scale [40, 146, 147]. The reaction passes smoothly through the benzyl acetate stage.

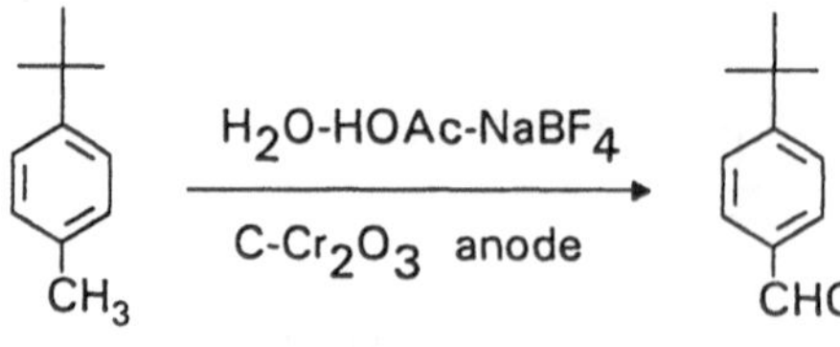

Conversion : quantitative;
yield: 87%; current efficiency: 67%

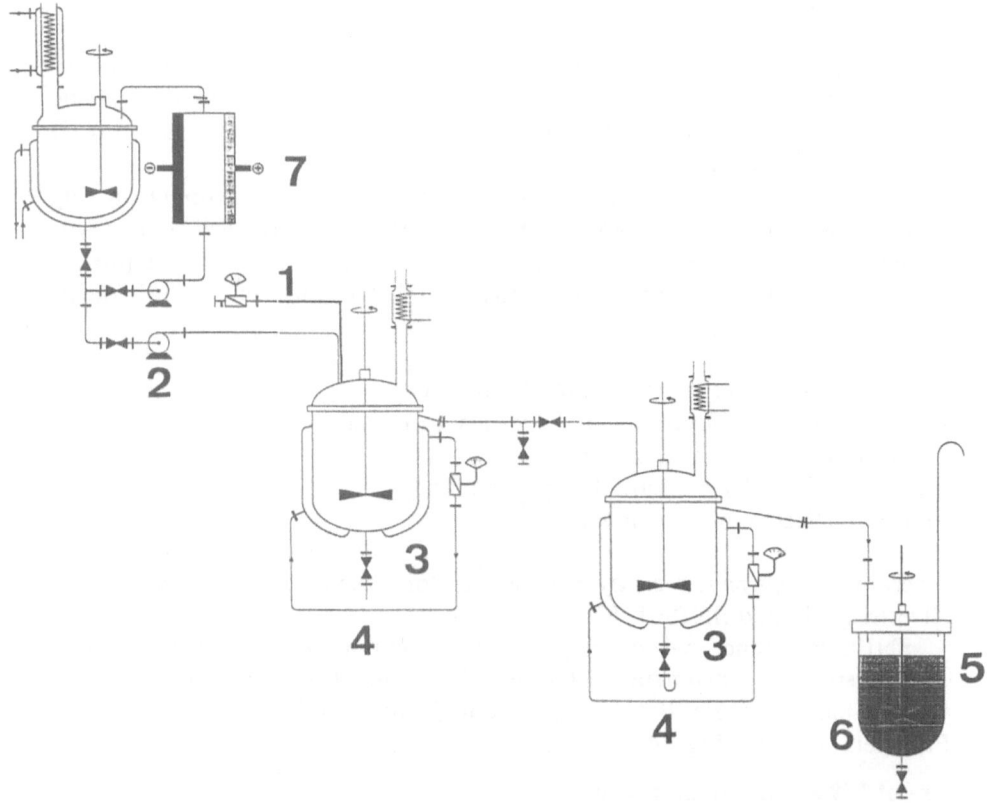

Fig. 13. Continuous indirect oxidation of toluene and toluene derivatives using a cascade process. Pilotstate [140]
(1) Dosage of TBT; (2) electrolyte; (3) chemical reaction (total = 3 vessels); (4) thermostate; (5) raw TBB; (6) electrolyte; (7) electrolysis cell

Table 12. Comparison of the continuous and batch oxidation process [140]

	Continuous	Batch
TBT	9.4 kg/h	80 kg/batch
Aquival. Mn^{3+}/1 kg TBT	8.3	9.0
Reaction volume	$0.034 \, m^3$	$0.4 \, m^3$
Conversion X_{TBT} (%)	28.8	24.8
Selectivity S_{TBB} (%)	83.6	83.8
Yield/Mn^{3+} (%)	85.1	67.0
Yield (kg TBB/h m^3)	72.2	15.3

TBT = *p-tert*-butyl toluene
TBB = *p-tert*-butyl benzaldehyde

3.1.2.2 Processes Using Consumable Electrodes

Electrochemical reduction of organic compounds in aprotic solvents is a very suitable way for preparing reactive anionic intermediates of great synthetic value [148]. Carbanions electrogenerated from organic halides readily react with various electrophiles.

Electrocarboxylation is carried out when CO_2 is used as electrophile offering an interesting alternative to organometallic synthesis. A prerequisite of this type of electroreduction in industrial scale is electrolytic cells especially adapted to use aprotic solvents. These cells must fullfil the following requirements [148, 149]:

– have a narrow and constant interelectrode gap;
– have consumable electrodes that are easy to replace;
– be run in explosive atmosphere;
– be operated under pressure (for gas handling);
– have an effective heat removal.

Several such cells have recently been devised for these reactions by CGE [150], Silvestri [151] and SNPE [152–155].

SNPE in France has developed a new technology specially adapted to electrolysis with consumable electrodes. SNPE has been operating for a few years a multi-purpose 10 t/year pilot unit located in ISOCHEM's plant in Pithiviers (France), Fig. 14.

Use of M9 sacrificial electrodes
The magnesium-based "pencil sharpner" electrolytic cell used is shown in Fig. 15. It has the following features:

– a low energy consumption due to a narrow interelectrode gap,
– a good selectivity of the electrolytic reaction due to a constant interelectrode gap and good agitation,
– use of a massive consumable electrode, e.g. 38 cm in diameter,
– easy replacement of consumed electrodes,
– efficient heat removal from the electrolytic cell,
– possibility of using gaseous reagents under pressure.

Mg is used as the consumable electrode. Industrial grade magnesium is of sufficient purity for these reactions (Pechiney manufactures 99.7% Mg costing about 4 dollars per kg). For most syntheses only about 0.2 kg of magnesium are consumed per kg of product which means about 0.8 dollars per kg. Magnesium salts are disposed of by standard procedure or can be transformed into valuable magnesium stearate [149].

The cell is suitable for the manufacture of (i) metal alkoxides, (ii) thiolates, (iii) carboxylates, (iv) hydroxides, (v) organometallics and chemicals such as diphenylacetic acid, (vi) *ortho-* and *para*-trifluoromethyl-benzoic acid, (vii) aryl-

Fig. 14. Pilot plant using consumable electrodes [115]

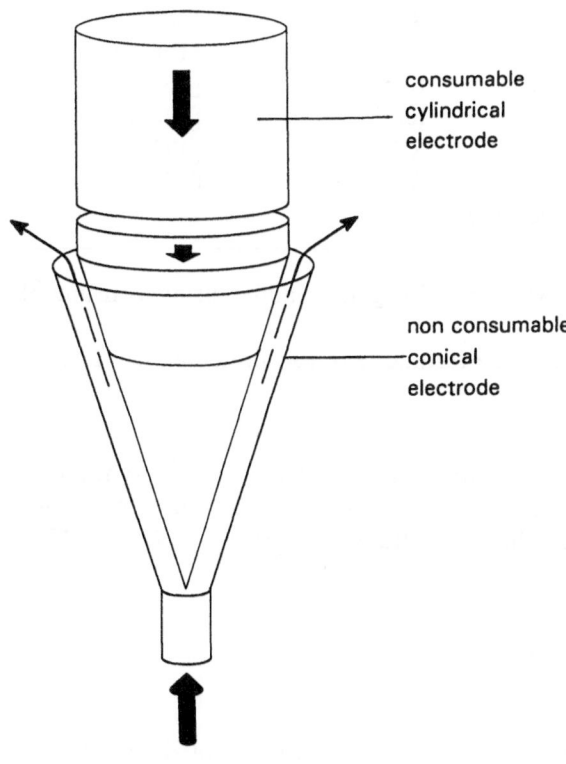

consumable
cylindrical
electrode

non consumable
conical
electrode

electrolyte

Fig. 15. Pencil sharpener electrolysis cell using consumable electrodes [115]

161

propionic acids or *meta*-trifluoromethylbenzyl-acetone, (viii) for electrocoagulation or (ix) for the manufacture of metal powders.

Electrochemical reduction of organo halides is a convenient alternative to organometallic synthesis, with the following advantages.

(1) no use of expensive, hazardous solvents,
(2) no use of dangerous reducing metals,
(3) safe operation at room temperature, easy control,
(4) single step reaction,
(5) improved selectivity.

The following diagram summarizes the possible products for different electrophiles

Organic halide	electrophile	product	
	H^+	RH	
	CO_2	RCOOH	[156–158]
	R'COX	RCOR'	[159, 160–163]
	R'COR''	RR'R''COH	[164]
$RX + e^-$	DMF	RCHO	[165]
	R'X'	RR'	[166–168]
	CO_2, cat.	RCOR	[169]
	Ph_2PCl	$RPPh_2$	[170, 171]
	SO_2	RSO_2H	[172]
	R'_3SiCl	$RSiR'_3$	[173]
	metal of the electrode	"R-Met"	

Examples for reductive carboxylations of organic halides using consumable anodes are given in Table 13

A list highlighting the scope of the technique is also presented in Table 9 b. See also Refs. [175, 176].

Use of Zn sacrificial electrodes

A typical example is the synthesis of oxalic acid. Electrochemical synthesis of oxalic acid by reduction of CO_2 in aprotic media with a Zn sacrificial anode was brought to pilot scale by the Dechema Institute some years ago (1981) [177]

$$CO_2 + e^- \longrightarrow (CO_2^- \cdot)$$

$$2CO_2^- \cdot \longrightarrow {}^-O_2C - CO_2^-$$

$$Zn \longrightarrow Zn^{2+} + 2e^-$$

The synthesis is performed with Zn anodes; after hydrolysis of Zn oxalate, the mother liquors are sent to recovery, from which the Zn can be recycled to the

Table 13. Electrochemical carboxylation of organic halides at dissolving metal anodes [174]

Overall reaction :

$$RX + Mg + CO_2 \text{---> } RCOO^- + X^- + Mg^{2+}$$

Medium : 0.02 M Bu_4NBr/DMF, CO_2 at 10°C, undivided cell

Entry	Halide	Acid	Isolated yield (%)
1	$CH_3(CH_2)_8CH_2Br$	$CH_3(CH_2)_8CH_2COOH$	75
2	$PhCH=CHCH_2Cl$	$PhCH=CHCH_2COOH$	80
3			70-90
4			50-80
5	$PhCF_3$	$PhCF_2COOH$	65
6	PhCl	PhCOOH	70
7	$CH_3COC_6H_4Br$	$CH_3COC_6H_4COOH$	90
8	$PhCH=CHBr$	$PhCH=CHCOOH$	80
9			70

electrosynthesis unit. Current and product yields of 85% and 90%, respectively, were claimed. The system, which poses no environmental problems, could be competitive with chemical methods should the latter incur higher costs as a result of more severe environmental restrictions [178].

Use of Al consumable electrodes [178, 179]

In the case of benzal chloride, the carboxylation in conventional diaphragm systems fails, leading to poor yields in phenylacetic and mandelic acid [180]. At an Al anode, carboxylation occurs because the self-esterification of the first carboxylate anion onto the second chloride group is hindered by the formation of Al complex salts [181]. Yields of phenylmalonic and chlorophenylic acetic acids up to 30% each have been obtained [178].

$$\text{C}_6\text{H}_5\text{-CHCl}_2 \xrightarrow[-\text{Cl}^-]{+\ 2e^-\ +\ \text{CO}_2} \text{C}_6\text{H}_5\text{-CH(Cl)-COO}^-$$

$$\xrightarrow{-\text{Cl}^-} \quad \text{(phenyl epoxide, } =\text{O)} \quad \xrightarrow{\text{OH}^-} \quad \text{C}_6\text{H}_5\text{-CH(OH)-COOH}$$

$$+\text{Al}^{3+} \longrightarrow \quad \text{Ph-CH(Cl)-C(O)-O}{\longrightarrow}\text{Al}^{3+} \xrightarrow[-\text{Cl}^-]{+\ 2e^-\ +\ \text{CO}_2} \quad \text{Ph-CH(COO}^-)\text{-C(O)-O}{\longrightarrow}\text{Al}^{3+}$$

Electrocarboxylation of ketones and aldehydes

Electrochemistry offers new routes to the production of several commercially relevant α-arylpropionic acids, used as non-steroidal anti-inflammatory agents (NSAI) [178, 182]. A preparative method based on sacrificial Al-electrodes has been set up for the electrocarboxylation of ketones [117, 183–187] and successfully applied to the electrocarboxylation of aldehydes, which failed with conventional systems. The electrocarboxylation of 6-methoxy-acetonaphthone to 2-hydroxy-2-(6-methoxynaphthyl)propionic acid, followed by chemical hydrogenation to 2-(6-methoxynaphthyl)-2-propionic acid – one of the most active NSAI acids – has been developed up to the pilot stage [184, 186].

In this case, the aluminium salt of the hydroxy acid is soluble in the reaction medium. It precipitates on addition of a number of solvents such as alcohols or ethers. 90% product-selectivity was observed in preparative syntheses (up to 40 g of pure 2-(6-methoxynaphthyl)propionic acid per run). The power consumption was given as about 2.7 kWhkg^{-1}.

Electrocarboxylation of imines

The first studies on the electrocarboxylation of the $> \text{C} = \text{N} -$ double bond of imines were performed on conventional systems [178, 188, 189]. The studies showed that in most cases the reaction is almost unselective, leading (i) not only to the desired monocarboxylation, and to (ii) the corresponding N-substituted amino acid with yields not exceeding 60% but also to (iii) products of monocarboxylation on the nitrogen atom, (iv) of dicarboxylation, (v) of carbon-carbon coupling, or (vi) of hydrogenation [178, 189–191]. Electrocarboxylation with Al sacrificial anodes [192, 193] in a diaphragmless cell appears to be more selective. The hydrogenation is almost totally suppressed and the dimerization substantially reduced. Monocarboxylation at the carbon atom and dicarboxylation are the main reactions [178]. An electrochemical plate and frame cell, with semi-continuous renewal of the particulated Al sacrificial bed, has been proposed for the scale-up of electrocarboxylation processes, and successfully used for the carboxylation of benzalaniline [178, 179].

$$>C = N-$$

$$+2e^- +2CO_2 \longrightarrow \underset{\underset{|}{COO'}}{\overset{COO'}{\underset{N}{\diagdown}}}\diagdown \quad \text{acid hydrolysis} \quad \downarrow \; -CO_2$$

$$\underset{\underset{H}{\overset{|}{N}}}{\overset{COO'-}{\diagdown}}\diagdown$$

$$+ 2e^- + CO_2 + H^+$$

For further details the reader is referred to the excellent paper by Chaussard et al. [149]: "Use of sacrificial anodes in electrochemical functionalization of organic halides" and the paper by Silvestri et al. [178]: "Use of sacrificial anodes in synthetic electrochemistry. Processes involving carbon dioxide".

Degner [194] and Couper et al. [75] have recently critisized the technology as it unavoidably produces, after the separation of the products, aqueous solutions containing stoichiometric amounts of salts of metals used as anodes. Solutions to this problem are possible as demonstrated in the case of Mg and Zn [177] electrodes. Al and Mg can easily be precipitated as hydroxides, recovered by filtration and dehydrated to the corresponding oxides, whereas Zn is recycled electrochemically.

3.1.2.3 Indirect Processes Using Modified Electrodes

The transfer of redox equivalents can be achieved by an electrocatalyst (mediator) or a modified electrode. Indirect electrolysis can lead to a better selectivity due to the specific interaction of the mediator with the substrate. However, low turnovers and the need to separate the mediator from the product are possible disadvantages, as mentioned above. The nickel hydroxide electrode [195, 196] is fairly free from these disadvantages. The following mechanism for the oxidation at the nickel hydroxide electrode has been proposed in the literature [195].

(a) $Ni(OH)_2 + OH^- \longrightarrow NiOOH + H_2O + e^-$

(b) $R-CH_2OH_{sol.} \longrightarrow R-CH_2OH_{ads.}$

(c) $R-CH_2OH_{ads.} + NiOOH \longrightarrow R-\dot{C}HOH + Ni(OH)_2$

(d) $R-\dot{C}HOH + H_2O \longrightarrow R-CO_2H + 3H^+ + 3e^-$

(e) $R-\dot{C}HOH + 3NiOOH + H_2O \longrightarrow RCO_2H + 3Ni(OH)_2$

At a nickel metal surface in alkaline aqueous medium a thin film of Ni(II) hydroxide is formed. At $+0.63$ V (vs NHE) the film is oxidized to Ni(III) oxide hydroxide (a). After adsorption of the substrate at this surface (b) hydrogen atom abstraction at the α-carbon of the substrate occurs in the rate determining step (c). The intermediate radical is then further oxidized either directly (d) or indirectly (e) to the product.

165

The nickel hydroxide electrode resembles in its applications and selectivity the chemical oxidant nickel peroxide. The nickel hydroxide electrode is, however, cheaper, easy to use and in scale-up, and produces no second streams/ waste- and by-products [196]. Nickelhydroxide electrode has been applied to the oxidation of primary alcohols to acids or aldehydes, of secondary alcohols to ketones, as well as in the selective oxidation of steroid alcohols, cleavage of vicinal diols, in the oxidation of γ-ketocarboxylic acids, of primary amines to nitriles, of 2,6-di-*tert*-butylphenol to 2,2′,6,6′-tetra-*tert*-butyldiphenoquinone, of 2-(benzylideneamino)-phenols to 2-phenyloxazols, of 1,1-dialkylhydrazines to tetraalkyltetrazenes. For details the reader is referred to Ref. [195].

The oxidation of carbohydrates at the nickel hydroxide electrode has been addressed in only a few papers [195]. Seiler and Robertson [197, 198] developed a technical process, which allows the oxidation of 2,3,4,6-di-*o*-isopropylidene-L-sorbose (**18**, DAS) to protected gulonic acid (**19**). The acid, an intermediate of the vitamin C-synthesis, can be produced in a scale of two tons per day at the nickel hydroxide electrode.

(18)   **(19)**

(DAS)   (93%)

The cell used (Fig. 16) is an undivided, so called "Swiss-roll" cell (Figs. 16 and 17) with a Ni mesh anode and a steel cathode.

3.2 Inorganic Synthesis

New electrochemical processes are also of interest in inorganic chemical synthesis.

"Electrochemical technology has had a long association, for instance, with the pulp and paper industry through the supply of classical bleaching chemicals like sodium hydroxide, chlorine, sodium hypochlorite and sodium chlorate/chlorine dioxide". In this context the reader is referred to an excellent review paper by Oloman "Electrochemical Synthesis and Separation Technology in the Pulp and Paper Industry" [129].

Ozone

Ozone is a powerful though relatively unselective oxidant which is being considered in several pulp mills as a replacement for chlorine in bleaching chemical pulp. The first installation of ozone bleaching equipment was in 1992. The ozone required for a large pulp mill is about 10 ton/day O_3 for a 1000 t/day pulp [129]. The phase volume ratio (gas/pulp) of about 8 wt% ozone by corona

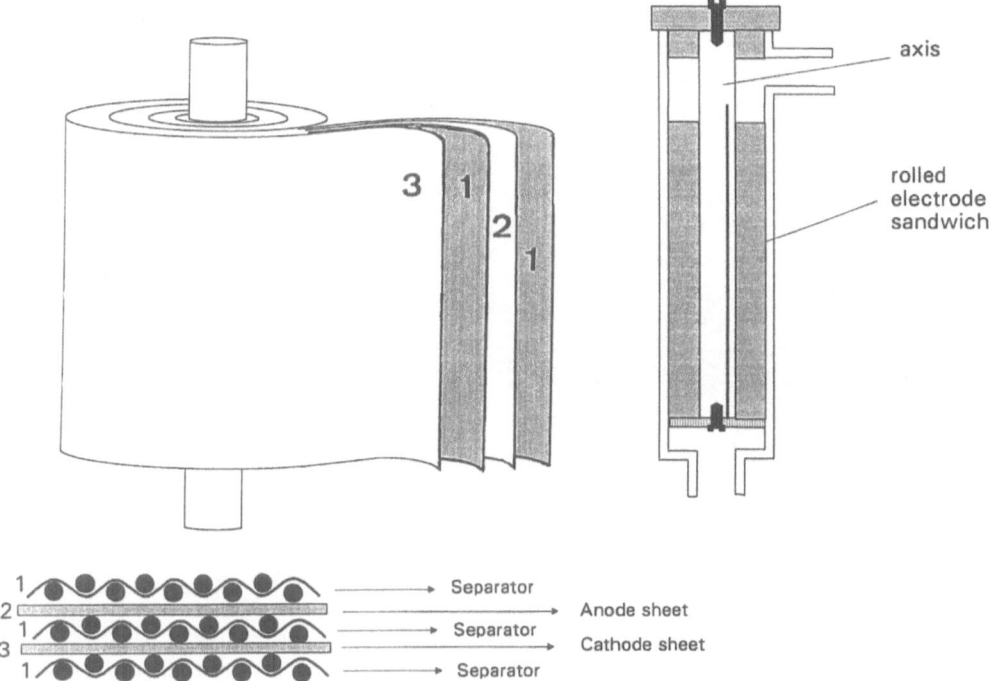

Fig. 16. "Swiss roll" electrolysis cell [198] (1) separator; (2) anode sheet; (3) cathode sheet

discharge is too high for conventional medium consistency bleaching equipment. Gas compressing and other methods have been proposed for overcoming this problem, however, they are costly. Foller [199] has reviewed electrochemical methods for on-site generation of ozone. Three ways have been seriously considered for the electrochemical production by which ozone can be prepared for commercial applications.

1) Electrolysis of water contained within a perfluorinated sulphonic acid membrane (Membrel water electrolysis) developed by ABB [133, 201–205] and Sasakura [200]. Current efficiencies reach 14–15% at current densities of 10 000 A m^2 and more. The cells are generally immersed directly in the water so that ozone is introduced directly into the water to be treated, Fig. 18.

$$3H_2O \longrightarrow O_3 + 6H^+ + 6e^-$$

The solid polymer electrolyte cells are viewed as being particularly appropriate for the treatment of high purity water systems, including the provision of ultra pure water for the pharmaceutical industry, cf. Ref. [205]. The process is often coupled with UV radiation which serves to decompose unwanted, residual ozone [133].

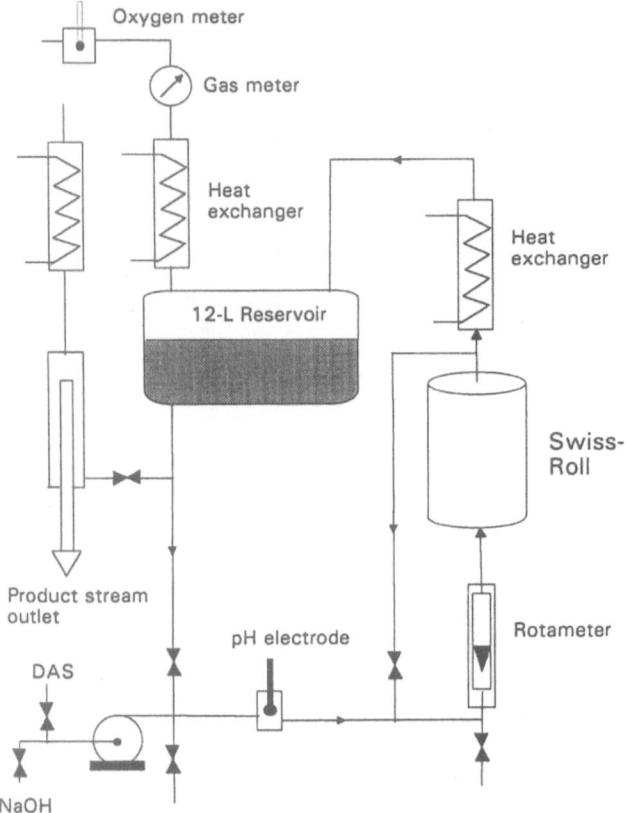

Fig. 17. Schematic diagram of the flow system used for batch and continuous oxidation of DAS (Diacetone-L-sorbose) with a "swiss roll" cell with an anode area of 3 m²

2) An alternative route to ozone is through the electrolysis of fluoroboric acid [129, 135, 199, 206–208].

The reaction is carried out in high concentration fluoroboric acid at a high overpotential refrigerated anode such as glassy carbon or lead dioxide to give ozone at a current efficiency around 40%. This electrosynthesis which produces ozone at concentrations up to about 40 wt% in oxygen is being developed by ICI/Oxy Tech Inc. for small scale commercial applications [129, 135, 199, 207, 209]. This approach is viewed as being particularly appropriate for small scale applications that require high concentrations of ozone, e.g. use in potable water supplies, for oxidizing cyanides and organics, such as phenols in industrial process liquors [199]. A modification of the process, operated at 30 °C and up to

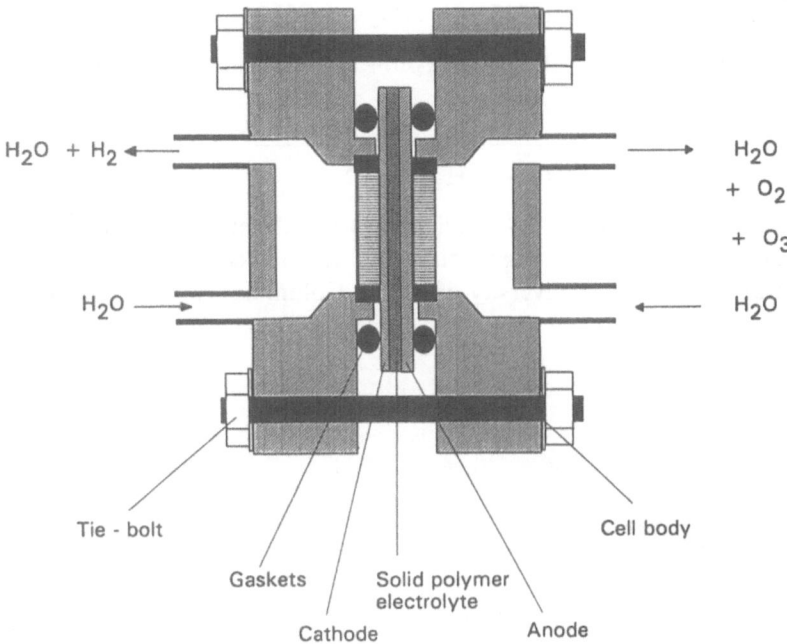

Fig. 18. Experimental cell for ozone generation in water. The hardware is made from Ti on the anode side and stainless steel on the cathode side [201]

10 kA/m^2 with an oxygen depolarised cathode, is being developed in Japan by Permalec Electrode Ltd. [129, 200].

3) Fischer GmbH (Meckenheim, Germany) has developed a cell involving the electrolysis in phosphate buffer that can produce up to 23 wt% ozone, roughly 1.3 g/h [48, 199].

Electrosynthesis gives a good concentration for pulp bleaching but the scale-up of electrosynthesis reactors to the range of 1 ton/day of ozone is problematic. The capital and operating costs probably make electrosynthesis uncompetitive with the corona discharge process [129].

H_2O_2

Hydrogen peroxide is an environmentally friendly reagent which is presently used in conventional chlorine/chlorine dioxide bleach sequences and is a major component of proposed sequences with oxygen, ozone and other non-chlorine reagents. A large chemical pulp mill consumes 10 to 20 ton/day H_2O_2. The high price of peroxide of about \$ 1/kg H_2O_2 is a major hindrance to its wide application in pulp bleaching and brightening [129].

The fact that most pulp and paper applications use dilute solutions of hydrogen peroxide in alkali has kindled interest in the electrosynthesis of alkaline peroxide solutions by cathodic reduction of oxygen.

Cathode reaction: $O_2 + H_2O + 2e^- \longrightarrow HO_2 \cdot + OH^-$

Anode reaction: $2OH^- \longrightarrow 1/2O_2 + H_2O + 2e^-$

Solution reaction: $HO_2 \cdot + H_2O \longrightarrow H_2O_2 + OH^-$

Net reaction $1/2O_2 + H_2O \longrightarrow H_2O_2$.

The electroreduction of oxygen in alkali occurs readily at carbon electrodes. Currently used electrodes are: a graphite fiber trickle-bed/perforated bipole reactor, operated at U.B.C (University of British Columbia) [210, 211]; a com-

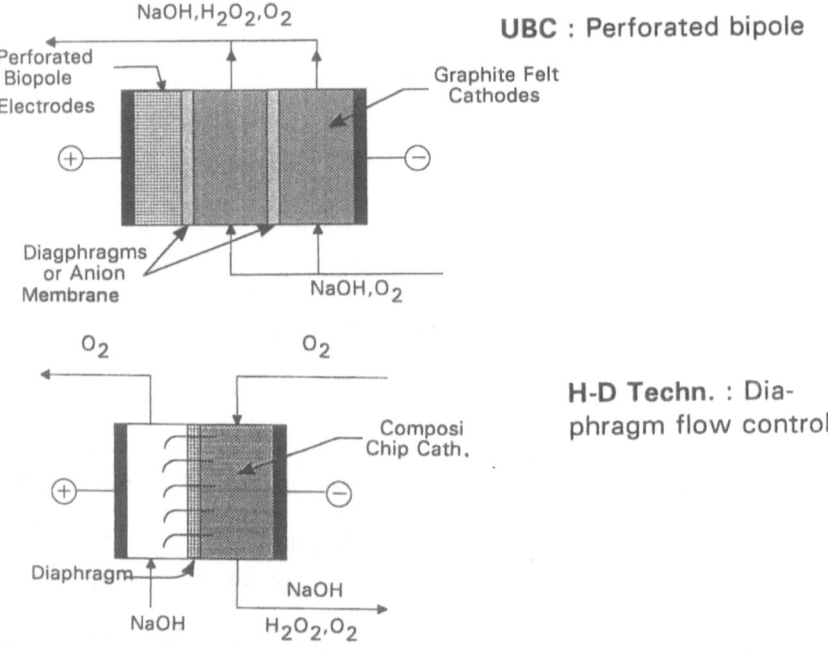

UBC : Perforated bipole

H-D Techn. : Diaphragm flow control

E-Tek : Gas diffusion

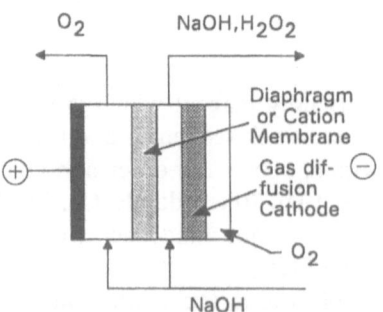

Fig. 19. Alternative cell designs for the electrosynthesis of alkaline peroxide solutions [129]

posite chip carbon/Teflon reactor (Dow Chemical/H-D-Tech Inc. [212]; and a carbon cloth reactor (E-TEK, Inc.) [199]. Figure 19 shows the principle features of cell constructions in the U.B.C., H-DTech and E-TEK peroxide reactors. The E-TEK technology is presently being tested in a FM21 commercial scale reactor.

H-D Tech Inc. installed a commercial electrosynthesis plant at the Muskogee mill of Fort Howard Paper in 1991. The H-D Tech reactor operates near atmospheric pressure at $0.6 \, kA/m^2/2V/30 \, °C$ to produce solutions containing up to about 4 wt% H_2O_2 in 6 wt% NaOH at about 85% current efficiency [129].

Co-generation of peroxide and chlorate in the same electrochemical cell is currently studied at the University of B.C. (British Columbia) as hydrogen peroxide and chlorate are often used in the same pulp bleaching sequence. A divided trickle-bed cell using a RAIPORE 1035 [213] anion exchange membrane has produced useful concentrations of H_2O_2 and $NaClO_3$ with current efficiencies of about 70% at $2.4 \, kA/m^2/4V$. The commercialization of this cell type, however, depends on a reasonably priced robust anion exchange membrane, which is not yet available. [129] Sale/manufacture of the RAIPORE membrane stopped at the end of 1992.

N_2O_5 [214, 214a]

ICI is focusing on the electrosynthesis of N_2O_5, an effective nitration agent which could be of interest for the fine chemicals and pharmaceutical industries. Dinitrogen pentoxide (up to 35 wt%) is almost 100% ionized in solutions of nitric acid, giving a high concentration of nitronium ion. The dinitrogen pentoxide system is less acidic than the mixed acid system (nitric acid/sulfuric acid 70–90%) widely used as nitrating agent. Due to the reduced concentration of free protons and water, N_2O_5 can be used to nitrate substances sensitive to free protons and water. Although it has been synthezised under laboratory conditions – for example by dehydration of nitric acid with phosphorus pentoxide – it is not widely used because of the difficulty of producing it on a commercial scale. Electrooxidation of dinitrogen tetroxide to dinitrogen pentoxide was reported back in 1910. ICI's Electrochemical Technology division now offers a small-scale production unit, based on its FMO1 laboratory electrolyser, and design packages for pilot and full-scale plants. A simplified diagram for the production of N_2O_5 in anhydrous nitric acid is shown in Scheme 8.

At the cathode nitric acid is reduced to dinitrogen tetroxide and water. The dinitrogen tetroxide is cycled to the anode chamber where it is oxidized to the dinitrogen pentoxide. The anolyte product contains up to 35 wt% N_2O_5 in nitric acid with less than 1 wt% N_2O_4.

Sodium dithionite (hydrosulphite)

$Na_2S_2O_4$ is usually produced by reduction of $NaHSO_3$ with reducing agents such as sodium borohydride, zinc or sodium formate. The electroreduction of sulfurous solutions as a source of dithionite has been extensively studied by Oloman [129, 215, 216]. Olin Corp. has recently brought electrosynthesis of dithionite to commercial scale with plants of about 12 ton/day operating in Charleston, USA

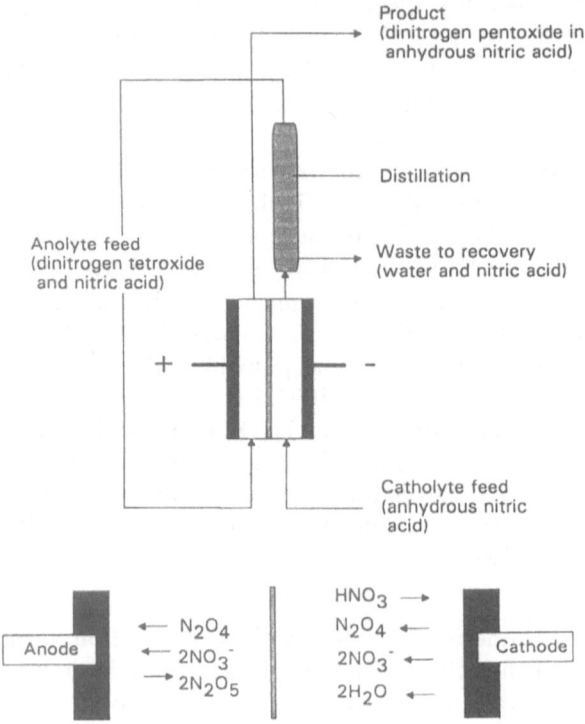

Scheme 8. Simplified diagram of ICI's technique for the production of nitrogen pentoxide in anhydrous nitric acid [214]

and Sao Paulo, Brazil [129, 217]. A flow-through cathode composed of metal fiber felt which operates at about 3 kA m^{-2} and 90% current efficiency is used in a divided cell with a cation exchange membrane. The catholyte solution containing $NaHSO_3$ and $Na_2S_2O_4$ is recycled through the cathode and enriched with sulphur dioxide in an external vessel. Olin has also investigated the removal of thiosulfate from dithionite solution by electroreduction [129, 218].

Hypochlorite and low-tonnage chlorine electrolysers

At one time, sodium hypochlorite was manufactured electrochemically on a substantial scale. Now it is regarded as a by-product of the chlor-alkali industry [10]. On the other hand, there are many situations where low volumes of hypochlorite may be required or the requirement is irregular. Aqueous solutions of hypochlorite are much safer than chlorine gas but contain $< 15 \text{ wt}\%$ of active chlorine. Hence, storage and transportation costs are relatively high. Often the most convenient and cost-effective solution is to electrolytically generate OCl^- in situ [10].

Cells for the hypochlorite production do not principally require a separator, since cathodically generated hydroxide

$$2H_2O + 2e^- \longrightarrow H_2 + 2OH^-$$

is immediately consumed in the electrolyte by hydrolysis of anodically produced chlorine.

$$2Cl^- - 2e^- \longrightarrow Cl_2$$

$$Cl_2 + 2OH^- \longrightarrow H_2O + OCl^- + Cl^-$$

Different factors reduce the current efficiency [133]:

(1) At high OCl^- concentrations, high temperature and high turbulent flow conditions, reduction of the ClO^- becomes important.

$$ClO^- + H_2O + 2e^- \longrightarrow Cl^- + 2OH^-$$

Also, overoxidation to chlorate may occur at the anode

$$6OCl^- + 3H_2O - 6e^- \longrightarrow 2ClO_3^- + 4Cl^- + 6H^+ + 3/2O_2$$

(2) Oxygen may occur: at low chloride concentrations, at low temperature or in the case of insufficient electrolyte motion.

$$2H_2O - 4e^- \longrightarrow O_2 + 4H^+$$

In the case of seawater electrolytes (scaling of the electrodes) deposition of metal hydroxides, e.g. $Mg(OH)_2$ due to localised increases in pH, are possible. These problems are overcome by adequate cell design maintenance programms [133]. The past 10 years have seen the introduction of a number of small electrolysis cells for the generation of either hypochlorite or chlorine gas. For a detailed discussion of the different cell types the reader is referred to Ref. [10]. ElectroCell Systems AB, Sweden, commercializes a small unit ("Chlor-o-Safe") [79] for on-site production of sodium hypochlorite at concentrations of only 1%. Designed for swimming pools, municipal water plants, food processors, industrial plants, wastewater treatment plants, cooling towers, power plants, chemical manufacturers et al. It eliminates the hazards and expenses of transporting, storing and handling chlorine and commercial hypochlorite. The ElectroCell skid, mounted, fully instrumented electrolytic system for on-site chlorine generation for use in water disinfection, is shown in Fig. 20.

Technical data of "Chlor-o-Safe" units are as follows: [99]

Size	S1	S2	S3	S4
Capacity Cl_2 g/h	80	160	320	640
Water consumption l/h	7	14	28	56
Salt consumption kg/h	0.2	0.4	0.8	1.6
Power consumption kW, DC	0.4	0.8	1.6	3.2
Rectifier V/A	6/100	6/200	10/200	10/400

Chlorine dioxide

Chlorine dioxide (ClO_2) is a powerful oxidizing agent. It has a selective reactivity that makes it useful in many applications such as water treatment, disinfec-

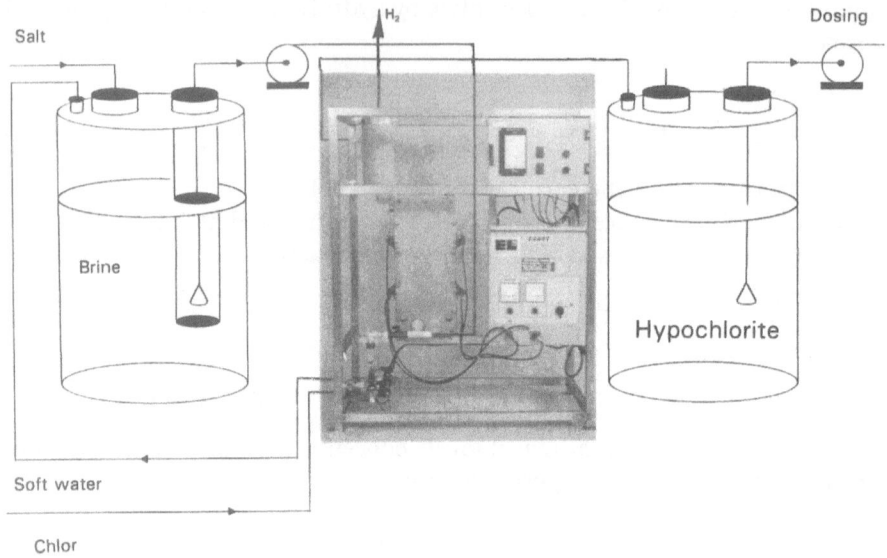

Fig. 20. Chlor-o-Safe: On-site production of sodium hypochlorite, Model S, ElectroCell Systems AB, Sweden [99]

tion/sanitization, bleaching in the pulp and paper industry, and chemical oxidation where chlorine or other oxidizing agents are unsuitable. ClO_2 is an extremely effective bactericide which is equal to or superior to chlorine on a mass-dosage basis. Unlike chlorine, ClO_2 does not hydrolyze in water and therefore its germicidal activity is relatively constant over a broad pH range. Reactions of organics also differentiate ClO_2 from Cl_2 [219]. Chlorine dioxide reacts by an oxidation mechanism with organics and does not predominently yield chlorinated organics as is the case with chlorine or hypochlorite [219, 220]. Proposed regulations on AOX will require that pulp mills substitute all chlorine with chlorine dioxide and thus reduce chlorinated organic compounds subsequently discharged to the environment [224].

The Olin Corporation has developed an electrochemical chlorine dioxide generator technology that safely and conveniently produces aqueous chlorine free solutions of chlorine dioxide at chlorite molar conversion efficiencies of 95% or better [221–224]. The chlorine dioxide is produced by direct oxidation of sodium chlorite, Fig. 21. The electrochemical ClO_2 generator system is shown in Fig. 22.

The electrochemical generator is designed for both small (0.136–4.5 kg ClO_2/day) and larger scale (0.5–27 kg/h range and more) chlorine dioxide production rates. The chlorine dioxide solution from this system is suitable for sanitizing and disinfection applications as well as waste water treatment.

Eka Nobel, a major supplier of sodium chlorate and chlorine dioxide has focussed on the possibilities to convert sodium chlorate and sodium sesquisulfate into products which can be used internally by pulp mills. A recently

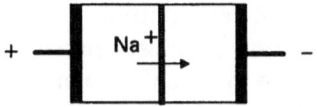

Anode	Cathode		
$NaClO_2 \longrightarrow Na^+ + ClO_2^-$	H_2O	$\rightleftharpoons$	$H^+ + OH^-$
$ClO_2^- \longrightarrow ClO_2 + e^-$	$H^+ + e^-$	$\longrightarrow$	$1/2\ H_2$
	$Na^+ + OH^-$	$\longrightarrow$	$NaOH$

$$NaClO_2 + H_2O + Energy \longrightarrow ClO_2 + NaOH + 1/2\ H_2$$

Fig. 21. Electrolytic (anodic) oxidation of chlorite [219]

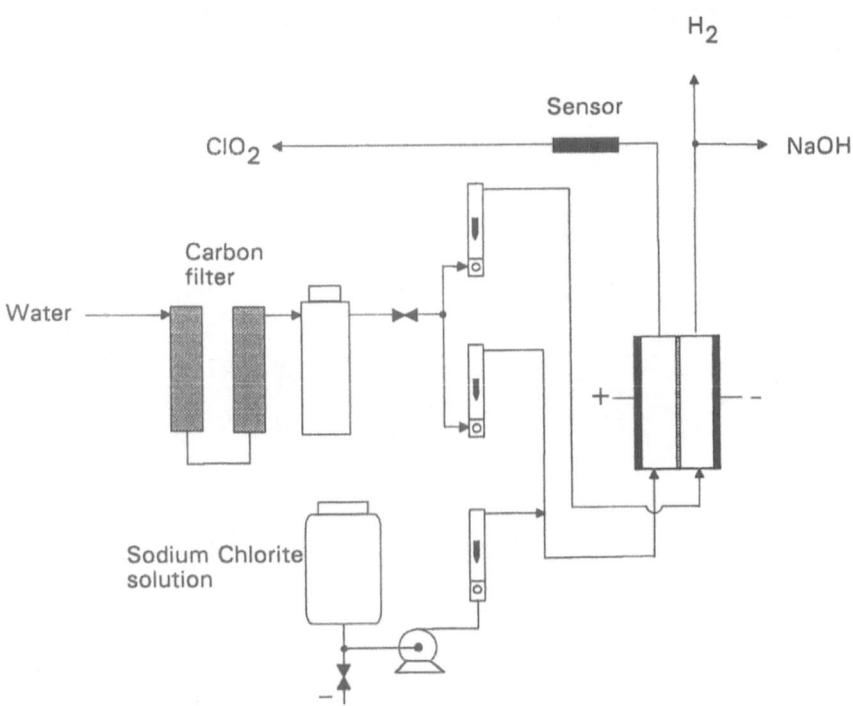

Fig. 22. Electrochemical ClO_2 generation from $NaClO_2$ [219]

P.M. Bersier et al.

developed process combines electrochemical salt splitting and generation of chlorine dioxide [225]. The chlorine dioxide processes based on hydrogen peroxide as reducing agent (SVP-HP) are due to formation of neutral salt cakes

$$NaClO_3 + 1/2H_2O_2 + H_2SO_4 \longrightarrow ClO_2 + 1/2O_2 + NaHSO_4$$
$$+ H_2O$$
$$NaClO_3 + 1/2H_2O_2 + NaHSO_4 \longrightarrow ClO_2 + 1/2O_2 + Na_2SO_4$$
$$+ H_2O$$

better suited for integration compared with systems based on methanol. Advantages of the H_2O_2 process are, apart from the formation of a neutral salt cake thanks to the more efficient use of protons, a faster reaction kinetics and an increase of the production capacity of a typically 40 tons per day (TPD) ClO_2 plant to 55 TPD. The electrochemical process involved for splitting the formed Na_2SO_4 saltcake into H_2SO_4 and NaOH – which are recycled to the pulp process – uses a two-compartment cell, incorporating special anodes. Hydrogen depolarized anodes (DPA = Hydrina) manufactured by DeNora Permelec are used. Hydrina anodes exhibit an energy consumption of 2800 kWh/ton NaOH as compared to 4800 for the DSA-O_2 anodes. The simplified flow scheme of an integrated process is shown in Scheme 9.

A demonstration plant based on SVP-HP and Hydrina technology with a ClO_2 capacity of 2 TPD and a production of about 1.2 TPD of NaOH started in 1993.

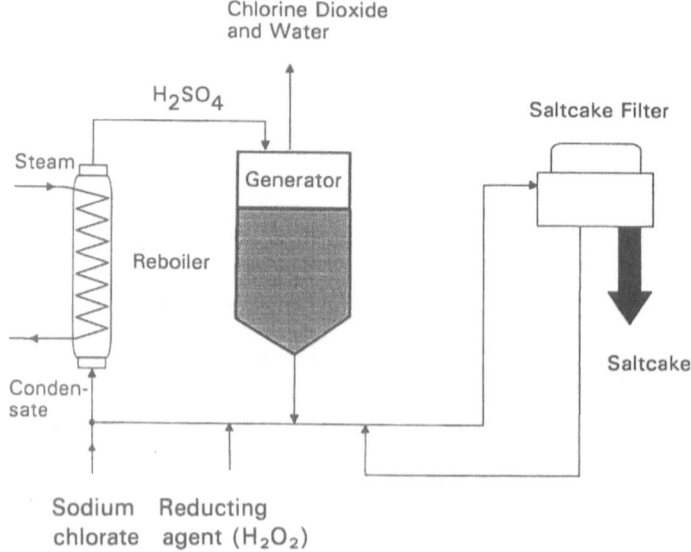

Scheme 9. Simplified flow scheme of an integrated SVP process, combining salt splitting and ClO_2 synthesis [225]

4 Electrochemical Treatment of Waste and Effluents

As a consequence of the activities of modern industries, many man-made organic compounds and inorganic pollutants, including, colorants, have found their way into the environment, without much knowledge being available on their possible harmful effects on it [226]. Toxicity is being perceived more and more by the public as a major threat to its well-being. In the light of the ever-more stringent environmental regulations and rising costs, the following requirements must be met by the (chemical) industry of the 1990s and beyond [46, 226]:

(1) Avoidance or minimization of wastes. Even with high yield methods, for one ton of finished dye, one ton of by-product may arise in addition to auxiliaries such as solvents, catalysts.
(2) Utilization of wastes.
(3) Ecologically acceptable waste disposal. A company's reputation can be severely damaged by mismanagement of waste.
(4) Avoidance of production processes or production of substances which give rise to insoluble problems of waste disposal.

Worldwide environmental regulations are setting limits on the discharge of a wide range of compounds, including heavy metals, cyanides, a large number of chlorinated organic compounds, aromatic unsaturated hydrocarbons and their derivatives, etc. [10]. Because many of these species are electrochemically active, electrolysis is a possible route for their elimination, recovery or recycling.

Electrochemical processes and technologies can contribute to a cleaner environment by two principle alternatives [23, 24, 227, 227a, –228]:

(1) Process integrated environmental protection which means the invention of new processes or the modification of known processes in such a way that less or no waste is produced by the process. cf. Sect. 3.
(2) The process is directed towards purification of different kinds of industrial wastes, cf. Kreysa, Ref. [228].

Advances in electrochemical engineering including cell design and construction materials (e.g. Ebonex, new types of ion-exchange membranes, etc.) and also a better understanding of the electrochemistry of electroactive species make electrochemical techniques increasingly more attractive for water and effluent treatment, air pollution abatement and control, recycling of chemical process streams, destruction of waste materials and toxic chemicals, nuclear decontamination by electrochemical techniques. Nowadays readily available off-the-shelf hardware (cells) can help to meet the ever-more demanding environmental legislation. An in depth discussion of currently used electrochemical technologies in the environmental field is beyond this chapter and must be restricted to a brief presentation of different cells on the market (Sect. 4.1) and to a number of selected successful applications to the treatment of inorganic (Sect. 4.2) and organic pollutants (elimination, decolorization of wastewaters and effluents (Sect. 4.3)).

For a detailed discussion the reader is referred to a number of recent monographs [9, 10, 23] and review articles [50, 226–231]. The role of electrochemistry in waste water and effluent treatment is still queried, as remarked by Pletcher and Walsh [10]. One answer would be, "relatively small" since there are many competitive methods which are cheaper on a large scale and use less energy. Principle types of processes used in local-authority sewage works are listed in Table 14.

Mechanical and biological methods are very effective on a large scale, and physical and chemical methods are used to overcome particular difficulties such as final sterilization, odor removal, removal of inorganic and organic chemicals and breaking oil or fat emulsions. Normally, no electrochemical processes are used [10]. On the other hand, there are particular water and effluent treatment problems where electrochemical solutions are advantageous. Indeed, electrochemistry can be a very attractive idea. It is uniquely clean because (1) electrolysis (reduction/oxidation) takes place via an inert electrode and (2) it uses a mass-free reagent so no additional chemicals are added, which would create secondary streams, which would as it is often the case with conventional procedures, need further treatment, cf. Scheme 10.

With the increasing costs of raw materials and the threat of depletion of world reserves of many resources, electrochemical processes should become more attractive to reuse and recycle wastes/materials. The recovery of metals in chemical solutions is very important from both the environmental and economical view points [232].

Electrolysis has the potential advantage that a metal can be recovered in its most valuable forms as metal film or powder and sold or recycled to the process. Cf. also Walsh, Ref. [133]. Heavy metals, such as copper from metal complex dyes, or from catalysts in industrial effluents, have become a problem in clarification plants because of their toxic effects on microorganisms. Their disposal through deposition after chemical or physical treatment is senseless,

Table 14. Types of processes used in local-authority sewage works [10, p.332]

Stage	BOD range mg cm^{-3}	Mechanical/biological processes	Physical/chemical processes
Primary	400–250	Comminution, sedimentation, sludge digestion or incineration	Flocculations by chemical additives, flotation
Secondary	250–40	Percolation/activated sludge filters, biolog. oxidation, nitrification	Flotation, coagulation by additives filteration (e.g. PO_4^{-3}, F^-, heavy metals), filtration
Tertiary	40– < 20	Filtration, oxidation, ponds desinfection	Treatment with Cl_2 or O_3, adsorption on high surface-area C, osmosis, UV sterilization

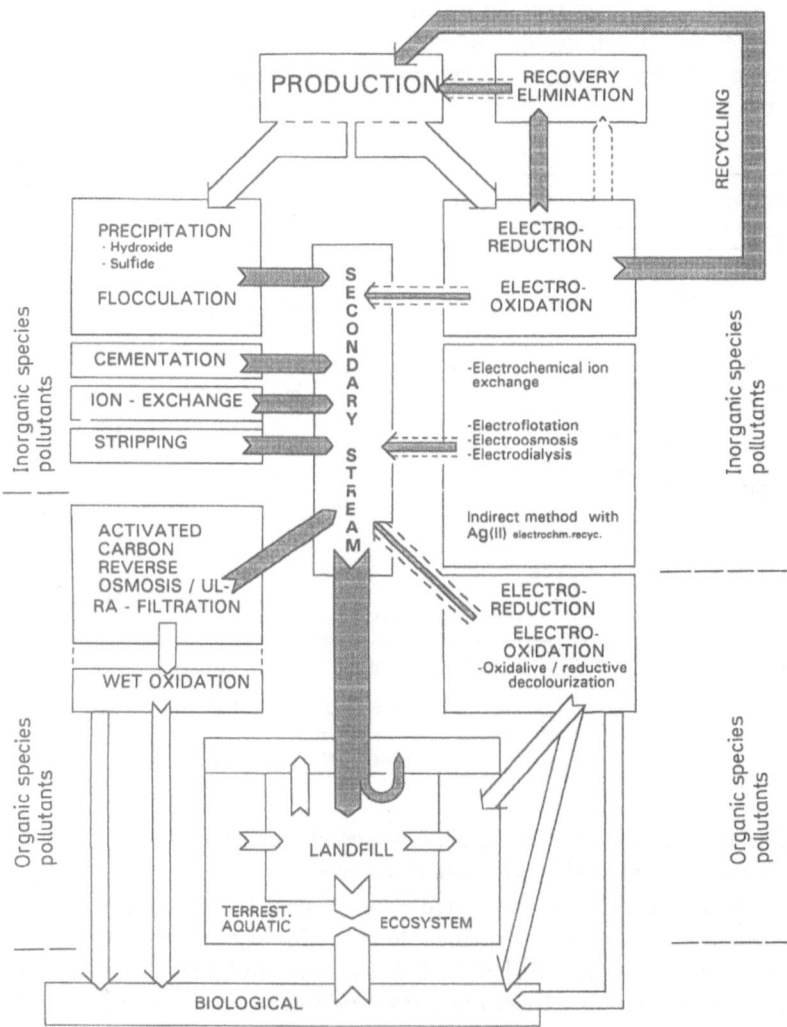

Scheme 10. Streams of conventional and electrochemical treatment of inorganic and organic pollutants in wastewaters and effluents [46, 231]

since this only results in a local change in the environmental problem. Therefore, the only long-term solution is the recycling (recovery) of metals, e.g. through electrochemical processing, and subsequent feedback into the process [226].

4.1 Electrochemical Hardware for Treatment of Wastewater and Effluents

The design of practical systems depends primarily on the objective of the process [233].

179

Properly designed cells can recover metals from solutions as dilute as 1 mg/l, or as concentrated as 200 g/l. In this context see also Ref. [13, 133, 234 et al].The following scheme shows 3 classes of cell types developed for various metal concentrations

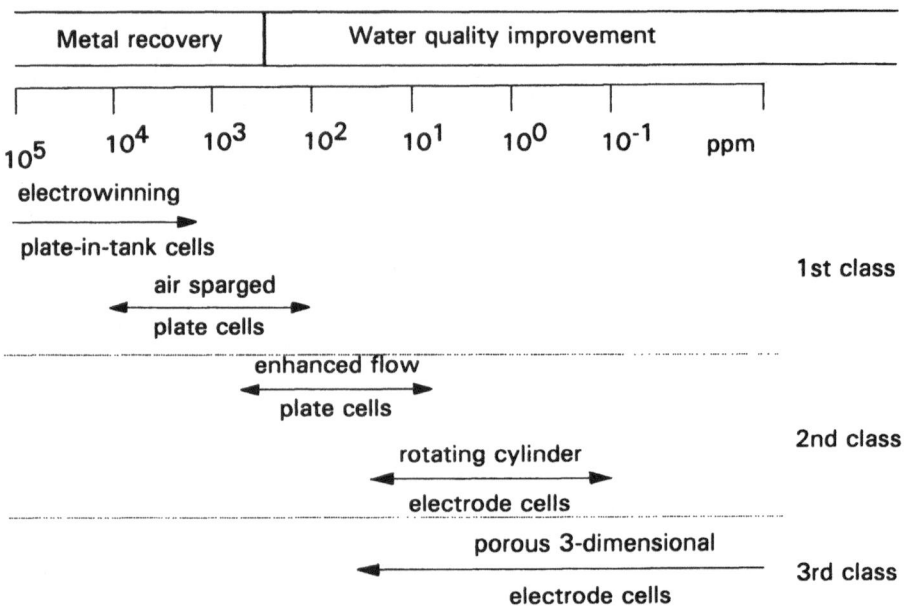

Characteristics and features of commercially successful systems:
1st Class: **Up-graded tank cells with plate electrodes**
Heraeus markets a multicathode metal depletion system. In the system, one or two anodes face a large number of permeable cathodes, e.g. in form of copper strips similar to expanded metal. [90, 234a]. Plattner and Comninellis have developed large scale reactors; total plate electrode surface: ca. 20 m², cf. ref. [140].

2nd Class: **Enhanced flow cells**
(1) Chemelec cell [13, 133, 233, 235–239, 247]
To enhance mass transport, the Chemelec Cell uses a partially fluidized bed. Closely spaced gauze or expanded metal electrodes consisting of alternating anodes and cathodes are separated by nonconducting spherical glass particles. Compared to laminar flow parallel to a plate electrode the mass-transfer is enhanced by a factor of up to 6 in the presence of the fluidized bed by acting as a turbulance promotor. This design is a compromise for ease of operation without the full advantages of three-dimensional electrodes. The limit for effective operation lies in the range 50 to 150 mg/l. This type is not suitable for the removal of low levels nor for the treatment of waters with high organic loads, such as effluents from chemical or from dyestuff production and processing

plants. The cathodes can readily be removed to recover the deposited metals. Chemelec Cells are mostly used to prevent building up of metal ions in recycled-water streams such as the washwaters from electroplating processes. Hence, most units are used in the plating industry to handle such metals as Cu, Ni, Au, Ag, Zn, Cd, Pb. Ni, Fe. The standard Chemelec Cell $(0.5 \times 0.6 \times 0.7 \text{ m})$ with a total electrode area of 3.3 m^2 (six cathodes and seven anodes) recovers $70\text{--}400 \text{ g h}^{-1}$ of metal. More than 500 units have been commissioned since 1982 [233].

(2) Another configuration that achieves high transfer coefficients is the rotating cylinder electrode (**RCE**) [13, 133, 240–246].

Robinson and Walsh have reviewed earlier cell designs. The performance of a 500 A pilot plant reactor for copper ion removal is described. Simplified expressions were derived for mass transport both in single pass [243] and batch recirculation [244]. For a detailed discussion of the principle and the role of the rotating cylinder electrode reactor in metal ion removal the reader is referred to Refs. [13] and [241] (46 references).

The Eco-Cell process for continuous removal and production of metal in powder form is summarized in Fig. 23.

An Eco-cell plant operating in Denmark contains a 4 kA Eco cell followed by a 1 kA Eco cascade cell and treats $200 \text{ m}^3 \text{ day}^{-1}$ of copper phthalocyanine pigment effluent, reducing the Cu^{2+} from 400 mg dm^{-3} to $2\text{--}3 \text{ mg dm}^{-3}$ and recovering about 80 kg of metal per day [10, p. 349].

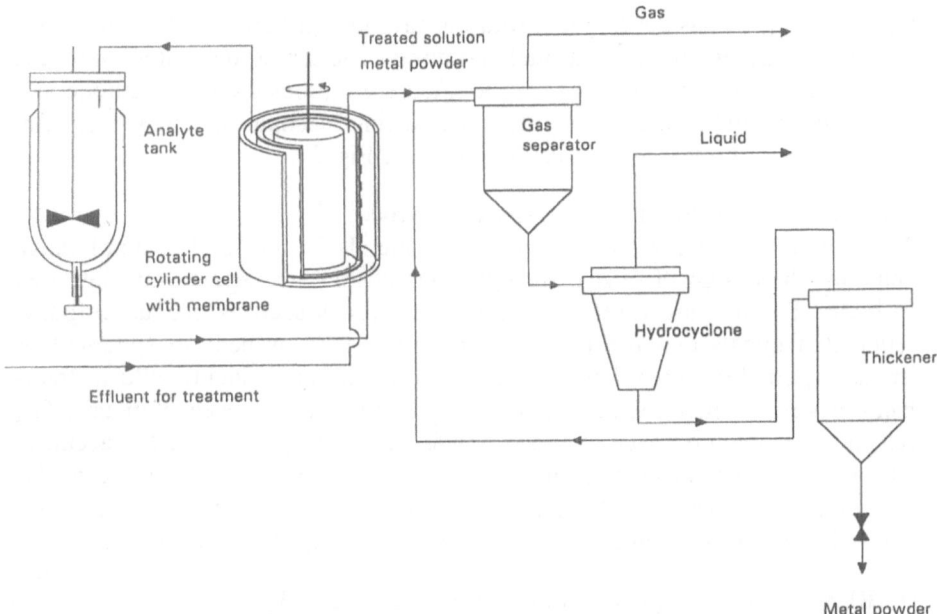

Fig. 23. The Eco-Cell process for continuous removal and production of metals in powder form

A typical 5 kA Eco cell has a cathode drum with a radius of 0.37 m, a height of 0.74 m and a cathode-membrane gap of about 1 cm. The cathode is rotated at 100–200 rev. min^{-1}. In rotating-cylinder electrode cells, high fractional conversion can be obtained by employing an Eco cascade cell.

The cascade configuration – partition walls subdivide the catholyte chamber into segments – eliminates back mixing effects and achieves the high degree of conversion to the required low concentrations. For an inlet concentration of 100 ppm outlet concentrations of 2 ppm Cu(II) and less have been reported.

(3) enViro-Cell [247–249]
In this cell, mechanical vibration is applied to the cell housing to enhance the transfer in the parallel plate tank cell [248]. The vibrations are transfered to the electrolyte resulting in an increase of the mass-transfer coefficient. The cell is extensively used in industry for the pretreatment of higher and high metal concentrations which is finally purified by a packed bed electrolysor if the required conversion is not too high [247].

(4) The cathodic system of the novel cell developed by Hertwig et al. [250, 251] consists of a rotating basket/drum filled with copper particles and of the current feeder mounted into the moving bed. The anode is a perforated plate or another structure suitable for gas-evolving reactions. The new cell allows the treatment of wastes with up to 20 g metal per liter. Advantages of this novel reactor are: a high degree of compactness and the possibility of cleaning electrolytes down to concentrations in the ppm and lower range allowed by laws for effluents. Cf. also Refs. [13, 133].

(5) Heraeus [90] has developed a rotating cathode Zn-band for the continuous removal of Zn without additional handling. The Zn is deposited onto the rotating cathode. The Zn band can be lifted continuously off the cathode and wound up or threaded into an automatic cutting mechanism. The principle of the rotating cathode Zn-band cell is shown in Fig. 24.

3rd Class: **Cells with three-dimensional electrodes** [233]:
The elimination of low concentrations of metals in the mg/l (1 ppm) range demands the design of cells with high rates of mass transport and high area cathodes. Suitable are three-dimensional electrodes such as porous graphite, reticulated metals or carbon, metal or carbon felts or beds of spherical or irregular particles. Three-dimensional electrode cells are suited to effluent treatment processes where the main objective is to remove low levels of metals. The cells may be operated over extended periods without significant metal accumulation. Two principal arrangements of cells with three-dimensional electrodes are possible with respect to the electrolyte and current flow directions. These are referred to as (a) flow-through and (b) flow-by arrangements, cf. Fig. 25.

(1) RETEC cell from Eltech Systems Corp [23, 252, 253].
Three-dimensional Cu or Ni foam and reticulated carbon (for the recovery of noble metals) cathodes are used. Between each pair of cathodes an inert,

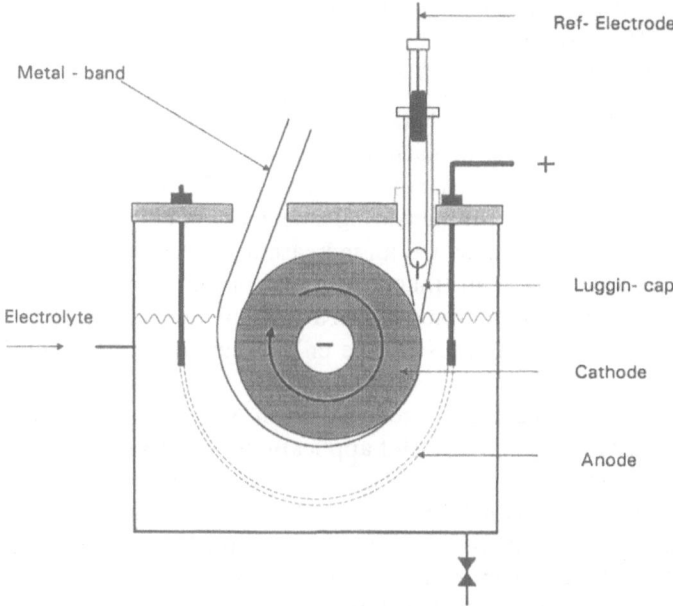

Fig. 24. Principle of a rotating cell with a Zn band, developed by Heraeus [90]

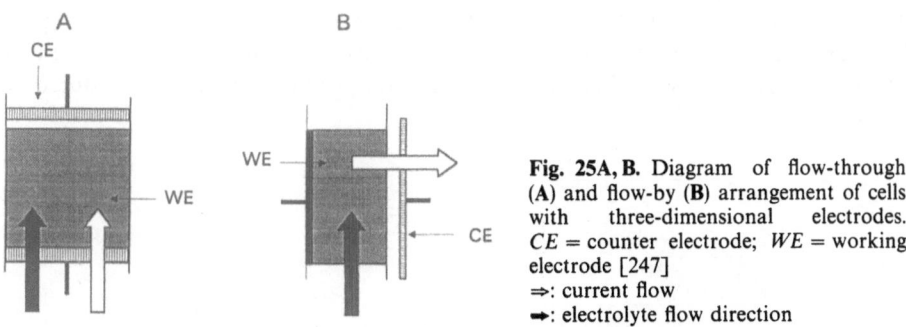

Fig. 25A, B. Diagram of flow-through (**A**) and flow-by (**B**) arrangement of cells with three-dimensional electrodes. CE = counter electrode; WE = working electrode [247]
⇒: current flow
→: electrolyte flow direction

DSA-O_2 oxide-coated Ti anode (DSA = Dimensionally Stable Anode) is placed, thus creating a large cathode volume. The effluent solution flows perpendicularly through the electrodes with a typical flow rate of 0.5 dm^3 s^{-1}. The flow-through metal electrodes have an active area approximately 15 times their geometric area. The cell allows air sparging to increase the mass-transfer. The current efficiency is about 40% when the inlet concentration of the metal ions is 150 to 1500 ppm and the concentration at the out-let is about 50 ppm. The cell is currently used for the treatment of recirculated wash-waters from acid copper, copper cyanide, zinc cyanide, zinc chloride, cadmium sulphate, cadmium cyanide and precious metal plating and washwaters from electroless copper deposition. Since the foam metal electrodes are relatively expensive the electrodes must be regenerated by dissolving the deposited metals. Electrode dimensions: 1.8 × 0.6 × 0.5 m. The 50 electrodes are operated with monopolar electrical connections (Rectifier supply = 200–800 A at 2–6 V).

(2) enViro-cell by enViro-Cell Umwelttechnik GmbH (Oberursel, Germany)

The cell consists of a packed bed of carbon granules contained in a cell compartment. The cell features a profiled compartment, Fig. 26, with increasing cathode area toward the cell outlet, thus, ensuring a decrease in current density along the direction of the flow. The result is an increase in current efficiency as well as conversion in a single pass of an effluent solution through the cell. In the common option, a single 1 × 1 m planar anode has a particulate bed, depth 0.15 to 0.26 m, on either side. Treated species by the enViro-cell system are summarized in Table 15.

The metal ion concentrations can be reduced from 80 ppm to less than 1 ppm with a single-pass operation. Successful applications on a technical scale are summarized in Table 16.

The regeneration of the bed can be performed by (1) replacement of the carbon granules plus deposited metal by a vacuum cleaner. Fresh packing granules may be added, most conveniently on a weight basis. Replacement is feasible in view of the low price of the material used as cell packing or (2) by the in situ regeneration by acid washing or current reversal.

(3) This type of electrode has also been successfully used for the decolorization of wastewaters and effluents from dyestuff production. Characteristic of such effluents are high metal loads, high loads of organics and of halogens [55, 226, 231]. The presence of halogens neccessitates the use of a divided cell to avoid the formation of halogenated organic compounds (AOX) in the course of

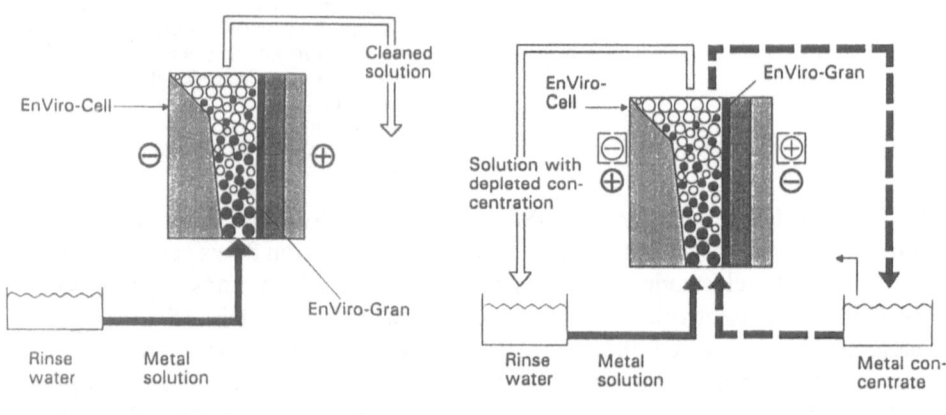

Continous mode of operation
Regeneration by exchanging
the granular graphite filling

Intermittent batch operation
with electrical regeneration
via pole reversal

Fig. 26. Mode of operation of en-Viro-Cell technology [10, 247]

Table 15. Treatment of effluents by the enViro-cell system

Reduction		Oxidation
Deposition	Change of valency	
Cadmium	$Fe^{2+} \longleftrightarrow Fe^{3+}$	Cyanid
Cobalt	$Cr^{3+} \longleftrightarrow Cr^{6+}$	Cyanid complexes
Nickel		Nitrite
Tin		EDTA
Lead		Chloride
Copper		Dyestuffs
Silver		CSB, (COD)
Mercury		

Table 16. Some industrial applications of the enViro-cell [249]; Packed bed electrolysis

Application	Element recovered	Flow (l/h)	C-in (ppm)	C-out	Energy consumption kWh/m^3
Chlor-alkali electrolysis	Hg	5000	1	0.05	0.16
Prod. of cell. acetate	Cu	20000	20	1.9	0.08
Dye stuff production	Cu	6000	400	2	4.0
Dye stuff production	Hg	2000	4	0.05	2.5
Plastic production	Cu	22000	6	0.5	0.15
Plastic production	Cu	3000	50	0.05	1.8
Wash weater gas purif.	Hg	1000	1	< 0.01	0.05
Rinse from fixing bath	Ag	1000	25	< 1	0.5

C-in: concentration before electrolysis
C-out: concentration after electrolysis at cell outlet

the treatment due to the formation of Cl_2, ClO^-, ClO_2 at the anode in a monocell. Cf. also Sect. 4.3.3., Table 26.

With the success of the three-dimensional electrodes, it has become commonplace for suppliers of plate and frame cells to offer designs that allow their operation using electrodes with a high surface area. Reilly Tar and Chemicals Corp. and ElectroCell systems AB supply systems that can utilize a packed bed electrode [75, 79, 254–256].

(4) ElectroCell Systems AB (Täby, Sweden)

MP, ElectroSyn and Electro Prod Cells can be modified to accomodate three-dimensional electrodes. The ER cell designed by Simonsson [79, 256] is based on a central catholyte compartment, 10.8 mm thick, filled with irregular graphite particles, (size 1–2 mm). Electrical contact to the bed is via a titanium grid and on either side of the bed are membranes and plate anodes. The packed beds are operated in the flow-by mode and the unusual ratio of height to width was designed to minimize channeling. Detailed mass studies demonstrate that there is a turbulent flow regime down to low flow rates. For larger scale operation a number of cells are mounted in a stack. Thus a cell for the treatment of 1 m^3/h in a single pass mode will have 4 stacks (10 cathodes and 5–11 anodes). Data for the removal of Cu, Zn and Ag ions from wastewater streams are summarized in Table 17.

Copper can be removed from solutions containing strong complexing agents, such as EDTA or Quadrol, or from acidic sulfate media. The technology can thus be applied to solutions from the electronics and printed circuit board industry as well as from washwaters from the electroplating industry. Silver removal was applied for the treatment of wastewaters from photographic fixing baths. The removal of Zn(II) was most effective from alkaline solutions but some acid solutions could also be handled. The reduction of metal ion concentrations below 1 ppm is feasible with reasonable current efficiencies and a cell voltage in the range of 2 to 3 V. Similar performance could be obtained on solutions of low ionic strength provided the specific conductivity is $> 10^{-6}$ Ohm^{-1} cm^{-1}. Up to 0.2 kg of metal can be loaded onto each electrode without a significant increase in pressure drop. The metal is recovered as a concentrated metal ion solution either by current reversal or acid washing [79]. An Electro Prod Cell with a packed bed of carbon particles is used commercially to control the Zn^{2+} ions in an acidic, ferric chloride solution used in pickling operations [10]. In order to maximize the cathodic current efficiency and to achieve a high conversion in a single pass, an analogous strategy to the enViro Cell is possible. The packed bed may be profiled by using decreasing grain sizes in the flow direction.

Characteristics of the cells 1 and 4 with three-dimensional electrodes are that the reactors must be leached to recover metals as concentrates [133].

Examples of electrochemical reactors for metal removal and recovery are summarized in Table 18.

Table 17. Data illustrating the operation of the ElectroCell ER cell for the removal of metal ions from waste solution streams [79]

Metal	Inlet conc/ ppm	Solution Characteristics	pH	Particle size/mm	Flow rate l/min	Cell voltage/ V	Cell current/ A	Outlet conc/ ppm	Current efficiency (%)
Cu	67	CuSO$_4$/H$_2$SO$_4$	1.4	1.0–1.4	2.0	2.2	13	0.03	52
Cu	26	Complexed with Quadrol	2.1	2.0–3.15	1.0	2.3	4	0.6	35
Zn	44.6	ZnSO$_4$ H$_2$SO$_4$/K$_2$SO$_4$	5	2.0–3.15	1.0	3.2	8	0.44	27
Ag	910	Thiosulfate fixed solution	3.6	2.0–3.15	1.0	0.5–2.0	2	0.7	20

Table 18. Examples of electrochemical reactors for metal removal and recovery [10]

Reactor	Company	Type of cathode	Frequency and method of product removal	Cell normally divided?	Performance enhancement mainly via:		
					Electrolyte movement	Electrode movement	High electroactive area
ER cell	ElectroCell AB	For example, packed bed of carbon particles within a parallel plate and frame	Discontinuous via anodic or open circuit leaching	Yes	✓		✓
Envirocell	enviro-cell Umwelttechnik 6 mbH	Contoured packed bed of carbon, possibly with non-conductive particles	Discontinuous via anodic or open-circuit leaching or vacuum removal of bed	Yes	✓		✓
FBE reactor	(Originally AKZO) Billiton Research b.v.	Fluidized bed of metal particles in a tube-and-shell type geometry	Continuous via withdrawal of grown particles	Yes	✓	✓	✓
Eco-cell	(Originally Ecological Engineering Steetley Engineering Ltd)	Outer surface of a rotating cylinder with powdered-metal deposit	May be continuous via automatic scrapping and fluidization of metal powder	Maybe		✓	✓
Chemelec cell	BEWT (Water Engineers) Ltd	Vertical mesh (or plate) in an electrolyte with fluidized glass beads	Discontinuous by manual scraping or reuse as anodes in plating	No	✓		

Table 18. (continued)

Reactor	Company	Type of cathode	Frequency and method of product removal	Performance enhancement mainly via:			
				Cell normally divided?	Electrolyte movement	Electrode movement	High electroactive area
Concentric cell	Wilson Process Systems	Inner surface of a cylindrical foil (copper or stainless steel)	Discontinuous: cathode may be furnace-refined in the case of gold	No	√		
Rotating electrode cell	Wilson Process Systems	Rotating cylindrical foil (usually stainless steel) or static cylindrical foil with rotating anode (larger cells)	Discontinuous by manual scraping or flexing	No	√	√	
Recowin	Eco-tec	Vertical plates in tank, air agitatation possible	Discontinuous via manual stripping of metal as sheet	No	√		
Geocomet reactor	Geoma AG	Rotating tubular bed or impact rod	Continuous as settled powder via a sludge cone	No			
Ketec cell	Eltech	Vertical metal (or carbon) foam electrodes in a tank	Discontinuous: cathode steel (chemically stripped for reuse carbon or precious metals (reused as anode in plating)	No			

4.2 Electrochemical Treatment of Inorganic Pollutants and Waste

Metalls and metalloids are characterized by special ecochemical features. They are not biodegradable, but undergo a biochemical cycle during which transformations into more or less toxic species occur. They are accumulated by organisms and cause increased toxic effects in mammals and man after long term exposure [55].

The elimination and recovery and the recycling of inorganic species are very well documented in the literature [28,133,228,229,231,247,257,258].

What follows are selected examples which have found pilot plant and technical applications.

4.2.1 Treatment of Liquors Containing Dissolved Chromium cf. Refs. [10,133]

Trivalent chromium is an essential element in mamallian systems, whereas hexavalent chromium is considered to be a moderate to severe industrial hazard [259].

Soluble chromium species, especially Cr(VI) are used in many sectors of industry, e.g.[10,260]:

(1) Electroplating of chromium from chromic acid leads to spent or contaminated baths and rinsewaters.
(2) Pickling, etching, stripping solutions in metal finishing.
(3) Etching of plastic surfaces prior to electroplating.
(4) Chromate conversion coatings for Al and Mg alloys.
(5) Sodium perchlorate production.
(6) Corrosion inhibition in cooling waters.
(7) In tannery, and
(8) as mediators in organic syntheses, cf. Sect. 3.

Although under constantly increasing pressure from legislation, (lawful maximum value of $Cr_{tot} = 20$ ppm, $Cr(VI) = 0.5$ ppm in Switzerland and Germany) chromium-containing liquors are often difficult to replace by alternative processes. Therefore it has become increasingly attractive to recycle these liquors or to find suitable treatments prior to disposal, cf. Ref. [10]. Numerous recent papers [90,234a,260–263] deal with this problem. Electrodialysis [133] and electrolytic treatment are providing an important contribution to the recycling and safe disposal chromium containing wastes, as illustrated by following two examples.

1) Chromium acid recovery [90]
The principle of a plant for Cr(III)/Cr(VI) oxidation and electrodialytic Fe removal developed by Heraeus Elektrochemie (Germany) is shown in Fig. 27A, a chromic acid recovery system in Fig. 27B.

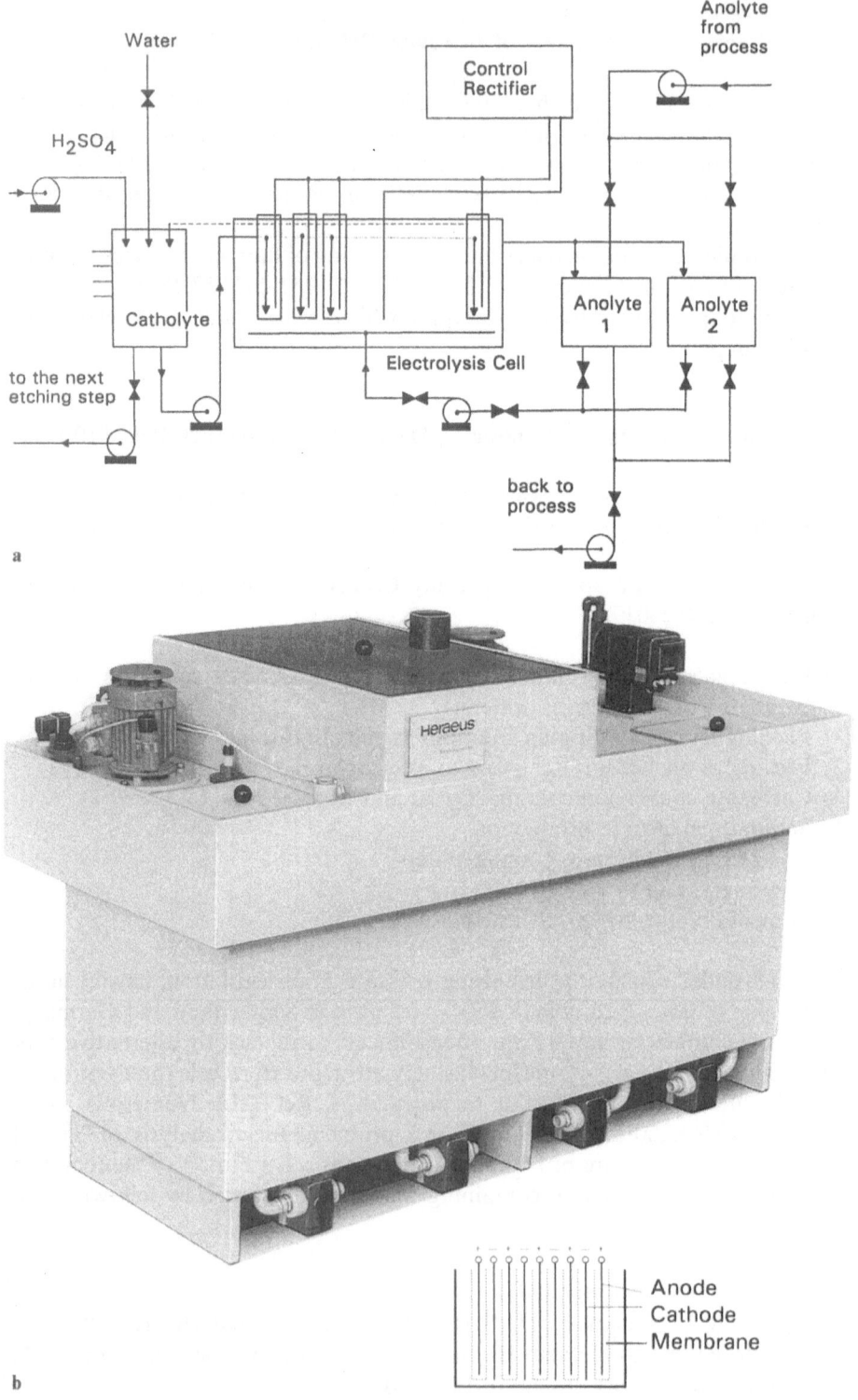

Anolyte
from
process

Water

Control
Rectifier

H_2SO_4

Catholyte

to the next
etching step

Electrolysis Cell

Anolyte
1

Anolyte
2

back to
process

a

Anode
Cathode
Membrane

b

Typical process conditions are:

Cr(VI)	200–300 g/l
Cr(III)	2–3 g/l
Fe_{tot}	10 g/l
F^-	600 ppm
Membrane area	2.5 m^2
Current density	0.5 kA/m^2

System duty:

1) Oxidize 0.1 kg/h Cr(III) to Cr(VI)
2) Remove 0.1 kg/h Fe from anolyte
3) Transfer catholyte as etchant to next process step.

During the process, a Cr(III) build-up often occurs which results in a serious reduction of the process performance when a certain Cr(III) concentration (typically a few grams/liter) is reached. In addition, contaminating metals, such as Fe are dissolved during the process from the work pieces or are carried over from other processes. The iron content below the required level is secured by electrodialytic transition of the Fe^{2+} through the cation exchange membrane from the anolyte to the catholyte. The liquor from the process plant is collected and circulated via a pump through the anode compartments of a parallel plate type electrolytic cell. Ebonex Ti_4O_7 (cf. Sect. 2.3) with special coatings anodes are used. Electrolytes used are 2 to 5% H_2SO_4.

2) An alternative to electrodialysis is electrocoagulation [133]
Chromium and other metals are precipitated by means of anodically dissolved Fe^{2+}.

$$Fe^0 \rightarrow Fe^{2+} 2e^- \text{ Anode}$$

$$3Fe^{2+} + CrO_4^{2-} + 4H_2O \rightarrow 3Fe^{3+} + Cr^{3+} + 8OH^-$$

$$Cr_2O_7^{2-} + 6Fe(OH)_2 + 7H_2O \rightarrow 2Cr(OH)_3 + 6Fe(OH)_3 + 2OH^-$$

In the USA Andco Chemical Corp. and Niagara Environmental Assoc., Inc. have widely used this procedure for the treatment of cooling tower blowdown waters and in treatment of metal finishing process streams. The undivided cell geometry makes use of cold-rolled steel plates as anodes. Disadvantages of this procedure: Formation of second streams, e.g. large amounts of sludges which must be dumped or treated.

3) Regeneration of chromic-sulfuric acid mixtures from wax bleaching processes [260]
Since 1927 chromic acid has been used for the oxidative bleaching of "Montan" waxes. Hoechst has been operating the electrochemical regeneration of chromic acid for more than 90 years. The newly developed electrochemical process (as enlargement and partial replacement for the existing plant) is based on the

Fig. 27. A Principle of a plant for Cr(III)/Cr(VI) oxidation and electro-dialytic Fe-removal [90]
B Compact electrolysis-membrane system (Photograph supplied by Heraeus Elektrochemie)

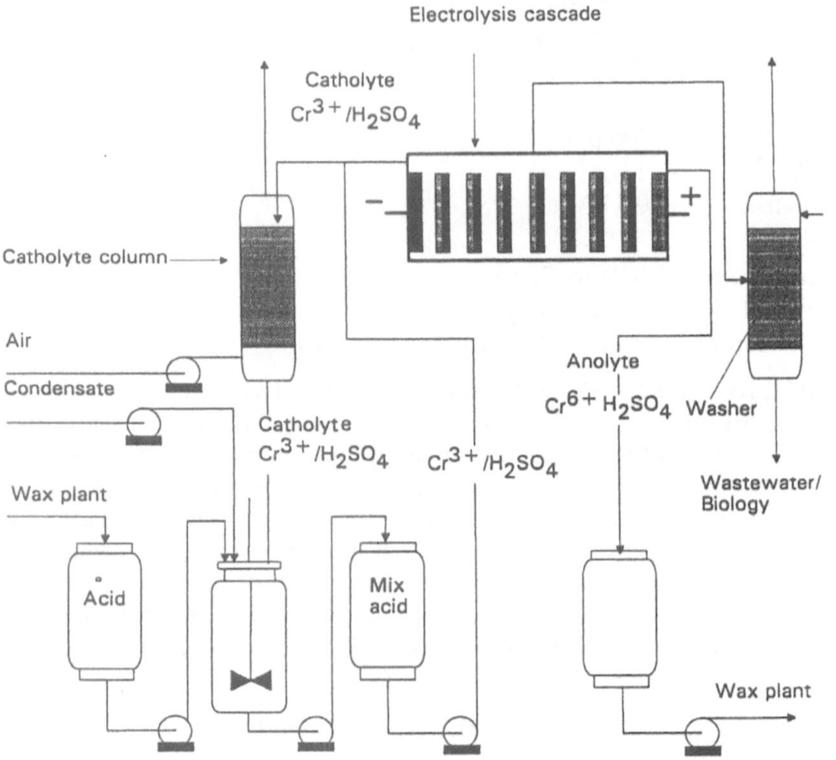

Fig. 28. Flow-sheet for the regeneration of chromium/sulfuric acid mixtures [260]

technology of the Hoechst-UHDE membrane-cell for chlor-alkali electrolysis using PbO_2/Ti anodes, Ni and Ni-alloys as cathode and a cation exchange, polyfluorinated polymer membrane. Figure 28 shows the lay-out.

A bipolar pilot membrane cell with six cell elements each of 1.8 m^2 has been installed and run with industrial electrolyte.

The following mechanism for the oxidation of Cr(III) to Cr(VI) on a PbO_2 surface has been postulated. PbO_2 operates as "oxygen" carrier anode

$$2Cr^{3+} + 3PbO_2 + 6H^+ \rightarrow 2Cr^{6+} + 3PbO + 3H_2O$$

Regeneration of the PbO_2:

$$3PbO + 3H_2O - 6e^- \rightarrow 3PbO_2 + 6H^+$$

Under the given operational conditions only β-PbO_2 is formed. [262a]

The current efficiency for pure Cr(III)-sulfuric acid is in the range of 90%. Organics, present in the industrial liquors, especially higher dicarboxylic acids, interfere, thus necessitating the use of divided cells. The scheme of the membrane cell used is shown in Fig. 29.

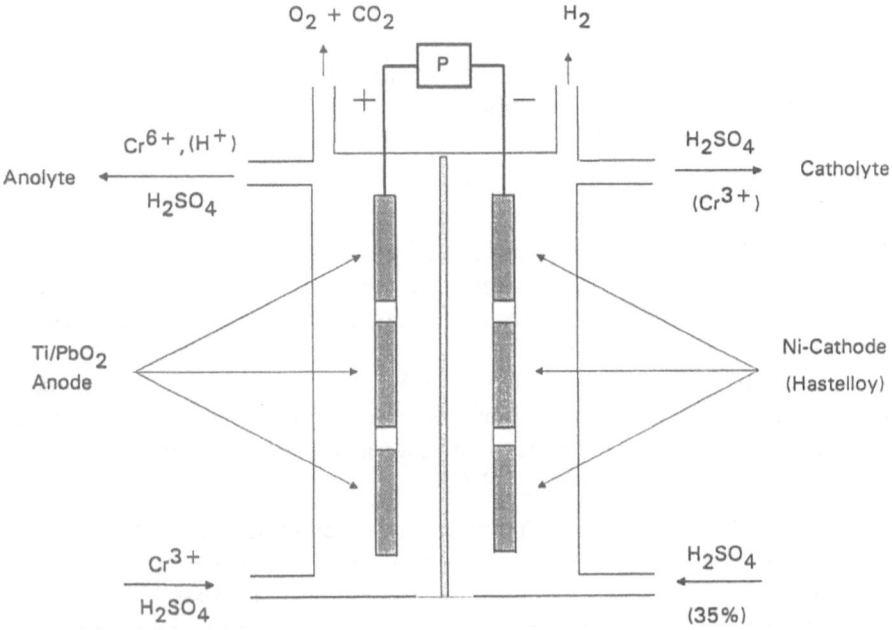

Fig. 29. Scheme of the membrane cell used for the chromium electrolysis [260]

4) In-situ generation of Cr(II)

Robinson et al. [263] have studied both at the laboratory and semi-technical scale the electrochemical generation of chromous ion (Cr(II)) for the reduction of sodium hydroxymandelate (SHM) to 4-hydroxy phenylacetic acid (HPA), a Tenormin intermediate, as a potential replacement for the existing Zn(Hg) reduction process.

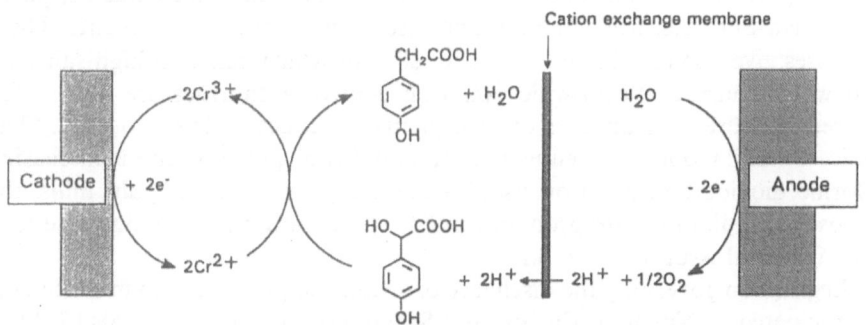

Process conditions using a FM 21 Cell are:

Current density   3 kA/m^2
Current efficiency   45%
Conversion   99%
Product yield   95%

The electrochemical generation of Cr(II) produces yields comparable to the conventional zinc powder route. Advantages of the electrochemical method include reduced revenue costs and a more environmentally acceptable process with respect to the heavy metal effluents.

4.2.2 Salt Splitting [264-269, 269a,270]

One of the most difficult wastes for industry to dispose of is that of solutions of inorganic salts [270]. Unlike organic salts which can be treated by various methods to convert the organic species into carbon dioxide and water, inorganic salts can rarely be broken down.

Large amounts of sodium sulfate are produced as a by-product of many diverse industries. Some of this material is recycled internally, some is upgraded and sold as a product. Most is disposed as a waste in landfills, or discharged to deep-wells, or bodies of water. [129,269]

Amount of sodium sulfate per annum generated by industry. (269)

Industry	kt/year
Chemicals and allied products	1400
Paper and allied products	200
Petroleum and coal products	75
Primary metals	60
Textile mill products	30

The mineral salt content of process effluent streams is rapidly becoming a focus of environmental concern.

Most attention is currently focused on the inorganic salt effluents, particularly Na_2SO_4, e.g. from pulping and viscose fiber production plants. These processes give rise to a large tonnage of effluent which deliver a high salt load into water courses. World-wide pollution control authorities are putting increased pressure on managers of such plants to reduce their salt output. "The choice is stark: reduce your effluent or be shut down" [135]. With the increasing chlorine dioxide consumption used in bleaching processes in pulp mills, the disposal of sodium sulfate produced in ClO_2 generation is becoming a serious environmental issue, cf. Sect. 3.2.

Legislation governing the discharge of sodium sulphate in pulp mill effluents is anticipated in Northern Europe and Scandinavia during the 1990s [271].

The keys to success are:

(1) Operable technology
(2) Robust technology
(3) Flexible technology
(4) Economic technology

If the Na_2SO_4 can be split into its component acid and base, the sulfuric acid and the caustic soda can be recycled to the processes. Thus a potential effluent processing charge can be seen as a raw material credit.

ElectroCell System AB [99], EL-TECH[269], ICI [271,272], and de Nora [129,273] are now developing electrohydrolysis of sodium sulfate for commercial applications. In the electrohydrolysis process sodium sulfate is fed as anolyte to an electrochemical cell divided by a cation specific membrane. Protons are generated in the anolyte, hydroxyl ions at the cathode. Sodium ions cross the membrane to produce a catholyte solution of sodium hydroxide. The net reaction is:

$$Na_2SO_4 + 3H_2O \rightarrow 2NaOH + H_2SO_4 + H_2 + 1/2O_2$$

Three different electrolytic approaches to salt splitting are competing with each other on the market [50], Fig. 30.

Two compartment cell

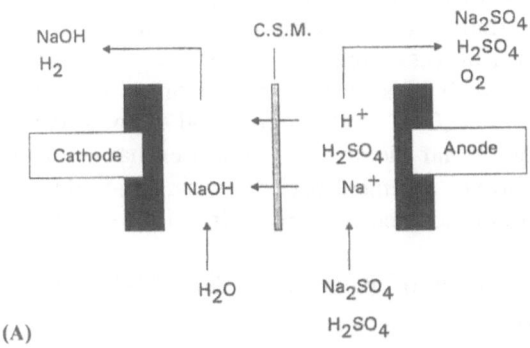

(A)

Three compartment cell

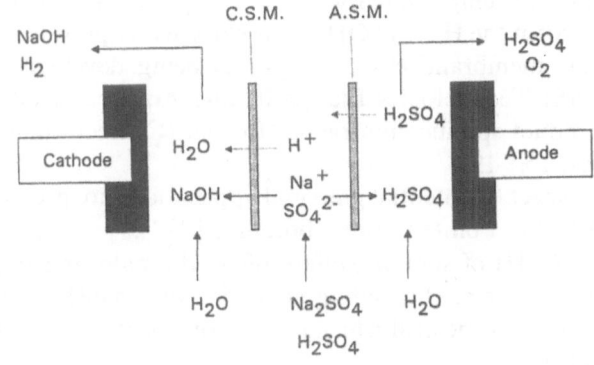

(B)

Fig. 30. Two compartment (**A**) and three-compartment (**B**) cell for salt splitting [135] A.S.M. = anion selective membrane; C.S.M. = cation selective membrane

(1) Two-compartment cells (Fig. 30A):
As acid is formed at the anode, the problem quickly becomes the competion between H^+ and Na^+ migrating through the single cation permeable membrane. The process operating at 2-5 kA/m^2 typically produces solutions of 20 to 25% sodium hydroxide along with 15% sulfuric acid – the latter containing up to 20% unconverted sodium sulfate [129,135].

(2) Three-compartment cells (Fig. 30B): [274]
This type of cell is limited by the performance of the anion-exchange membrane. The membrane can tolerate only a limited concentration of acid in the anolyte before backdiffusion of protons through the anion membrane becomes rather significant, causing a decrease in the cell's current efficiency, and acid gets into the salt stream. Unconverted salt in the product solutions can be eliminated by the use of the three compartment device, but this option substantially adds to the capital costs and operating complexity of the plant.

System schematics for a two- and three-compartment cell are shown in Fig. 31A und Fig. 31B.

(3) Electrodialysis by use of bipolar membranes [129,270,271]
The bipolar membrane is an alternative cell arrangement which can act as a direct source of acid or base for a process stream.

A bipolar membrane is a sandwich of a cation and an anion exchange membrane which splits H_2O to H^+ and OH^- under a potential of about 0.9 V. The reactor consists of a stack, of bipolar membranes, cation exchange membranes and anion exchange membranes arranged between a single anode and a single cathode with parallel hydraulytic circuits for the salt, acid product and alkali product, Fig. 32.

The net reaction uses only one mole of water per mole of NaOH as no oxygen and hydrogen is generated

$$Na_2SO_4 + 2H_2O \rightarrow H_2SO_4 + 2NaOH$$

The process operates at current densities of about 1 kA/m^2 and unit cell voltage of 1.5 V. The specific energy consumption is about 2 kWh/kg NaOH. Under the influence of the electric gradient the H^+ and OH^- ions emerge on opposite faces of the membrane. Bipolar membrane electrodialysis is being developed by several companies, e.g. WSI Technologies Inc. [270] and Aquatech Systems [129,275,276]. Typical product specification ranges for the ICI electrodialysis process is summarized in Table 19.

Pilot plant runs with respect to potential pulp mill applications are presently being made at the PAPRICAN Pointe Claire laboratory [277].

Electrolytic hydrolysis (E-H) of sodium sulfate offers the pulp and paper industry the capability of reducing or eliminating plant discharge while generating valuable caustic soda and sulfuric acid which can be sold or returned to the process to close the loop [269].

The feasibility of electrochemically recovering pure sodium hydroxide and lignin from kraft black liquor has been explored in several studies. Electrodialy-

Two compartment cell

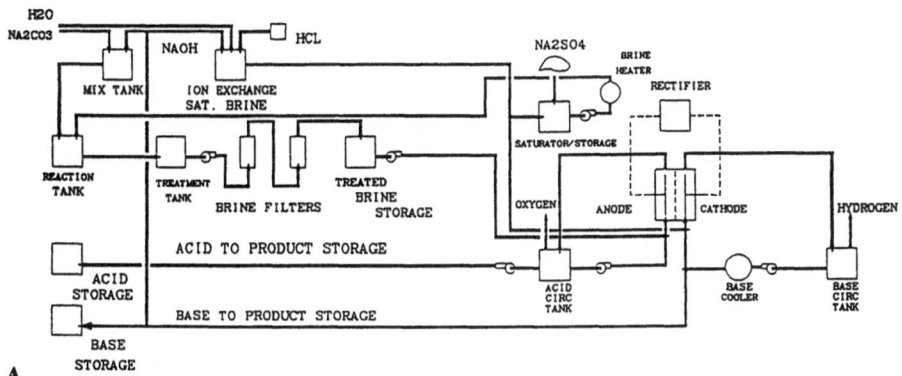

Three compartment cell

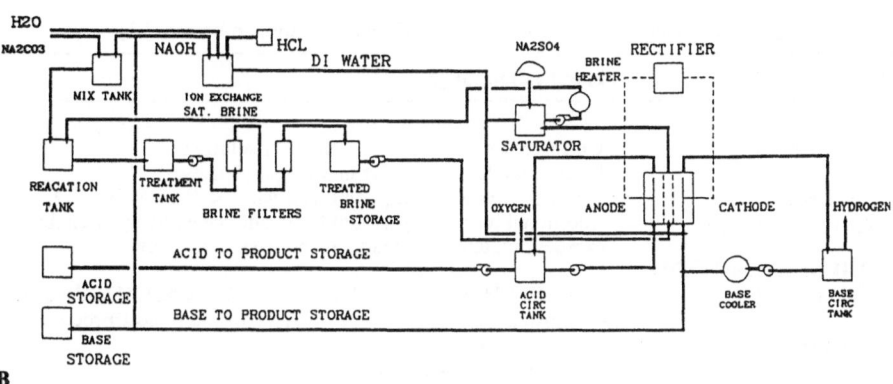

Fig. 31A, B. System schematics for a two (**A**) and a three-compartment (**B**) cell [269]

sis of weak black liquor (WBL) was studied as a process to recover the sodium from the process liquor and precipitate lignin [278,279]. Bipolar membrane technology is not suitable for liquors containing organics because of the severe fouling of the anion-selective membranes. The recovery of sodium as caustic soda using electrodialysis would thus be limited to the sodium present as free NaOH, which is only a minor fraction of the total sodium in WBL. Moreover, with an electrodialysis process, the maximum current density is low (100 mA/cm^2) compared to electrolysis processes (400–500 mA/cm^2). This means that the caustic soda production rate for an electrodialysis system would be less than one fourth of the rate provided by an electrolytic system of the same effective membrane area. Electrolytic routes for processing WBL to increase kraft mill production have also been suggested [278,280,281,282]. Lignin is

197

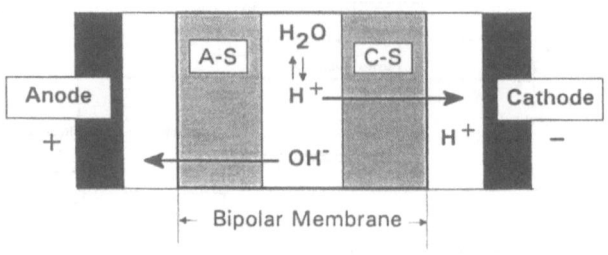

Bipolar Membrane

Construction and Operation

A-S = Anion-selective membrane
C-S = Cation-selective membrane

Fig. 32. Bipolar membrane electrodialysis of NaA to HA and NaOH ("salt splitting") [270]

Table 19. Typical product specification ranges for ICI electrodialysis process [271]

Product	Specification range
ACID:	
-Ex electrolyser	Up to 15% w/w in aqueous Na_2SO_4 solution
	Direct recycle possible for many host processes
-Ex ICI electro- hydrolysis process	Up to 50% w/w H_2SO_4 combined with < 5% w/w Na_2SO_4
	Additional technology available outside the electrolyser enables an excellent match to be made at the E-H/host interface
SODIUM HYDROXIDE:	
-Ex electrolyser	Typically 20% w/w NaOH but up to 32% w/w NaOH under special circumstances. Direct recycle possible for many host processes
OXYGEN:	100% v/v (dry basis)
	High purity gas with a very low N_2 content
HYDROGEN:	100% v/v (dry basis)
	High purity gas. A valuable "green" fuel substitute or reducing agent

usually precipitated and sodium recovered by conversion of the sodium salts to caustic soda. Clouthier et al. [278] have studied the electrolysis of weak black liquor. Experiments run with an ElectroCell MP cell in the batch mode showed that electrolytic processing of WBL could reduce more than 75% of the lignin content while converting up to 80% of the sodium salts to a high quality caustic soda solution. The schematic diagram of the experimental apparatus in the batch mode operation is shown in Fig. 33.

Operation conditions

Application of a divided cell containing one pair of electrodes (Pt-coated Ti anode; 316 type stainless steel cathode) with an effective area of 100 cm². Nafion-324 was used as the membrane. Two 8 l tanks contained anolyte (feed) and catholyte (caustic). A coil-type heat exchanger was used to maintain the heat

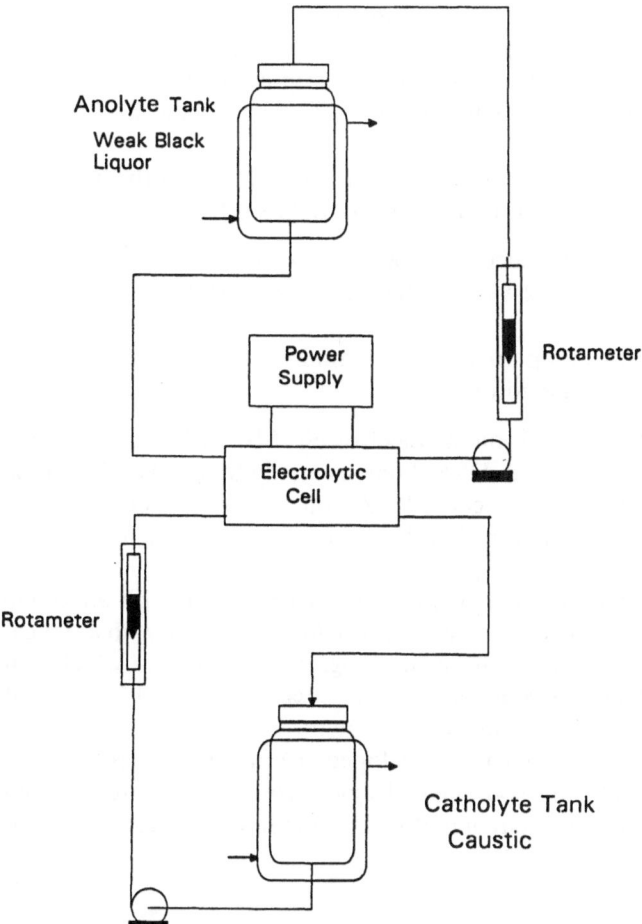

Fig. 33. Schematic diagram of the experimental apparatus for WBL electrolysis under batch mode operation [278]

of the two solutions at 55°C. The system offers the flexibility and ease of operation of a self-contained modular system with the number of units in operation adapted to the mill's needs. Thus the proposed electrolytic process could represent an attractive alternative for reducing the organic load on a recovery system thereby increasing the production capacity of an existing kraft mill. It offers numerous benefits such as: off-loading the recovery furnace, the lime kiln and the causticizing plant; recovering sodium as valuable caustic soda; producing oxygen for bleaching uses and recovering hydrogen from water electrolysis for energy uses [278].

Azarniouch and Prahacs have patented an electrochemical recovery procedure for NaOH and other valuable substances (lignin, organic compounds, H_2SO_4, H_2, O_2 or O_2/Cl_2 mixtures) from spent liquors and bleach plant effluents [282].

Genders et al. (Electrosynthesis Co Inc.) in conjunction with the Ormiston Mining Company [264–266,282a] have developed a new technology which converts sodium sulfate into two useful products: ammonium sulfate (used extensively as a fertilizer) and sodium hydroxide, The scheme of the modified cell is shown in Fig. 34.

To maintain the pH at 2, ammonia is continuously added to the anolyte.

Advantages of the ammonium modified process: [282a]

(i) NaOH and $(NH_4)_2SO_4$ produced can be recycled or marketed.
(ii) Product concentrations > 32 w% NaOH and $> 40\%$ $(NH_4)_2SO_4$ are obtained.
(iii) The current densities of up to 2.5 kAm^2 minimize the membrane area required for a given production rate.

The laboratory unit was based on a ElectroCell MP cell with an electrode/membrane area of 0.01 m^2. A DSA (Dimensionally Stable Anode) anode served as oxygen electrode, Ni as cathode. Anion exchange membranes = Neosepta ACM (Tokuyama Soda); AMH; cation ex-change membranes = Nafion 324; Nafion 902.

The scale-up was performed with a single ElectroProd Cell with an anode area of 0.4 m^2, with an interelectrode gap of approximately 8 mm, with a typical cell voltage of 5.3 V. A further development program is underway which will extend the scale-up to a 20 cells stack (membrane area > 8 m^2) of ElectroProd cells. [282a] The details are found in the US Patent, Ref. [265].

Osmotek Inc. (Corvallis, Ore.) has developed a different electrolytic approach to the recovery of pure sodium hydroxide from "black liquor" from paper mills. The black liquor is circulated to the anode of an electrolysis cell

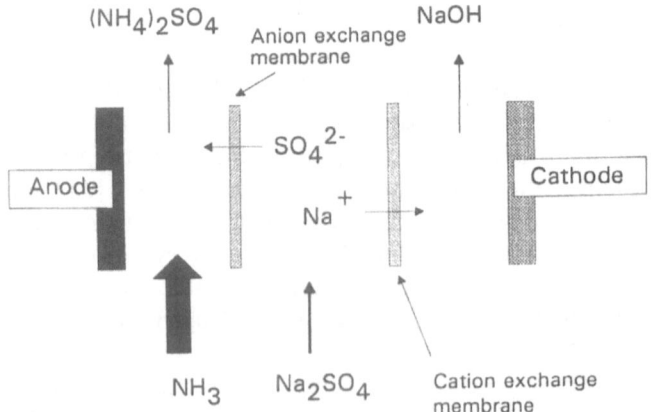

Fig. 34. Modified sodium sulfate salt splitting process for the production of both ammonium sulfate and sodium hydroxide [13, 264–266]

with a cation permeable membrane. The oxidation of the water leads to the neutralization of the impure sodium hydroxide. Pure caustic soda is formed at the cathode cf. Ref. [50].

Lantagne and Vélin [267] have reviewed the application of dialysis, electrodialysis and membrane cell electrolysis for the recovery of waste acids. Because of the new trends governed by environmental pressures, conventional treatment methods based on neutralization and disposal are being questioned. Membrane and electromembrane technologies are considered to be potential energy-efficient substitutes for conventional approaches. Paper mills will focus on the application of ion-exchange membranes namely dialysis, electrodialysis and membrane cell electrolysis for recovery of waste acids.

4.3 Electrochemical Treatment of Organic Pollutants and Destruction of Organic Pollutants

Not all environmental contamination is due to heavy metals or purely inorganic species, much is due to organic spills of petroleum, products, chemicals, solvents, pesticides, etc. Many industrial processes generate extremely toxic residual wastewaters, hardly biodegradable and requiring a chemical or physicochemical treatment. Incineration and landfill (sometimes after chemical treatment or encapsulation) have been the first resort for disposal of toxic industrial wastes [283]. Incineration of organic wastes can pose emission problems, in that discharge of very toxic materials in the offgas can occur if the combustion conditions are not carefully controlled. Thus burning of 2,4,5-trichlorophenol and its derivatives at 600 °C produced up to 5000 ppm (0.5%) 2,3,7,8-tetrachlorodibenzodioxin (2,3,7,8-TCDD = Seveso Dioxin), whereas 2,4,6-trichlorophenol forms, under the same conditions, only the by far less toxic 2,4,7,9-TCDD [284]. Worthy of notice is that the discharge limits for combustion products such as dioxins and dibenzofurans have to some extent followed analytic detection limits downwards and are often in ppb.

Electrochemical destruction of organics can be an economically viable alternative to incineration, carbon beds, bioremediation, deep well disposal and other methods as destruction to very low acceptable levels is possible [227a]. Electrochemical techniques are in fact superior to incineration or deep well disposal as it is a final solution and not a transfer of a toxic material from one environment to another, e.g. to the groundwater or the atmosphere [285]. Common destruction pathways include both *direct* and *indirect* electrolysis. Many electrochemical degradation pathways remain unclear and may be a mixture of direct and indirect processes depending on the pollutant and its intermediates [84, 285a].

What follows is the discussion of a number of selected examples of applications that tackle environmental pollution. For a detailed discussion the reader is refered to recent monographs [23] and a number of excellent reviews such as Refs. [84, 227a].

4.3.1 Direct Pathways in Electrochemical Destruction of Organics

Complete oxidation of organics to CO_2 and water requires the transfer of a large number of electrons (commonly $4e^-$ for every C), and hence a very high consumption of energy. The destruction of aniline, for instance, requires 28 electrons per molecule

$$C_6H_5NH_2 + 13H_2O \rightarrow 6CO_2 + NH_4OH + 28H^+ + 28e^-$$

If the reaction were carried out in an electrochemical cell, assuming a cell voltage of 3 V and a current efficiency of 95%, the cost would be $1.78/kg at $0.07/kWh [285]. Energy costs for electrochemical oxidation of pollutants are listed in Table 20 [84].

This is acceptable only for very toxic materials and radioactive wastes from the nuclear industry.

Table 20. Energy costs for electrochemical oxidation of pollutants [84]

Examples	Energy Cost ($/lb) Assumption: 100% CE, 5 V, 0.06 $/kWh
$CH_2O + H_2O \longrightarrow CO_2 + 4H^+ + 4e^-$ MW30	0.49
$H_2S + 4H_2O \longrightarrow H_2SO_4 + 8H^+ + 8e^-$ MW34	0.86
(a) $NH_3 + 3H_2O \longrightarrow HNO_3 + 8H^+ + 8e^-$ MW17	1.72
(b) $2NH_3 \longrightarrow N_2 + 6H^+ + 6e^-$	0.65
$C_6H_6 + 12H_2O \longrightarrow 6CO_2 + 30H^+ + 30e^-$ MW78	1.41
$C_6H_5NH_2 + 15H_2O \longrightarrow 6CO_2 + HNO_3 + 36H^+ + 36e^-$ MW93	1.42

EC = energy consumption (kWh/kg) = nFV/MW.CE
CE = current efficiency

4.3.1.1 Direct Processes

Anodic destruction of CN^- in parallel to the cathodic dechlorination of ArylCl [84, 286]

At the anode:

$$CN^- \rightarrow CN\cdot + e^-$$

$$2CN\cdot \rightarrow NCCN$$

$$NC\cdot + H_2O \rightarrow CNO^- + 2H^+ + e^-$$

$$2CNO^- + 4H_2O \rightarrow N_2 + 2CO_3^{2-} + 8H^+ + 6e^-$$

At the cathode:

$$Aryl\text{-}Cl + 2H^+ + 2e^- \rightarrow ArylH + HCl$$

(a) The direct oxidation of CN^- has been reported as being kinetically hindered on Pt and graphite anodes, probably due to absorbed species [286].

Other routes to the destruction of CN^-:
(b) Indirect cyanide oxidation with electrogenerated hypochlorite, ClO^- being generated in an undivided cell

$$CN^- + ClO^- \rightarrow CNO^- + Cl^-$$

(c) Photoelectrochemical cyanide oxidation at irradiated semiconductors is another route.

$$2CN^- + 2H_2O + 2h_{VB}^+ \rightarrow 4H^+ + 2CNO^-$$

$$4H_2O + 6h_{VB}^+ \xrightarrow{\quad} N_2 + 2CO_3^{2-} + 12H^+$$

$$O_2 + 4H^+ + 4e^- \longrightarrow 2H_2O$$

4.3.1.2 Combination of an Electrochemical Treatment with Other Methods

Combination of electrochemistry with other methods for a complete destruction, e.g. especially the eletrochemical pretreatment as precursor to biological degradation is gaining much interest. Numerous examples are to be found in the modern literature [135, 287, 288, 289a,b,c 290].

a) Reduction processes
Wash effluents from nitration installation can cause problems due to the high nitrophenol concentrations, their high toxicity to sewage bacteric and their general resistance to biodegradation. Electrochemical pretreatment tests in the laboratory showed a reduction in toxicity, and improvement of the color smell, etc. The electrochemical pretreatment is also attractive because of the absence of solid waste coproducts and its ease of operational control, [135].

The toxicity of chlorinated compounds is connected with the chlorine content. The decrease of toxicity of chlorophenols with decreasing number of chloro substituents, expressed as EC_{50}, using the Microtox method [291], is shown in Table 21a. [290], cf. also Refs. [292, 293].

The biodegradability of organic compounds is strongly affected by chlorine substitution in the molecule as shown by the DDT-related compounds, Table 21b.

Many wastewaters from the chemical process industry contain organochlorine compounds which are, amongst others, responsible for the toxicity of such wastes. The electrochemical reduction of chlorine atoms from organic molecules prior to their discharge may greatly facilitate the subsequent biological treatment of toxic wastewater [290]. Schmal et al. of TNO Delft (the Netherlands) have examined the electrochemical dechlorination of eight priority

Table 21a. Toxicity of chlorophenols [290]

Compound	EC_{50} (mg/dm^3)
Pentachloro	0.1–1
Tetrachloro	0.1
2,4,6-trichloro	7
2,4-dichloro	4
2-chloro	22
Phenol	22–42

Table 21b. Effect of substitution on the biodegradability of compounds of the type $(C_6H_4R)_2CHCCl_3$ [290]

R	Biodegradability
Cl(DDT)	1
OCH_3	60
CH_3	500
SCH_3	3000

compounds from the EC-list: [294] 2-amino-4-chlorophenol; 4-chloronitrobenzene; dichlorvos (DDVP); hexachloroethane; pentachlorphenol; 2,4,5-T; tetrachloroethene; 1,2,4-trichlorobenzene, using a MP-Cell (Eletro Cell Systems AB). TNO and Electrosynthesis Co (Lancaster USA) used cathodes based on bundles of carbon fibers and carbon felt [290, 295]

The overall reaction for the cathodic dehalogenation is:

$$R - Cl + 2e^- + H_2O \longrightarrow RH + Cl^- + OH^-$$

Inherent advantages of the cathodic dehalogenation are:

(1) Treatment at ambient temperature;
(2) No additional chemicals;
(3) Selective removal of chlorine while the organic skeleton remains (cf. Fig. 35) to be digested by the biological route, which is still the cheapest way.

Figure 35 shows the decay of PCB as well as the rise and fall of the four intermediate chlorophenols, which confirms that the organic skeleton remains intact during the cathodic treatment. See also Table V, Ref. [290].

The expected total costs of the process are of the order of DM 10/m^3 wastewater which is comparable with the cost of adsorption on activated carbon, and about 1/10 of the costs of incineration.

The water insoluble, highly chemically and thermally stable PCBs used as insulating fluids for transformers and capacitors, in paints, copy paper, etc., are extremely toxic, persistent in the environment and bioaccumulating. PCBs are currently destroyed by incineration of concentrates at high temperatures or chemically with sodium metals or organosodium. Both processes are costly. The cathodic reduction/elimination of the chlorine from polychlorinated biphenyl

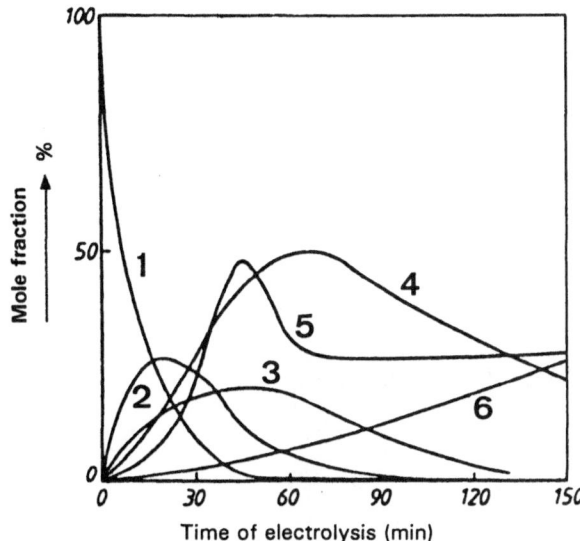

Fig. 35. Mole fractions of intermediates and product during electrolysis of PCP (290) (*1*) PCP, (*2*) tetrachloro-, (*3*) trichloro-, (*4*) dichloro- (*5*) monochloro- (*6*) phenol

wastes [84] according to the following equation

$$C_{12}H_{10-n}Cl_n + 2ne^- + nH^+ \longrightarrow C_{12}H_{10} + nCl^-$$

greatly reduces the toxicity and is less energy intensive than complete oxidation. A novel process developed by Electrosynthesis Co Inc. and PCB Sandpiper (Columbus, Ohio) [295–298] and commercialized by ElectroCell Systems AB [73] can be used to reduce large volumes of mineral and silicone oils containing about 1000 ppm polychlorinated biphenyls from transformer washing. Figure 36 shows the rig with an MP Cell.

Electrolysis conditions: divided cell, 25 °C; flow through carbon felt cathode (SGF-2300); graphite anode; current density, 15 mA/cm²; catholyte 1.0 M Et₄NCl/PC (propylene carbonate)/mineral oil (1:1) as an emulsion.

An emulsion of the oil and propylene carbonate (imiscible with the oils) is pumped through the cell with a carbon-felt cathode. The polychlorinated biphenyls are continuously extracted into the electrolysis medium. 30–50% current efficiencies for the decrease in the polychlorinated biphenyl concentrations from 700 to less than 1 ppm are possible.

(b) Oxidation processes

The price of the oxidation equivalent varies with the reagent [288]. Oxygen in air is the cheapest, followed by chlorine, electricity, hydrogen peroxide and finally ozone. Oxydation with oxygen at low temperature is only feasible biologically. Chlorine often forms very stable, highly toxic chlorinated compounds, which often limits its use.

Electrochemical oxidation is clean, and is, in the presence of activating groups, efficient. The lack of stability of commonly used anodes (except Ebonex, Sect. 2.3) often limits its use. Hydrogen peroxide is an ad hoc expensive oxidant

Fig. 36. ElectroSynCell (with systems including pumps, tanks, heat exchanger, instrumentation and piping) used for the elimination of PCBs from mineral and silicon oils

which can be used where small quantities of wastewater have to be detoxified. Ozone which can be used in all situations is the most powerful but also the most expensive oxidant.

Features of the various chemical processes are summarized in Table 22.

Electrochemical oxidation is appreciated as a very powerful tool for breaking up most resistant organic compounds. Anodic oxidation of organic contaminants is, thus, a potentially powerful method of controlling pollutants in wastewaters provided the process can be run with high current efficiency for the oxidation reaction. The current efficiency mainly depends on the reactivity of the organic pollutants with respect to the oxidants and their concentration. Electrochemical oxidation is due either to (1) direct oxidation of the molecule adsorbed at the surface of the electrode by withdrawal of electrons, or (2) indirectly by oxidation of the molecule by atomic oxygen or hydroxyl radicals, formed by electrolysis of water. Result: reduction of the COD (chemical oxygen demand) and an enhanced biological degradability measured as BOD (biological oxygen demand). Comninellis and co-workers have extensively studied the

Table 22. Comparison of the various chemical processes vs electrochemical oxidative destruction for the destruction of organic compounds [288]

Oxidant	SFr/kg equi.	Average Yield	Invest Cost for 100 kg COD/h (M SFr)	Treatment Cost SFr/kg COD
O_2^*	< 1		2–6.0	1–3
Cl	18	50%	3.0	6.50
H_2O_2	45	55%	3.0	12
O_3^{**}	60	100%	24	14
Kwh	10	56%	10	6.1

*Biological treatment
**Only two equivalents/mole
SFr = Swiss francs

electrochemical pretreatment of phenolic wastewaters [83a, 83b, 299, 300a–f] and of 1,4-benzoquinone – often present in wastewaters of various industries [301] – before biological treatment. During the electrolysis of the phenol the DOC (Dissolved Organic Carbon), TOC (Total Organic Carbon) and chemical oxygen demand (COD) were monitored. At the same time, the oxidation intermediates were analyzed by HPLC and the evolution of the toxicity was studied by the Microtox [291] and the activated sludge by respirometric [302] methods. The biodegradibility [303] for the pure intermediate substances was also measured. The toxicity and the biodegradability tests give an indication of the state where the biological treatment can take place.

The extremely toxic 1,4-benzoquinone can harm biological wastewater plants [301]. The benzoquinone formed during the phenol electrolysis thus necessitates a very severe electrochemical treatment before the biological treatment.

Comparison of the oxidation products of model phenol and benzoquinone compounds obtained with Pt and SnO_2 anodes show two main differences:

(1) At the SnO_2 anode only a very small amount of highly toxic intermediates (hydroquinone, catechol, benzoquinone) is formed. These intermediates are formed in large amounts on the Pt anode probably by chemical reaction of adsorbed hydroxyl radicals with phenol.
(2) Aliphatic acids (fumaric acid, maleic acid) are rapidly oxidized at the SnO_2 anode, whereas they are practically electrochemically inactive at the Pt anode. The process at the SnO_2 is able to completely oxidize phenol, which is quite unique for a low temperature process.

In this context see also Refs. [83a, 83b]. Comninellis and Plattner [287, 287a, 288] have developed a simple method for estimating the facility of the electrochemical oxidation of organic species based on a newly defined "electrochemical oxidizability index" (EOI) and the degree of oxidation using the "electrochemical oxygen demand" (EOD). Electrochemical oxidizability index for various benzene derivatives obtained at Pt/Ti and SnO_2-ABB-anodes are listed in Table 23.

Table 23. Comparison of the Electrochemical Oxidation Index (EOI) at Pt/Ti and SnO_2- Anodes (ABB-Anode) with organic substances [45a, 287, 288]

Compound	Pt anode	ABB-Anode
Ethanol	0.02	0.49
Acetone	0.02	0.21
Acetic acid	0.00	0.09
Formic acid	0.01	0.05
Tartaric acid	0.27	0.34
Oxalic acid	0.01	0.05
Malonic acid	0.01	0.21
Maleic acid	0.00	0.15
Benzoic acid	0.10	0.79
Naphthalene-2-sulfonic acid	0.04	0.51
Naphthalene-1-sulfonic acid	–	0.41
Phenol	0.15	0.60
Aniline	–	0.43
Benzene sulfonic acid	–	0.28
5-Methyl-3-aminoisoxazole	–	0.25
Orange II		0.58
Anthraquinone sulfonic acid		0.18
Nitrobenzene		0.80
Nitrobenzenesulfonic acid		0.46
Triaminotriazin		0.02
EDTA	0.30	0.30
p-NDMA	0.30	0.37
4-Chlorphenol	–	0.35
	Av. 0.05	Av. 0.34

4.3.2 Indirect Pathways/Electrode Processes in Electrochemical Destruction

Indirect pathways (cf. Ref. [84]) are based on

(1) generation of short lived reactive species

$(e)_{solv}$, $O\cdot$, $OH\cdot$, $O_2^{-\cdot}$, $\cdot HO_2$, O_3, ClO_2

(2) regenerable redox couples in solution:

At anodes: Cr(III)/Cr(VI); Ag(I)/Ag(II); Co(II)/Co(III); Cl^-/OCl^-

At cathodes: V(II)/V(III); Cr(II)/Cr(III); Cu(I)/Cu(II)

(3) heterogenous oxidations at anode oxides:

PtO_x, PbO_2, FeO_x, IrO_x, SnO_2,

MnO_x, AuO_x, TiO_x CrO_3

A novel complete indirect electrochemical destruction of organic chemicals of the types found in industrial wastes to CO_2 and H_2 using metallic mediators such as Ag(I)/Ag(II), Co(II)/Co(III) has been described [13, 133, 283, 304–308].

(a) Silver (II) ions can be generated electrochemically at the anode in a divided cell be oxidation of Ag(I) in acid solution. Oxidation of organic species by Ag(II) is well documented in the literature [309, 310]. The redox couple Ag(II)/Ag(I) has a very high $E°$ of $+ 1.98$ V. Only O_3, $S_2O_8^{2-}$ and F_2 have higher $E°$ values. $Ag^+ \rightarrow Ag^{2+} + e^-$ has a very low activation energy and therefore fast electrode kinetics, and a very fast solution kinetics; a high current density is possible without activation polarization (ca 120 mV at 0.5 A cm^{-2}) [304].

Different oxidation pathways for the oxidation of organics by Ag(I) are possible:

(1) Oxidation of Ag(I) to Ag(II), followed by reaction of Ag(II) directly with the organics
(2) Oxidation of Ag(I) to Ag(II), followed by reaction of Ag(II) with water/nitric acid to produce radicals which further react with organics
(3) Direct oxidation of the organics at the anode

It is thought that all three mechanisms can occur at the same time, each contributing to some extent, but mechanism (2) is thought to be dominant.

Following mechanisms are based on reaction (2).

At the anode (platinized titanium)

$$Ag^+ \longrightarrow Ag^{2+} + e^-$$

In nitric acid, Ag^{2+} forms a brown coloured complex $(AgNO_3)^+$ via

$$NO_3^- \longrightarrow NO_3\cdot + e^-$$

followed by

$$Ag^+ + NO_3\cdot \longrightarrow (AgNO_3)^+$$

This complex stabilises the Ag^{2+}, and allows it to travel through the anolyte. The $(AgNO3)^+$ carries the oxidation reaction into the bulk solution, and thus removes the limitation of the reaction having to happen on the electrode surface and the surrounding boundary layer. Reactions in the bulk solution are:

$$(AgNO_3)^+ + H_2O \longrightarrow HO\cdot + H^+ + Ag^+ + NO_3^-$$

followed by

$$\text{organics} + HO\cdot \longrightarrow CO_2 + CO + H_2O \text{ etc.}$$

The overall reaction can be written as

$$\text{organics} + nH_2O \longrightarrow (n/2)CO_2 + 2nH^+ + 2ne^-$$

Some CO can also be formed.

At the cathode the main reaction is the formation of nitrous acid:

$$NO_3^- + 3H^+ + 2e^- \longrightarrow HNO_2 + H_2O$$

In order to minimize operating costs of the process, nitric acid is recovered within the catholyte. A side stream is heated to boiling and refluxed into a packed column, into which air or O_2 is injected.

$$HNO_2 + 1/2O_2 \longrightarrow HNO_3$$

The overall cathode process can be summarized as:

$$4H^+ + 4e^- + O_2 \longrightarrow 2H_2O$$

Assuming little nitric acid loss, the *overall process* now becomes

$$\text{Organics} + O_2 \longrightarrow CO_2 + CO + H_2O + \text{inorganics}$$

Figure 37 shows a simplified block diagram of the process currently studied in a pilot plant rig. The process uses off-the-shelf cells (ICI FM 21) and plant technology.

Typical electrical power consumption values are [133]:

Kerosene: 35 kWhkg^{-1} for complete oxidation, based on 100% current efficiency and a cell voltage of 3 V.

CCl_4:   2.1 kWhkg^{-1} for oxidation to CO_2 and Cl_2.

Whereas kerosene is easy to burn CCl_4 requires fuel addition.

The low temperature/low pressure operating process developed by AEA (Dounray, Scotland), known as "Silver bullet", which allows one to process a wide range of waste types and compositions, cf. Table 24, makes it an attractive alternative to incineration for the more toxic and troublesome types of industrial organic wastes.

The movement of the mediated electrochemical oxidation technique using metallic mediators, e.g. Ag(II) as an alternative to incineration (being developed at the Lawrence Livermore National Laboratory (LLNL) [306,a,b, 307,a] primarily for the United States Department of Energy) into other fields such as destruction of infectious agents (i.e. AIDS contaminated blood, etc.), listed organic hazardous wastes, chemical warfare agents, and pesticides would be relatively easy. At the LLNL, the oxidation of several real and surrogate organic wastes, including ethylene glycol (EG) [306a, 307a], isopropanol, 2-mono-chloro-1-propanol, 1,3-dichloro-2-propanol, benzene, and Trimsol cutting oil have been studied in detail [306, 307, 311]. Mediators such as Ag(II), Co(III), and Fe(III) have been used in supporting electrolytes of HNO_3 or H_2SO_4.

In contrast to Ag(I), the ions Ce(III), Co(II) and Fe(II) remain soluble in the presence of halide ions liberated during the destruction of halogenated organics. Thus Co(III) is a halide-tolerant mediator. Destruction rates greater than 3L/d have been achieved with EG. Treatability studies of Trimsol cutting oil contaminated with various radionuclides were discussed by Hickman et al. [306].

(b) Destruction of carbonaceous waste by iron (III): [50, 300b, 311a, 311b]

Fe(III) in acidic media at 120–150 °C is a very efficient redox reagent for the oxidation of organics, polycyclic aromatics, coal, coke, etc. Graphitic type organic materials such as tarry and oily wastes, still bottoms, high sulfur coke, etc. can be rendered benign or be converted to humic acid without proceeding to complete combustion. The environmentally benign humic acid can be used in adhesives, fertilizers and lubricants.

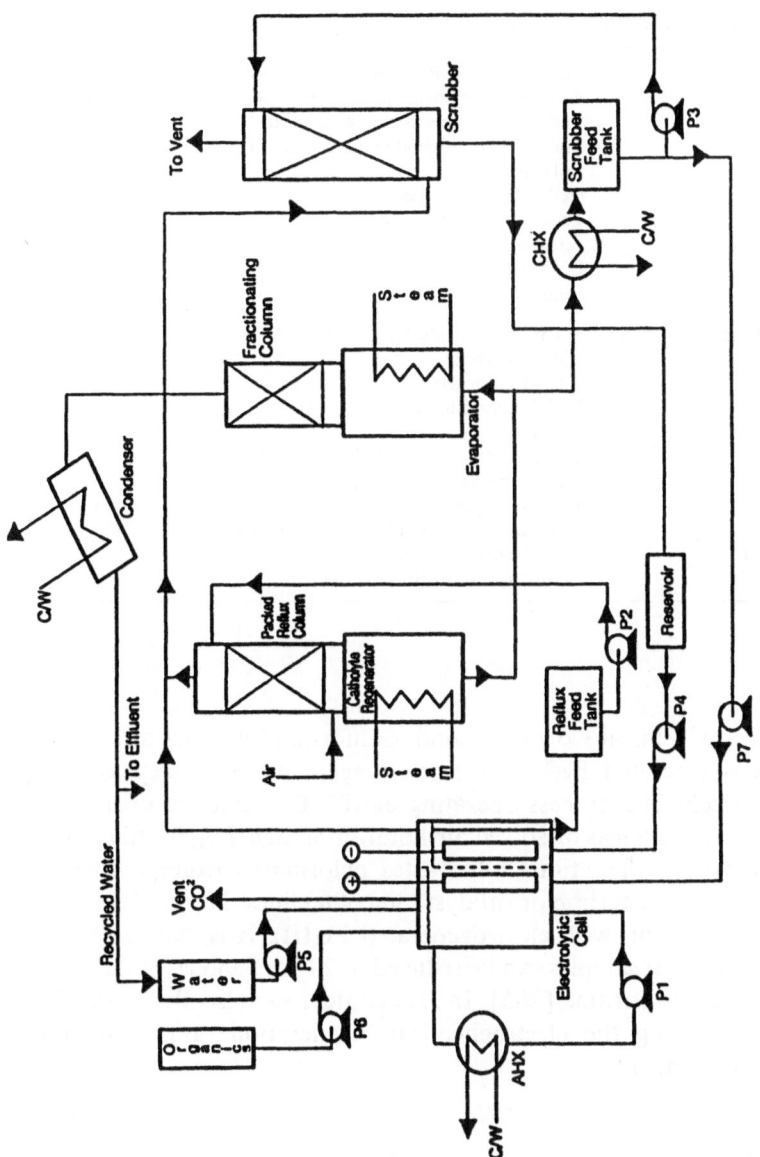

Fig. 37. Schematic diagram of the Ag(I/II)/nitric acid (DNE) electrochemical oxidation process [283, 304]

Table 24. Toxic wastes to which the "Silver Bullet" technology has been applied [304]

Dodecane	Epoxy resin (Tufnol, Bakelite, Epophen)
Kerosene	Styrene divinylbenzene ion exchange resins
Benzene	Reillex HPQ anion exchange resin
Toluene	Phenol/formaldehyde cation exchange resin
Octanoic acid	Polyurethane
Phenol	Cellulose (as tissues)
Rubber gloves	Tri-n-butyl phosphate
Detex	Lithium based grease
Cutting oil	Lubricating oil
Easing oil	Pump oil
Hydraulic fluid (Shell D 46-probably tritolyl phosphate)	
Xylene	Chlorobenzene
Acetic acid	1,2,4-trichlorobenzene
Dinitro stilbene	1,1,2,2-tetrachloroethane
m-nitro p-toluidine	PCBs
p-aminophenol	(2-chloroethyl) ethyl sulfide
p-nitrosophenol	various mixed organic amines
DMF	3-chloro-1-propanol
n-butanol	Irradiated polythenethyle (dose 1E9 R)
2-methoxyethanol	Irradiated PVC (dose 1E9 R)
Triethanolamine	2-chloro 4-fluorobenzoic acid

The Chevron Chemical Co. (Richmond, California) [50] uses a technology based on electrogenerated Fe(III) as redox reagent which is recycled to the electrochemical cell. The process operating at 180 °C requires a pressure cell. Hazardous materials such as dissolved air-flotation sludge (DAF) gathered from refinery runoff process operations, oily wastes, chlorinated aromatic hydrocarbons are adsorbed onto carbon or catalyst (Silicolite) fines. The adsorption step is followed by oxidation with electrogenerated Fe(III). It is claimed that the carbon content of DAF samples can be reduced to 20 ppm and less total organic carbon in the aqeous streams [285]. In this context see also Ref. [300b] and [311a, 311b] discussing the electrochemical regeneration of metallic redox couples (Fenton's reagent).

4.3.3 Electrochemical Treatment of Colored Effluents

The fact that dyes are extensively scattered throughout the environment, and that dye degradation products can exhibit genotoxic and ecotoxic properties has intensified the need for sufficiently sensitive and selective (analytical) methods and reliable destruction methods [226]. An important degradation pathway of water-soluble azo dyes (azo dyes makeup about 30% of all dyes used) is a reductive cleavage to aromatic amines. Strongly hydrophylic amines, e.g. sulfonated amines can be regarded as being detoxified, lipophilic aromatic amines can cause problems, cf. also Sect. 4.3.3.2.

Characteristics of wastewaters and effluents from dye production and processing are:

(1) often intensively coloured solutions, and
(2) high loads of (i) organic compounds/pollutants,
              (ii) "heavy/toxic" metals, and
              (iii) halogens.

The environmental tasks of the dye industry have been defined by the ETAD (Ecological and Toxicological Association of the Dye Manufacturing Industry) as

(1) monitoring of trace amounts of harmful inorganic compounds and impurities such as (i) heavy metals: As, Cd, Co, Cr, Cu, Hg, Ni, Pb and Zn, (ii) amines (educts, intermediates, byproducts or degradation products), (iii) PCBs, (iv) chlorinated compounds (AOX)
(2) control of coloured effluents within specific limits of color in effulents and wastewaters.

Are dyes environmental chemicals? Dyes enter the environment mainly in wastewaters from their production, from dying and recycling processes and from domestic sources. While colored organic substances generally contribute only a small part of the total organic load in a wastewater, their high degree of colour is easily detectable. 1 ppm (1 mg/l) of a dye is likely to cause visible coloration of the water. Besides being an aesthetic pollutant they may impede light penetration in the receiving body of water, thus interfering with biological process [312, 313]. As far as the public is concerned, the removal of color from wastewater is often more important than the removal of soluble, colorless organic substances which usually contribute to the major fraction of the biochemical oxygen demand (BOD). Methods of dye/textile effluent treatment may be classified broadly into three main categories: physical, chemical and biological, cf. Table 25 [314].

Table 25. Classification of methods of dye and textile effluent treatment [314]

Physical	Chemical	Biological
Sedimentation	Neutralization	Stabilization ponds
Filtration	Reduction	Aerated lagoons
Flotation	Oxidation	Trickling filters
Foam fractionation	Catalysis	Activated sludge
Coagulation	Ion exchange	Anaerobic sludge
Reverse osmosis	Electrolysis	Fungal treatment
Solvent extraction		
Ionizing radiation		
Adsorption		
Incineration		
Freezing		
Distillation		

Commercial dyes derive their colour from the relative complex chromophore systems which they contain. All chromophore moieties (except indanthrene dyes) are electrochemically active. Polarographic and voltammetric reduction ranges are summarized in Fig. 38.

Electrochemical methods have been used in several ways for the treatment of effluents. A distinction must be made between:

(1) *Indirect* electrochemical procedures such as:
   (i) Electrocoagulation, (ii) electrodialysis, (iii) electroflotation, and (iv) processes with electrogenerated agents, i.e. "active" chlorine, generated at the anode by electrolysis of (added) NaCl.
(2) *Direct* electrochemical methods:
   Oxidation/reduction of the coloured species, involving a direct electron transfer,
(3) Combined electrochemical techniques:
   involving direct and indirect processes. Very promising are processes combining electrochemical techniques which need no additional chemicals, with the cheap biological process.

4.3.3.1 Indirect Electrochemical Procedures
For a detailed discussion, cf. Ref. [226]

(a) Processes based on sacrificial (Fe, Al) electrodes
The process is based on the simple reaction

Anode: $Fe \longrightarrow Fe^{2+} + 2e^-$

Cathode: $\dfrac{2H_2O + 2e^- \longrightarrow H_2 + 2OH^-}{Fe + 2H_2O \longrightarrow Fe(OH)_2 + H_2}$

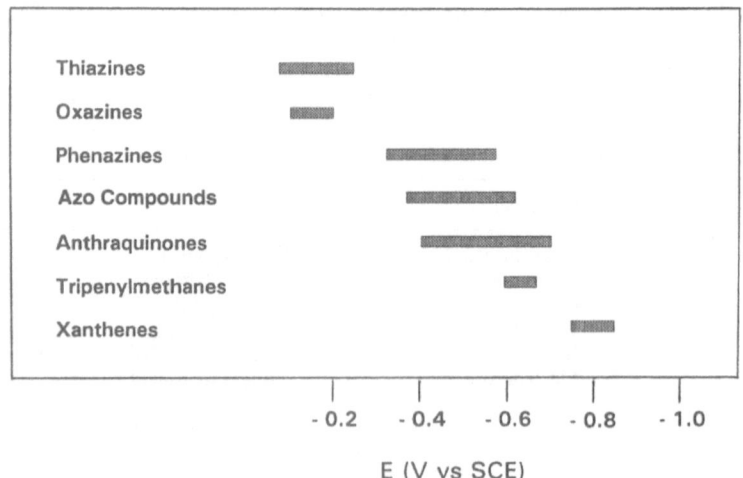

Fig. 38. Polarographic ranges of different dye classes, measured in 0.06 M phosphate buffer, pH 7, containing 1% ethanol, at 25 °C [315, 316]

Insoluble $Fe(OH)_2$ is formed which precipitates soluble and insoluble inorganic and organic pollutants from the solution. The mechanism for removing contaminants from wastewater is not yet fully understood [312, 317–319]. The process has proved itself to be highly effective in the removal of the color and heavy metals. Disadvantage: The iron sludge must be dumped (landfills are scarce) or dried, pelleted and sintered for disposal in iron mills (very costly).

(b) Processes with active chlorine generated at the anode
The active agent for decolorization is the active chlorine formed by electrolysis of added NaCl to the wastewater in a monocell or divided cell. The dyes are thus mainly oxidized in the bulk solution and not at the surface of the electrode. Conceptually the simplicity of on-site electrochemical generation of hypochlorite is very attractive. It also offers decided advantages from the safety standpoint [320]. The disadvantages are: The hypochloric acid formed during electrolysis acts as an oxidizing and chlorinating agent. Due to dissolution of the chlorine, formed during electrolysis – in water – and oxidation of organic compounds by active chlorine, the acidity of the anolyte increases. Below pH 3 the oxidation of organic compounds is accompanied by chlorination of organic compounds, with formation of poorly soluble, generally toxic organochlorine substances, such as AOX (adsorbable organic halogenated compounds). The prerequisite for the purification of wastewaters is thus a pH of 3–7 for the anolyte, which is best achieved by diluting the wastes with NaCl solution, or by addition of alkali to the wastes to neutralize the HCl.

4.3.3.2 Direct Electrochemical Procedures

Direct electrochemical reduction and oxidation treatment of pollution involving a mass-free reagent – the electron – is a very attractive idea, because it is a uniquely "clean" process, as (1) the reduction and oxidation take place at inert electrodes and (2) there is no need to add chemicals. The techniques of cathodic reduction/anodic oxidation of wastewaters containing dyes are relatively new and have drawn the attention of investigators in Japan, China, USA and Russia [55].

Reduction mode
Numerous commercial dyes are metal chelate complexes. These metals form pollutants which must be eliminated. One of the strongest points in favour of electrochemical reduction/removal of metal ions and metal complexes – the metal ions and weakly complexed ions form the toxic species – and of the metals from the metal-complex dye is that they are eliminated from the solution into the most favorable form as pure metal, either as films or powders. Polyvalent metals and metalloids can be transferred by reduction or oxidation treatment to one valency, or regenerated to the state before use, e.g. Ti(III)/Ti(IV), Sn(II)/Sn(IV), Ce(III)/Ce(IV), Cr(III)/Cr(VI), and can be recycled to the chemical process. Finally, they can be changed to a valence state better suited for separation, for instance, for accumulation on ion exchangers, etc. Parallel to the

reduction of the inorganic species, electroactive organic species, including the electroactive chromophores, are reduced.

Oxidation mode

As mentioned in Sect. 4.3.1, electrolytic oxidation is considered to be a powerful tool for breaking up most resistent organic compounds.

Toxic compounds are frequently rendered less toxic by electrochemical treatment, for instance dehalogenation of chlorinated derivatives such as PCBs or AOX (performed in a divided cell or in the absence of Cl ions in a monocell); odors are eliminated or greatly reduced, i.e. reduction of nitrotoluene which can be a serious odor nuisance; biodegradability is improved. Elimination of colloids and surfactants is possible.

The high organic loads differentiate the wastewaters and effluents from dye production and processing from inorganic/mineral acid wastes (galvanic, pitting, etc.) which have been treated very successfully by electrochemical means in the past. The fate of the organic compounds, including dyes, present in the samples treated electrochemically deserves much attention because formation of more hazardous, "toxic" products is possible, thus enhancing the toxicity of the treated wastes, a fact which has long received little or no attention in communal and industrial effluent treatment.

Well known are:

(i) formation of lipophilic (carcinogenic) amines due to the reduction of $-N = N-$ and NO_2-groups or due to desulfonation of aromatic amines. Strongly sulfonated hydrophilic amines can be regarded as non-toxic whereas many lipophilic amines are known or suspected to be carcinogenic to humans,
(ii) formation of phenols,
(iii) formation of chlorinated compounds (AOX) in the presence of halogens when working in monocells,
(iv) formation of surfactants,
(v) formation of polymers, poisoning the electrodes, or clogging the diaphragms.

The toxicity (before and after treatment) of solutions subjected to a chemical or electrochemical oxidation/reduction treatment should always be tested.

Papers and patents on the direct electrochemical decolorization of colored wastewaters and effluents are scarce [226, 318, 319, 322–325]. Thus a carbon felt electrode was used in a flow-through cell for treatment of dyeing wastewaters. COD removal rates of > 60% in a single stage and up to 80% by a two-stage treatment were reported. Anodic oxidation of Direct Blue 21 was performed in a divided cell with a PV porous diaphragm in flow-through mode. At the bottom of the anode compartment a stainless steel screen is laid to support lead spheres which act as the anode. The cathode was a stainless steel sheet [324, 325]. A bench scale set-up and flow diagram for direct electrochemical treatment of wastewaters using a fiber-optic guided-wavemeter for monitoring the color change (cf. also Ref. [326]) is shown in Fig. 39 [226].

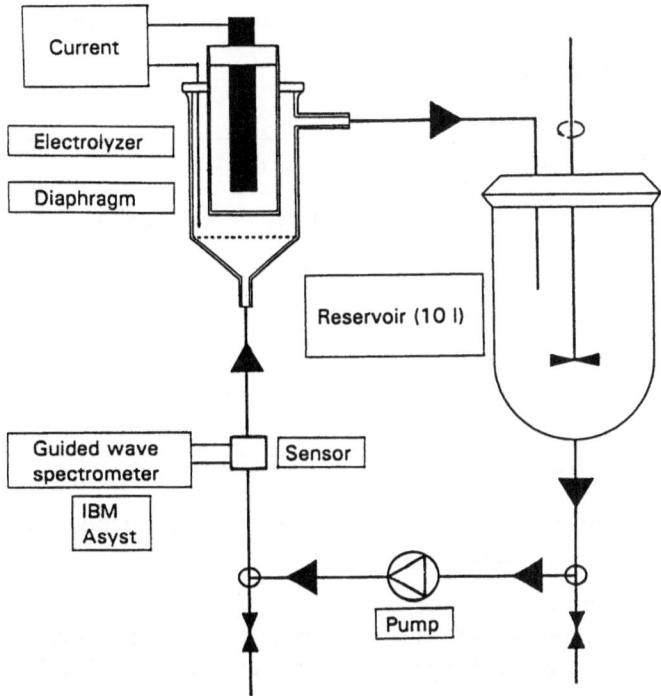

Fig. 39. Bench-scale set-up and flow diagram for removal of metals and dyes (decolorization) from wastewaters [226]

The tests are run in two parallel streams: the first stream is electrolyzed in a divided cell, the second stream in a monocell. The control of AOX formation or reduction during the progress of the electrolysis is thus possible, Table 26.

After reductive treatment the test solutions are purged with oxygen or air. The progress of the chemical and electrochemical decolorization is monitored:

(1) by simple "stain" tests (application of 30–50 μl of the test solution on a filter paper);
(2) spectrochemically using an on-line guided-wave meter with fiber optics [326];
(3) electrochemically by polarography, voltammetry and tensammetry; after HPLC separation with electrochemical and UV detection in the tandem mode.

The elimination of metals is monitored electrochemically and spectrometrically:

(1) the labile metal concentration is determined polarographically or voltammetrically (inverse stripping techniques),
(2) the total metal concentration after digestion by polarography, or voltammetry and by AAS and ICP-MS.

Table 26. Cu and AOX contents before and after reduction of Formazan Blue wastewater

| | Replicate measurements Cathode = Cu grid Anode = Ti / Ti coated | | | |
| | Divided Cell | | Mono Cell | |
	Cu mg/l	AOX mg Cl/l	Cu mg/l	AOX mg·Cl/l
Before	560	47	560	47
After	14	80	30	308
	62	76	24	310
IC50 value: Before	1 : 20		1 : 20	
After	1 : 20		1 : 20	
	1 : 10		1 : 10	

When discussing the treatment of wastewaters samples from production and pilot plant, the following distinction must be made:

(1) colored wastewaters with no or low metal loads,
(2) colored wastewaters containing high metal loads due to the presence of metals used as catalysts or contained in metal-complex dyes, cf. Table 26.

5 References

1. Kalhammer FR (1991) Energy, electrochemistry and electrochemical synthesis: Exploratory themes at the EPRI research 5th International Forum on Electrolysis, Ford Lauderdale, FL, Nov. 10–14
2. Trasatti S (1992) Electrochemistry and environment: The role of electrocatalysis, Abstract, PL-4. 43rd Meeting ISE, Cordoba, Argentina, Sept. 20–25
3. Weinberg N (1992) Electrolysis In: Genders and Weinberg, Ref [23]
3a. Grimm JKP (1801) Gilberts Ann Phys 7: 348
3b. Arnim LAv (1801) Gilberts Ann Phys 8: 257
4. Fichter F (1942) Organische Elektrochemie, Steinkopff, Dresden und Leipzig
4a. Zanetti R (1991) Chem Engineer, May, p 5
5. Steckhan E (ed) (1987) Electrochemistry I, Springer, Berlin Heidelberg New York (Topics in Curr Chem, vol 142)
6. Steckhan E (ed) (1988) Electrochemistry II, Springer, Berlin (Topics in Curr Chem, vol 143)
7. Steckhan E (ed) (1988) Electrochemistry III, Springer, Berlin (Topics in Curr Chem, vol 148)
8. Steckhan E (ed) (1990) Electrochemistry IV, Springer, Berlin (Topics in Curr Chem, vol 152)
8a. Pletcher D (1982) Industrial electrochemistry, Chapman and Hall, London
9. Genders JD, Pletcher D (1990) Electrosynthesis; from laboratory to pilot to production, The Electrosynthesis Co Inc, Buffalo
10. Pletcher D, Walsh FC (1991) Industrial electrochemistry, 2nd edn. Chapman & Hall, 1990, London; Chem Abstr 115: (1991) 37456x
11. Baizer MM, Lund H (eds) (1991) Organic electrosynthesis, 3rd ed, Marcel Dekker, New York
12. Pletcher D (1991) A first course in electrode processes, The Electrochemical Consultancy Ltd, Romsey, England

13. Walsh FC (1993) A first course in electrochemical engineering, The Electrochemical Consultancy Ltd, Romsey, England
14. Walsh FC (1991) Trends in electrochemical engineering, Bull Electrochem 7(5): 210
15. Mills GA, Walsh FC (1992) Electrochemical engineering widens its scope, Professional engineering, Nov 29; through (133)
16. Talbot JB, Fritts SD (1992) Report of the electrolytic industries for the year 1991, J Electrochem Soc 139: 2981
17. Electrochemical Technology in Industry – A UK Status Report, SJD Tait (ed), SCI Electrochemical Group, London, 1991; cited in Ref [133]
18. Coeuret F, Storck A (1984) Elements de Génie Electrochimique, Techniques at Documentation, Lavoisier, Paris
19. Heitz E, Kreysa G (1986) Principles of electrochemical engineering, VCH, Publ, Weinheim, New York
20. Rousar I , Micka K, Kimla A (1986) Electrochemical engineering, Vol 1 and 2, Elsevier, Amsterdam
21. Scott K (1992) Electrochemical reaction engineering, Academic Press, London, cited in Ref [13]
21a. Scott K (1992) Industrial waste water treatment: an electrochemical perspective, Symp Pap-Inst Chem Eng North West Branch 1992, 3(3, Integr Pollut Control Clean Technol) 6.1; Chem Abstr 117: (1992) 156885n
22. Coeuret F (1992) Introduccion a la Ingenieria electroquimica, Editorial reverté, SA, Barcelona
23. Genders JD, Weinberg NL (1992) Electrochemistry for a cleaner environment, The Electrosynthesis Co Inc, Buffalo
24. Pletcher D, Walsh FC, Whyte I (1990) The application of electrochemical techniques to the treatment of industrial process liquors, I Chem E Symp Ser, No 116 195, 35 refs; Chem Abstr 114: (1991) 170419h
25. Inst Chem Eng Symp Ser 1992, 127 (Electrochem Eng Environ 92), Loughborough Univ Technology
26. Lapique F, Valentin G, Storck A (1993) Electrochemical engineering for environmental protection, Electrochem Processing, Innovations and Progress, April 21–23, Glasgow
27. Millington JP (1993) Electrochemical technology for a cleaner environment, Electrochem Processing, Innovations and Progress, April 21–23, Glasgow
28. Sequeira CAS (ed) (1993) Environmentally oriented eletrochemistry, Elsevier, Amsterdam, to be published
29. Beck F (1974) Elektroorganische Chemie, Gundlagen und Anwendungen Verlag Chemie, Weinheim
30. Beck TR (1987) Industrial electrochemical processes. In: Yeager E, Salkind AJ (eds) Techniques of electrochemistry, Wiley, New York
31. Nohe H (1979) AIChE Symposium Series, 75 (185) 109; Chem Abstr. 91: 80881e
32. Roberts R, Ouelette RP, Cheremisinoff PN (1982) Industrial applications of organic synthesis, Ann Arbor Science Publ
33. Wagenknecht JH (1983) Industrial organic electro synthesis, Chem Educ 60(4)(1983) 271; Chem Abstr 98: 214827x
34. Weinberg NL (1983) Electrosynthesis technology, J Chem Educ 60(4)(1983)268; Chem Abstr 98: (1983) 197140y
35. Jansson R (1984) Organic electro synthesis, Chem Eng News 62 Nov 19: 43
36. Shono T (1984) Electroorganic chemistry as a tool in organic chemistry, Springer, Berlin Heidelberg New York
37. Torii S (1985) Electroorganic syntheses, Methods and applications Part I, Oxidations, Kodansha-VCH
38. Torii S (ed) (1987) Recent advances in electroorganic synthesis, Kodansha-Elsevier, Amsterdam
39. Hughes D (1987) Reviving electrochemistry, Chem Engineer Sept 17
40. Degner D (1988) Organic electrosyntheses in industry. In: Steckhan E (ed) Electrochemistry III. Springer, Berlin Heidelberg New York, p 1 (Topics in Curr Chem, vol 148)
41. Toomey JE Jr, Yu JC (1989) Making organic chemicals via electrochemical processing, Chem Engineer, 140–147, June
41a. Toomey JE Jr (1985) The electron as commercial reagent, CHEMTECH, 15, 738; Chem Abstr 104: (1986) 41820b

42. Shono T (1990) Electroorganic Chemistry, Academic Press
43. Weinberg NL, Mazur DJ (1990) Industrial electroorganic synthesis in the United States and Canada, Kagaku to Kogyo (Tokyo), 43 2002 (Japan), a review with 23 ref; Chem Abstr 114: (1991) 110525e
44. Little RD, Weinberg NL (eds) (1992) Electroorganic synthesis, Festschrift for Manuel M Baizer, (Held in Montreal, May 6–11, 1990); Marcel Dekker, New York, 1991; Chem Abstr 116: (1992) 161157k
45. Electrosynthesis in the Chemical Industry, 5th International Forum, Fort Lauderdale, FA, Nov. 10–14, 1991 Electrosynthesis Co. Inc., 72 Ward Road, Lancaster, NY, 14086-9779, USA
45a. Electrosynthesis in the Chemical Industry, 6th Int Forum, Fort Lauderdale, FA, Nov. 8–12, 1992
46. Bersier PM, Bersier J, in ref (45), and ref therein Electrosynthesis Co. Inc., 72 Ward Road, Lancaster, NY, 14086-9779, USA
46a. Swenton JS, Morrow GW (eds) (1991) Synthetic applications of anodic oxidations, Tetrahedron 47(4/5). Tetrahedron Symp in Print, No 42 (37 papers)
47. Utley JHP, Electrochemical methods for organic synthesis: Fundamentals and applications, in ref (45)
48. Samdani G, Gilges K (1991) Chem Engineer, May 37
49. Pletcher D, Weinberg NL (1992) The green potential of electrochemistry, Chem Engineer 99, August 98
50. Pletcher D, Weinberg NL (1992) The green potential of electrochemistry, Chem Engineer 99, November 132
51. Pletcher D (1993) Progress in organic electrosynthesis, Electrochem Processing, Innovations and Progress, April 21–23, Glasgow
52. ICI, Applied Electro Technology
53. Bersier PM, Howell J (1994) Electroanalysis, in preparation
54. BASF, Intermediates, cf Table 9a
54a. Krumpelt M, Weissman EY, Alkire RC (1979) AIChE Symp Ser 185, Vol 75, p iii
55. Bersier PM, Bersier J (1990) In: Ivaska A, Lewenstam A and Sara R (eds), Contemporary electroanalytical chemistry, Plenum, New York, p 109
56. EWG Nr C 176/3, 14.7.1982. Kommission, Mtlg der Kommission an den Rat über die gefährdenden Stoffe im Sinne der Liste I der Richtlinien des Rates 76/464/EWG (E_1)
57. Sanchez-Cano G, Montiel V, Aldaz A (1991) Tetrahedron, 47: 877
57a. Park K, Pintauro PN, Baizer MM, Nobe K (1985) J Electrochem Soc 132: 1850
58. Scott K (1991) Electrochim Acta 36(1991)1447; Chem Abstr 115: (1991) 168940y
59. Bersier PM, unpubl results
60. Weinberg NL, Genders JD, Mazur DJ (1990) US US 4,950,368; Chem Abstr 114: (1991) 255808j
61. Bersier J, Jäger H, Schwander H (1981) Eur Pat Appl 22 062, Ciba-Geigy; Chem Abstr 94: (1981) 158 314f
62. Jäger H, Plattner E, Bersier J, Comninellis C (1981) Eur Pat Appl EP 60, 437, Ciba-Geigy; Chem Abstr 98: (1983) 181 139u
63. Danly D In: Emerging opportunities for electroorganic processes, Marcel Dekker, New York, p 229; through (40)
64. Degner D, in ref (40), references 106–114
65a. Pistorius R, Millauer H, p-Benzoquinone diketals, Ger Off 2,547,463; Chem Abstr 87, (1977) 134621h
65b. idem, Ger Offen 2,547,386; Chem Abstr 87, (1977) 134622j
65c. Cramer J, p-Benzoquinone diketals, Ger Offen 2,739,315; Chem Abstr 90, (1979) 186593g
65d. Pistorius R, Millauer H, p-Benzoquinone diketals, Ger Offen 2,547,383, Chem Abstr 87, (1977) 134623k
66. Ito S, Iwata M, Sasaki K (1991) Tetrahedron, 47: 841
66a. Ito S, Kuroda J, Iwata M, Sasaki K, Okataki A (1993) J Appl Electrochem 23: 677, and references therein
67. Ito S, Yamasaki T, Okada H, Okino S, Sasaki K (1988) J Chem Soc, Perkin Trans 2 285
68. Ito S, Kunai A, Okada H, Sasaki K (1988) J Org Chem 53: 296
69. Chaussard and Lahitte, through (35)
70. Danly DE (1981) Scale-up of electro-organic processes, Lecture notes, UCLA short course in electroorganic chemistry/through ref (32)

71. Danly DE, in ref (45)
72. Carlsson L (1984) Proc Journée d'information sur l'électrosynthèse organique, EDF, SORAPEC, 29 Oct
73. Bersier PM, Carlsson L, Bersier J (1992) 43rd ISE Meeting, Cordoba, Argentina, September, 20–25, Electrochemical hardware for a better environment, Poster 3–22
73a. Bersier PM, Carlson L, Bersier J (1993) 44th ISE Meeting, Berlin, September 5–10, p 680, Poster V 7.13
74. Brooks WN (1986) The ICI (Mond) FM21 cell as a multi-purpose electrolyzer, Inst Chem Eng Symp Ser 1986, 98 (Electrochem Eng) 1–12, 320–1; Chem Abstr 105: (1986) 180432k
75. Couper AM, Pletcher D, Walsh FC (1990) Chem Rev 90: 837
75a. Meli G, Leger J-M, Durand R (1992) J Appl Electrochem Soc 23: 197
75b. Beden B, Leger J-M, Lamy C (1992) In: Modern aspects of electrochemistry, 22: 97
76. Millington JP, Dalrymple IM (1983) (Electricity Council) Eur Pat Appl 64417, 1983; Chem Abstr 98: 62142p
77. Reilly, Tar, Chemical, through ref (10), p 326
77a. Toomey JE, Filter press electrochemical cell with improved fluid distribution system, (Reilly Tar and Chemical Corp), US Pat 4,589,968
77b. Toomey JE (1985) Eur Pat Appl EP 122,736; Chem Abstr 102: (1985) 14136k
78. Carlsson L, Sandegren B, Simonsson D, Rihovsky M (1983) J Electrochem Soc 130: 342
79. Carlsson L, in ref (23) p 251
80. Carlsson L, Holmberg H, Johansson B, Nilsson A (1982) In: Weinberg NL and Tilak BV (eds), Techniques of electroorganic synthesis, J Wiley, New York, Vol 3, p 179
80a. Carlsson L, Sandegren B, Simonsson D, Rikovsky, M, Proc Electrochem Soc 83–6 (Proc Symp Electro- Chem Process Plant Der 1982) 8; Chem Abstr 99, (1983) 183980f
81. Bjareklint A, Carlsson L, Sandegren B (1984) Design of EC modularized electrochemical cells, In: White RE (ed), Electrochem Cell Design, Plenum, p 197
82. Stucki S, Kötz R, Suter W, Carcer B (1991) J Appl Electrochem 21(1991)99; Chem Abstr 114: 173750v
83. Kötz R, Stucki S, Carcer B (1991) J Appl Electrochem 21 14; Chem Abstr 114: 71061d
83a. Comninellis C, Pulgarin C (1993) J Appl Electrochem 23: 108
83b. Etwaree T, Savall A, Comninellis C (1992) Recent Progr Génie Proc (6)20, 215–224
84. Weinberg NL, in ref (45a)
85. Clarke R, in ref (45a), Ceramic electrodes in environmental electrochemical technology, and references therein
86. Hayfield PCS, Clarke RL (1989) Proc-Electrochem Soc Meeting, Los Angeles
idem (1989) Proc Electrochem Soc 1989, 89–10, 87–100; Chem Abstr 111: (1989) 203912g
87. Clarke R, Pardoe R, in ref (23), p 349, and ref therein
88. Weinberg NL, Genders JD, Clarke RL (1990) Eur Pat Appl EP 369,732; Chem Abstr 113: (1990) 122587u
89. Clarke RL (1993) Applications of Ebonex electrode materials in environmental processes, Electrochem Processing, Innovations and Progress, April 21–23, Glasgow
90. Mayr M, Blatt W, Busse B, Heinke H, in ref (23), p 365
91. Langlois S, Coeuret F, J Appl Electrochem 19(1989)51 Société Sorapec, 192 rue Carnot, 94120 Fontenay Sous Bois, France
92. Langlois S, Coeuret F (1990) J Appl Electrochem 20 740; Chem Abstr, 113: (1990) 199957x
93. Idem (1990) J Appl Electrochem 20 749; Chem Abstr 113: (1990) 199999n
94. Tentorio A, Casolo-Ginelli U (1978) J Appl Electrochem 8(1978)195; Chem Abstr 89(1978)83309u
95. Wang J, Electrochim Acta 26(981)1721
96. Davis ThA, in Ref (45)
97. Davis ThA, in Ref (23) p 173
98. Electrosynthesis Co, Inc, 72 Ward Road, Lancaster, NY 14086-9779, USA
99. ElectroCell Systems AB, Box 7007,S 18307 Täby, Sweden
100. EA Technology Service to Business, Industrial Applications of Electrochemistry, EA Technology, Capenhurst, Chester CH1 6ES
101. EDF (Electricité de France), Direction des Etudes et Recherches, St Denis, France
102. Baizer MM (1979) Chem and Ind, 435
103. Danly D (1981) Adiponitrile via improved EHD, Hydrocarbon Process, Int Ed 60(4) 161; Chem Abstr 94: (1981) 199649n

104. Danly DE (1979) Discovery, development and commercialization of the electrochemical adiponitrile process, Amstrong Lecture, Part 2 Chem and Ind (London) (13)439; Chem Abstr 92: (1980) 42379k
104a. Bard AJ (1993) EIRELEC'93, September 11–15, Adare, Ireland
104b. Harrison S, Labrecque R, Théorêt A, in Ref [45]
105. Weinberg NL (1985) Electrosynthesis of polyols, US US 4,478,694; Chem Abstr 102: (1985) 35368c
106. Weinberg NL, Lipsztajn M, Mazur DJ, Reicher M, Weinberg HR, Weinberg EP (1988) Stud Org Chem (Amsterdam) 30(Recent Adv Electroorg Synths) 441(1987); 108: (1988) 74781a
107. Weinberg NL, Mazur DJ (1991) J Appl Electrochem 21 895; Chem Abstr 115: (1991) 265322f
108. Weinberg NL, Mazur DJ (1992) in ref. (44), p 55; Chem Abstr 116(1992)223556a
109. Genders JD, in Ref [45]
110. BASF, Direct indirect electrochemical oxidation of alkoxytoluenes
111. Otsuka Chemical through Ref [35]
112. Genders JD, Weinberg NL, Zawadzinski C (1991) The direct electrosynthesis of L-cysteine free base, Electroorg Synth (Manuel M Baizer Meml Symp) 1990, (Publ 1991) 273; Chem Abstr 116: (1992) 138606x
113. Genders JD, Mazur DJ, Weinberg NL (1991) High yield methods for electrochemical preparation of cysteine and analogs, Eur Pat Appl EP 436055, 10 July, 1991; Chem Abstr 115: (1991) 169016p
114. Hitchman ML, Millington JP, Ralph TR, Walsh FC (1989) The electrochemical reduction of cystine to L-cysteine, I Chem E Symp Ser 112: 222
115. SNPE, pamphlet
116. Hoechst, through Ref [48]
117. Wagenknecht, Monsanto, through, Ref [48]
118. Evans J, St Smedley, in Ref [45]
119. Tuffrey NE, Jiricny V, Evans JW (1985) Hydrometallurgy, 15: 33, through Ref [118]
120. Takenaka S, Oi R, Ch Shimakawa, Shimokawa Y, Tateyama N, in Ref [45]
121. Oi R, Shimakawa C, Shimokawa Y, Takenaka S (1987) Bull Chem Soc Jpn 60: 4193
122. Oi R, Shimakawa C, Takenaka S (1988) Chem Lett 899
123. Simonet J (1983) In: Baizer MM, Lund H (eds.) Organic electrochemistry, An introduction and a guide, 2nd Ed, Marcel Dekker NY, p 843
123a. Konno A, Fukui K, Fuchigami T, Nonaka T (1991) Tetrahedron 47: 887
123b. Steckhan E (1987) Organic syntheses with electrochemically regenerable redox systems, Top Curr Chem 142: 1
124. Simon H, Bader J, Günther H, Neumann S, Thanos J (1984) Biohydrogenation and electromicrobial and electroenzymatic reduction methods for the preparation of chiral compounds, Ann NY Acad Sci 434: 171, and refs therein
125. Bersier PM (1983) J Pharm & Biomed Anal 1:475
126. Bersier PM (1993) Electrochem Processing, Innovations and Progress, April 21–23, Glasgow
127. Pletcher D (1991) Indirect electrolysis involving phase transfer catalysis, Electroorg Synth (Manuel M Baizer Meml Symp) 1990 (Publ 1991) 255; Chem Abstr 116: (1992) 115434a
128. Bersier PM, Werthemann D (1983) J Wood Chem Technol 3: 335
129. Oloman C, Electrochemical synthesis and separation technology in the pulp and paper industry, in Ref [45a]
130. Oehr KH, Ceric Sulphate, US, US 4,313,804, 1982; Chem Abstr 96: (1982) 94013p
130a. Oehr KH (1983) Electrolytic oxidation of solutions containing cerous ion and cobalt, Can CA 1,153,332; Chem; Abstr 99: (1983) 183996r
131. Kreh RP, Spotnitz RM, Lundquist JT (1989) J Org Chem 54: 1526
132. Harrison S, Velin A, Théorêt A (1993) Electrochem Processing Innovation and Progress, April 21–23, Glasgow
133. Walsh FC (1993) Recycling and process streams: The scope for electrochemistry, Electrochem Processing, Innovation and Progress, April 21–23, Glasgow
134. Kramer K, Robertson PM, Ibl N (1980) J Appl Electrochem 10 29; Chem Abstr 92: (1980) 215013s
135. Couper AM, in Ref [23] p 237
136. Bersier PM, Bersier J (1989) Analyst, 114: 1531
137. Bersier PM, Bersier J, Sedelmeier G, Hungerbühler H (1990) Electroanalysis 2: 373
138. Ibl N, Kramer K, Ponto L, Robertson P (1979) AIChE Symp Ser, 75 (185) 45; Chem Abstr 91: (1979) 80879k

139. Walsh FC, Worked examples, in Ref [13], p 339, Chapter 10
140. Vaudano P (1992) Chimia 46(1992)103; Chem Abstr 117: (1992) 29180d
140a. Comninellis C, Plattner E (1983) (to Givaudan & Cie), EP 131001; See PCT Int Appl. W08402, 522; Chem Abstr. 102: 59345
140b. Vaudano P, person commun
141. Comninellis Ch, Plattner E (1982) J Electrochem Soc, 129: 749
142. Comninellis Ch, Plattner E (1987) J Appl Electrochem, 17: 1315
143. Comninellis Ch, Plattner E (1986) Chimia 40: 413
144. Comninellis Ch, Plattner E (1987) In recent Adv in electroorg synthesis, Proc 1st Int Symp on Electroorganic Synthesis, S Torii (ed), Kodansha, Ltd, Tokyo, p 463
145. Brenet J (1987) Bull Soc Chim Fr, 1
146. Degner H, Siegel H (1980) Ger Offen 2,855,508, BASF, AG; Chem Abstr 93: (1980) 140 061r
147. Degner H, Roos H, Hannebaum H (1983) Ger Offen DE 3,132,726, BASF; Chem Abstr 98: (1983) 169382k
148. Robin Y, in Ref [45]
149. Chaussard J, Folest J-C, Nédélec JY, Périchon J, Sibille S, Troupel M (1990) Synthesis 5: 369
150. Fauvarque JF, Jutand A, François M (1988) J Appl Electrochem 18(1988)109; Chem Abstr 109: (1988) 13588s
151. Silvestri G, Filardo G, Gambino S (1988) Eur Pat Appl EP 283 796; Chem Abstr 110: (1989) 65 730x
152. Chaussard J (1987) EP 219 367; Chem Abstr 107: 66612
153. Jud JM, Research and development related to electrochemical processes, (25 references), in Ref [45]
154. Troupel M Nedelec JY and Périchon J, in Ref [44], p 335
155. Chaussard J, Storck A, Lapique F, Hornut JM (1988, SNPE) Fr Pat 2617197; Chem Abstr 111: 86198r
156. Moingeou MO, Chanssard J, Travpet M, Saboueau C (1988, SNPE) Fr Pat 2609474 Eür Pat Appl EP 277, 048; Chem Abstr 110: 65729d
157. Dunach E, Périchon J (1988) J Organomet Chem 352: 239
158. Silvestri G, Gambino S, Filardo G (1986) Tetrahed Lett 27: 3429*
159. Moingeau MO, Chanssard J (1986, SNPE) Fr Pat 257926, Eur Pat Appl EP 198, 743; Chem Abstr 106: 223025v)
    Fr Pat dem 8804254
160. d'Incan E, Sibille S, Périchon J (1986) Tetrahed Lett 27: 4175
161. Saboureau Ch, Troupel M, Sibille S, Périchon J (1989) J Chem Soc Chem Commun, 1138
162. Durandetti S, Sibille S, Périchon J (1989) J Org Chem 54: 2198
    *through Ref [115]
163. Sibille S, Mcharek S, Périchon J (1989) Tetrahedron 45: 1428
164. Périchon J, Rabemanantsoa A, Sibille S, d'Incan E (1986, SNPE) Fr Pat 2579627; Chem Abstr 106: 223026w
    Fr Pat dem
165. Fr Pat dem 88/15235 (cited in Ref [115])
166. Fr Pat dem 87/15975, 89/05626 (cited in Ref [115])
167. Nédélec JY, Ait-Haddou-Mouloud H, Folest JC, Périchon J (1988) Tetrahed Lett 29: 1699
168. Idem (1988) J Org Chem 53: 4720
169. Fr Pat dem 87/17671 (cited in Ref [115])
170. Folest JC, Périchon J (1988, SNPE) Fr Pat 2606427, Eur Pat Appl EP 268, 526: Chem Abstr 110: 47414t
171. Folest JC, Nédélec JY, Périchon J (1987) Tetrahed Lett 28: 1885
172. Folest J-C, Nédélec JY, Périchon J (1988) Syn Commun 18: 1491
173. Pons L, Biran C, Bordeau M, Dunogués J (1988) J Organomet Chem 358: 31
174. cf Ref [45]
175. Sock O, Troupel M, Périchon J (1985) Tetrahedron Lett 26: 1509
176. Heintz M, Sock O, Saboureau C, Périchon J (1988) Tetrahedron 44: 1631
177. Fisher J, Lehman Th, Heitz E (1981) J Appl Electrochem 11: 743; cited in Ref [178]
178. Silvestri G, Gambino S, Filardo G (1991) Acta Chem Scand, 45: 987
179. Silvestri G, Gambino S, Filardo G, Tedeschi F (1989) J Appl Electrochem 19: 946; Chem Abstr 112: (1990) 65357x
180. Wawzonek S, Shradel JM (1979) J Electrochem Soc 126: 401; Chem Abstr 91: (1979) 4801p
181. Silvestri G, Gambino S, Filardo G, Greco G, Gulotta A (1984) Tetrahed Lett 25: 4307

182. Rieu J-P, Boucherle A, Cousse H, Mouzin G (1986) Tetrahedron 42: 4095
183. Silvestri G, Gambino S, Filardo G (1986) Tetrahed Lett 27: 3429
184. Idem (1987) US Pat 4,708,780; cited in Ref [178]
185. Wagenknecht JH (1986), Monsanto Col US Pat 4,582,577; Chem Abstr 105: 50861p; continuation: US Pat 4,601,797; Chem Abstr 106: 24982e
186. Maspero F, Piccolo O, Romano U, Gambino S (1988) Eur Pat Appl EP 286,944; Chem Abstr 110: (1989) 75087g
187. Mcharek S, Heinz M, Troupel M, Périchon J (1989) Bull Soc Chim Fr 95
188. Weinberg NL, Kentaro-Hoffmann A, Reddy TB (1971) Tetrahed Lett 25: 2271
189. Hess U, Thiele R (1982) J Prakt Chem 324: 385
190. Hess U (1980) Z. Chem 20: 148
191. Hess U, Ziebig M (1982) Pharmazie 37: 107
192. Silvestri G, Gambino S, Filardo G (1987) In: Torii S (ed) Recent advances in electro-organic synthesis, Kodanska, Tokyo, p 287
193. Idem (1989) Gazz Chim Ital 118(1988)643; Chem Abstr 111: (1989) 58288b
194. Degner D (1988) Top Curr Chem 148: 1
195. Schäfer HJ, Schneider R (1991) Tetrahedron 47: 715, and references therein
196. Schäfer HJ (1987) Top Curr Chem, 142 (Electrochemistry I) 101; Chem Abstr 108: (1987) 45737k
197. Seiler P, Robertson PM (1982) Chimia, 36: 305
198. Robertson PM, Berg P, Reimann H, Schleich K, Seiler P (1983) J Electrochem Soc 130: 591
199. Foller PC, Allen RJ, Bombard RT, Vora R, The use of gas diffusion electrodes in the on-site generation of oxidants and reductants, in Ref [45]
200. Katch M, Nishiki Y, Nakamatsu (1992) A study on electrochemical ozone generator using gas diffusion cathode, Japan Electrochem Soc Meeting, April, cited in Ref [129]
201. Stucki S, Theis G, Kötz R, Devantay H, Christen HJ (1985) J Electrochem Soc 132(1985)367; Chem Abstr 102: (1985) 102388z
202. Stucki S (1983) Reaction and process technology of membrel water electrolysis, Dechema-Monogr 94(1983) 1932–1947 Reactiontech Chem Electrochem Processes; Chem Abstr 102: (1985) 86449n
203. Baumann H, Stucki S (1986) Swiss Chem 8(10a) 31; Chem Abstr 106: (1987) 72423v
204. Stucki S, Baumann H, Christen HJ, Kötz R (1987) J Appl Electrochem 17 773; Chem Abstr 107: (1987) 104976s
205. Baumann H, Stucki S (1988) Process for decomposing organic matter and/or microorganisms in pretreated feed water for ultrapure water circuits Eur Pat Appl EP 281, 940; Chem Abstr. 110: (1989) 44706d
206. Foller PC, Tobias CW (1982) J Electrochem Soc 129: 506
207. Foller PC (1993) Applications of gas diffusion electrodes in prospective electrolytic processes, Electrochem Processing, Innovations and Progress, April 21–23, Glasgow
208. Couper M (1993) Electrochemical generation and application of ozone at high concentrations, Electrochem Processing, Innovations and Progress, April 21–23, Glasgow
209. Couper AM, Bullen S (1992) The electrochemical generation of ozone at high concentrations, Inst Chem Eng Symp Ser 1992, 127 (Electrochem Eng Environ 92)49–58, Chem Abstr 117:(1994) 194570p
210. Oloman C, Watkinson AP (1986) Electrolytic generation of alkaline peroxide solutions, Can CA 1,1214,747; Chem Abstr 106: (1987) 127953a
211. Oloman C (1988) Perforated bipole electrochemical reactor, US Pat 4,728,409; Chem Abstr 108: (1988) 158067k
212. Clifford A, Dong D, Giziewicz E, Rogers D (1990) Electrosynthesis of alkaline hydrogen peroxide, Proc Electrochem Soc 1990, 90–10, 259–74; Chem Abstr 114: (1991) 131743q
213. Kalu EE, Oloman C (1990) J Appl Electrochem, 20 932; Chem Abstr 114: 31813u
214. ICI, pamphlet
214a.Kargin, Yu N Kargin-Power, O, Electrosynthesis of some phosphorous and nitrogen containing organic compounds 7th Int Forum on Electrolysis, No 7–11, 1993, Lake Buena Vista, FL
215. Oloman C (1970) J Electrochem Soc 117: 1604
216. Oloman C, Lee B, Leyton W (1990) Electrosynthesis of sodium dithionite in a trickle-bed reactor, Can J Chem Eng (1990)1004; Chem Abstr 114: (1991) 51972y
217. Bolick RE, Cawfield DW, French JM (1988) Electrochemical process for producing sodium hydrosulfite solutions Eur Pat Appl EP 257,815; Chem Abstr 108: (1988) 158089u

218. Stubbs JL Jr, Bolick II RE, Hauser EF Jr (1992) Electrochemical removal of thiosulfate from hydrosulfite solutions, US, US Pat 5,112,452; Chem Abstr 117: (1992) 71946h
219. Kaczur JJ, Cawlfield DW, New electrochemical chlorine dioxide generation technology, in Ref [45]
220. Cawlfield DW, Kaczur JJ (Olin Corp) (1991) Electrochemical chlorine dioxide generator, PCT Int Appl WO 91 09,990; Chem: Abstr 115: (1991) 169055a
221. Cawlfield DW, Kaczur JJ (1992) US Pat 5,084,149; Chem Abstr 116: 223787b
222. Lipsztaijn M (1988) Electrolytic production of chlorine dioxide Eur Pat Appl EP 293,151; Chem Abst 110: (1989) 123911v; (1989) US Pat 4767510, Eur Port Appl EP 266, 127: Chem Abstr 108: 223911t
223. Twardowski Z (1989) Integrated process for the manufacture of chlorine dioxide and sodium hydroxide, US, US 4,806,215; Chem Abstr 111: 176579v
224. Cawlfield DW, Mendiratta SK (1993) Chloric acid electrochemically for chlorine dioxide, Electrochem Processing, Innovations and progress, April 21–23, Glasgow
225. Landfors J (1993) Alternative on-site caustic soda production in the paper and pulp industry, Electrochem Processing, Innovations and Progress, April 21–23, Glasgow Sunblad B (1992, Eka Nobel) Can CA 2,071,810; Chem Abstr 118: 194663f
225a. Coin, RJ and Niksa, MF., ref [214a] [214a] Safe and economical production of chloric acid and chlorine dioxide from sodium chlorate solution
226. Bersier PM, Carlsson L, Bersier J, in Ref [45a]
227. Pletcher D, Electrochemical technology for a cleaner environment, Fundamental considerations, in Ref [23], Chapter 2
227a. Weinberg NL, in Ref [23]
228. Kreysa G (1992) Electrochemical technologies for clean environment, PL-1, 43rd ISE Meeting, Cordoba, rept 20–25, Argentina
229. Walsh FC, Reade GW (1993) Electrochemical techniques for the treatment of dilute metal-ion solutions, in CAS. Sequiera (ed), Environmentally oriented electrochemistry, Elsevier, Amsterdam
230. Walsh FC (1992) The design and performance of electrochemical reactors for environmental treatment and efficient synthesis, 43rd ISE Meeting, Cordoba, Argentina, Sept 20–25, IL-3-09
231. Bersier J, Bersier PM (1991) Electrochemical treatment of industrial wastewaters, scope and limitations, 42nd ISE Meeting, Montreux, Switzerland, August 25–30, 5–11
232. Froment M (1992) Tin oxide: A key material in the field of electrochemistry and the environment, PL-6, 43rd ISE Meeting, Cordoba, Argentina, September 20–25
233. Pletcher D, Walsh FC, in ref (23), p 51
234. Leclerc O (1993) The "Priam" process, Electrochem Processing, Innovations and Progress, April 21–23, Glasgow
234a. Mayr M, Blatt W, Ströder U, Heinke H (1991) Galvanotechnik (6), Sonderdruck
235. Lopez-Cacicedo CL (1975) Brit Pat 1423369(1973); through (247); Ger Offen 2,445,412; Chem Abstr 83: 1(1975) 102918m
236. Lopez-Cacicedo C (1974) Recovery of metals from rinse waters in Chemelec electrolytic cells, Trans Inst Met Finish 55, Pt 2 74; Chem Abstr 84: (1976) 10365w
237. Lopez-Cacicedo CL (1975) Inst Chem E Symp Ser No 42 29 (Hydrometallurgy); Chem Abstr 85: (1976) 164097t
238. Lopez-Cacicedo CL (1981) J Sep Process Technol 2 34; Chem Abstr 96: (1981) 26321n
239. EA Technology, Capenhurst UK, CH 1 6ES, Licensee: BEWT (Water Engineers) Ltd, Tything Road, Arden Forest Industrial Estate, Alcester, Warwickshire B49 6ES, England
240. Gabe DR (1974) J Appl Electrochem 4 (review with 312 refs) Chem Abstr 81: (1974) 130142f
241. Walsh FC, The role of the rotating cylinder electrode reactor in metal ion removal in ref (23), p 101, and references therein
242. Holland FS, Rolskov H (1978) Proc Effluent Water Treat Conv 1978, Paper No 9; Chem Abstr 92: (1980) 185300p
243. Robinson D, Walsh FC (1991) The performance of a 500 Amp rotating cylinder electrode reactor, Part 1. Current-potential data and single pass studies, Hydrometallurgy 26 93; Chem Abstr 114: (1991) 194767w
244. Idem (1991) Batch recirculation studies and overall pass transport, Hydrometallurgy 26 115; Chem Abstr 114: (1991) 194768x
245. Holland FS (1978) Chem Ind (London) (13)453; Chem Abstr 90: (1979) 13929t
246. Ricci LJ, Chem Eng (1975)29; through ref (247)

247. Kreysa G (1988) In: Stucki S (ed) Process technologies for water treatment, Plenum Press, New York, p. 65
248. Bruhn D, Dietz W, Müller K-J and Reynvaan C, EPA 86109265.8(1986); through ref (247)
248a. Dietz W (1986) Electrodeposition of metals from an electrolysis bath, Eur Pat Appl EP 171,647; Chem Abstr 104: (1986) 158198w
249. Müller KJ (1993) Recent applications of packed bed electrolysis, Electrochem Processing, Innovations and Progress, April 21–23, Glasgow
250. Hertwig K, Bergmann H, Nieber F (1991) Electrochemical treatment of metal containing water in particle electrode cells, 5–20, 42nd ISE Meeting, Montreux, Switzerland, August 25–30
251. Bergmann A, Hertwig K, Nieber F (1992) Experimentelle und theoretische Untersuchungen an einem neuen Reaktortyp für elektrochemische Abwasserreinignung, Dechema-Monographien, Vol 125, VCH, 1992, 65; Chem Abstr 117: (1992) 120361z
252. Konicek MG, Platek GF (1983) New Mater New Processes, 2(1983)232; Chem Abstr 98: (1983) 219531f
253. Marketed by Eltech Systems Corp., 12850 Bournewood Drive, Sugar Land, Texas, 77478, USA; through ref (233)
254. Chaney GA, Wilcox M, Goss D, Toomey JE (1985) Génies des procédés d'électrosynthèse organique, Perpignan, France, through (75)
255. Toomey JE, Chaney GA, Wilcox M (1987) In: Torii S (ed) Recent advances in electroorganic chemistry, Elsevier, p 245; through ref (75)
256. Simonsson D (1984) J Appl Electrochem 14 595; Chem Abstr 101: (1984) 200190c
257. Ford WPJ, Walsh FC, Whyte I (1992) Simplified batch reactor models for the removal of metal ions from solution Inst Chem Eng Symp Ser, 1992, 127(Electrochem Eng Environ 92)111; Chem Abstr 117: (1992) 197712x
258. Kreysa G (1991) Electrochemical recycling and treatment process for contaminated wastewaters, Abstr KL-5-1, 42nd ISE Meeting, Montreux, Switzerland, Aug 25–30
259. Bersier PM (1987) Anal Proc 24: 44
260. Herbst H, Petz K, Stenzel J (1992) Die elektrochemische Regenerierung von chromschwefelsäure in einer modifizierten Hoechst-UHDE -Membranzelle, Dechema Monographien, Vol 125, VCH, p 243
261. Bishara JI, Brannan JR, Dickie DK, Horvath RJ, Recovery of hexavalent chromium from plating rinse waters, in ref (45)
262. Dalrymple I and Sunderland G, The electrochemical recovery of metals and chemicals from waste sources, in ref (45a)
262a. Burbank J (1957) J Electrochem Soc 104: 693
263. Robinson D, Electrochemical generation of chromous ion for the reduction of hydroxymandelate, in ref (45a)
264. Genders D, Hartsough D, Thompson J, Novel approaches to salt splitting, in ref (45a)
265. idem, US 5,098,532, 1992
266. Genders D, Hartsough D, Thompson J (1993) Electrochem Processing, Innovations and Progress, April 21–23, Glasgow
267. Lantagne G, Vélin AP, Overview of the application of dialysis, electrodialysis and membrane cell electrolysis for the recovery of waste acids, in ref (45a), and references therein
268. Mani KN (1991) Electrodialysis water splitting technology, J Membr Sci 58 117; Chem Abstr 115: (1991) 52408b
269. Niksa MJ, Acid/base recovery from sodium sulphate, in ref (45a)
269a. Jorissen J, Simmrock KH (1991) J Appl Electrochem 21: 869
270. Johnson W, Bipolar membrane technology and applications, in ref (45)
271. Martin AD, Electrohydrolysis, A means for the recovery of the sodium hydroxide from sodium containing process streams, in ref (45a)
272. Martin AD (1992) Sodium hydroxide production by the electrohydrolysis of aqueous effluent streams containing sodium salts, Inst Chem Eng Symp Series, 1992, 127 (Electrochem Eng En 92), 153; Chem Abstr 117: (1992) 194581t
273. DeNora Permelec SpA (1992) Hydrina membrane electrolysers; through (129)
274. Millington JP (1983) An electrochemical unit for the recovery of sodium hydroxide and sulfuric acid from waste streams, Ion Exch Membr 1983; 195; Chem Abstr 101: (1983) 93624y
275. Mani KN, Chlanda FP, Byszewski CH (1988) Aquatech membrane technology for recovery of acid/base values from salt streams, Desalination 68 149; Chem Abstr 108: (1988) 189072n

276. Mani KN (1988) Proc-APCA Annu Meet 1988, 81st (1) Paper 88/6A 7, Aquateck membrane technology for regeneration of hazardous waste/acid effluent streams; Chem Abstr 110: (1989) 198503x

277. Paleologou M, Wong P-Y, Berry RM (1992) A solution to the caustic/chlorine imbalance: bipolar membrane electrodialysis, J Pulp and Paper Sci, 18(4) J138; Chem Abstr 117:(1991) 253647u

278. Clouthier J-N, Azarniouch MK, Callender D (1992) Electrolysis of weak black liquor, Part I, Laboratory study, Paper presented at the 1992 Int Chem Recovery Conf for Pulp and Paper Ind, Seattle, June, and ref. therein (16–21)

279. Ref 16–21 in ref (278)

280. Dyfverman A (1942) An attempt at electrolytic treatment of black liquor, Sven Papperstidn, 45: 535; cited in Ref [278]

281. Beaudry EG, Electrolytic separation technology for treating ionic wastes, slurries and process streams 4th Int. Forum on Electrolysis in the chem Jud, For Lauderdale, FL,. 1990 through (278)

282. Azarniouch MK, Prahacs S (1991) Recovery of NaOH and other values from spent liquors and bleach plant effluents, US 5,061,343 Chem Abstr 116: 23226f

282a. In Ref [13], p 332

283. Steele DF, Richardson D, Craig DR, Quinn JD, Page P in Ref [23], p 287

283a. Steele DF, Richardson D, Campbell JD, Craig DR and Quinn JD (1990) The low temperature destruction of organic waste by electrochemical oxidation, Trans I Chem E 68B: 115

283b. Steele DF (1989) A novel approach to organic waste disposal, ATOM, 393, July 10; cited in Ref [133]

283c. Steele DF (1990) Electrochemical destruction of toxic organic industrial waste, Platinum Met Rev 34(1)(1990)10; Chem Abstr 113: (1990) 157987b

283d. Steele F (1989) Electrolytic decomposition of organic liquid radioactive waste in presence of oxidizing species, Eur Pat Appl EP 297,738; Chem Abstr 110: (1989) 201272d

283e. Steele DF (1989) Oxidizing electrodecomposition of organic radioactive waste, Brit UK Pat Appl GB 2,206,341; Chem Abstr 110: 201273e

284. Bahadir M (1991) Chemie in unserer Zeit 25: 239

285. Clarke RL, in Ref [23], p 271

285a. Meissner W and Härtel G (1992) Zerstörung organischer Abwasserinhaltsstoffe durch Electrolyse, Dechema Monographien, Vol 125, p 523 VCH-Verlagsgesellsch

286. Kelsal GH (1991) SCl, Electrochemical techniques for a cleaner environment, April 19, London

287. Comninellis Ch (1988) Process technology for water treatment, plenum press In: Stucki S (ed), p. 71

287a. Comninellis Ch (1992) Trans I Chem E 70, Part B, 219

288. Plattner E, Comninellis Ch (1988) In: Stucki S (ed), Process technicology for water treatment, Plenum Press, p 205

289a. Comninellis Ch, Pulgarin C (1991) J Appl Electrochem 21: 703

289b. Ho CC, Chan CY, Khoo KH (1986) Electrochem treatment of effluents: A preliminary study of anodic oxidation of simple sugars using lead dioxide coated titanium anodes, J Chem Tech Biotechnol 36: 7

289c. Mill Th, Yao CCD, Smedley St, Dougherty B, Cox Ph, SRI international

290. Schmal D, van Duin PL, de Jong AMCP, (1991) Dechema-Monographien, Vol 124, VCH Verlagsgesellsch, p 241 and references therein

291. Bulich AA, Isenberg DL (1981) The use of luminescent bacterial system for the rapid assessment of aquatic toxicity, ISA Trans 20(1) 29; Chem Abstr 95: (1981) 12110k

292. Schmal D, van Erkel J, de Joung AMCP, van Duin PJ, Report R 87/037(1987) of MT-TNO to the Commission of EC

293. Schmal D, van Erkel J, van Duin PJ (1986) Electrochemical reduction of halogenated compounds in process wastewaters, Inst Chem Eng Symp Ser 1986, 98 (Electrochem Eng) 281; Chem Abstr 105: 158252e

293a. Schmal D, van Erkel J, van Duin PJ (1986) J Appl Electrochem 16: (1986) 422

294. EC list of 129 compounds; EC publications No. C 176.7 (1982)

295. Mazur DJ, Weinberg NL (1990) US, US 4,968,393, Nov 6, Chem Abstr 114: (1991) 131819u, and ref therein

296. Mazur DJ, Weinberg NL (1987) Methods for the electrochemical reduction of halogenated organic compounds, US, US 4,702,804; Chem Abstr 108: (1988) 175949s

297. Abel AE, Mazur DJ, Weinberg NL (1988) Processes for decontaminating polluted substrates, Eur Pat Appl EP 288,408; Chem Abstr 110: (1989) 63196d
298. Mazur DJ, Weinberg NL, Aurnou EA, Liolios EA, Kendall PM, The electrochem Soc Extended Abstr, Vol 87 (1987)1857; through (79), and ref therein
299. Comninellis Ch, Plattner E (1987) Symp Process technology for water treatment, BBC Research Center Baden (Switzerland), Sept 21/22
300. Seignez C, Pulgarin C, Comninellis Ch, Péringer P (1991) Pre-treatment of wastewater containing phenol before biological treatment, 42nd Meeting ISE, Montreux, p 5–01
300a. Seignez C, Pulgarin C, Péringer P, Comninellis Ch, Plattner E, Swiss Chem 14(1992)(1)25
300b. Savall A, Comninellis Ch (1992) Recents Prog Génie Proc 6, 20, Technol Innovates Epur Eaux 207
300c. Comninellis C (1992) Gas, Wasser, Abwasser, 72(11)791
300d. Comninellis Ch, Pulgarin C (1991) J Appl Electrochem 21: 1403
300e. Comninellis Ch (1992) Electrochemical treatment of wastewater containing phenol, Inst Chem Eng Symp Series, 1992, 127 (Electrochem Eng En 92)189; Chem Abstr 117: 177627v
300f. Comninellis Ch (1992) Electrochemical oxidation of organic pollutants for wastewater treatment, Proceed Electrochem Soc Symp, St Louis, May 17–22, St Louis, Missouri
301. Pulgarin C, Seignez C, Comninellis C, Plattner (1991) Electrochemical detoxication of waste water, 42nd Meeting ISE, Montreux, Switzerland; August 25–30 pp 5–06
302. Suschka J, Ferreira E (1986) Activated sludge respirometric measurements, Water Res 20 137; Chem Abstr 104: (1986) 55663q
303. OECD Guideline for testing of chemicals, Inherent biodegradability: modified Zahn-Wellens-test 302B; cited in Ref [300, 300a]
304. Steele DF, Richardson D, Campbell JD, Craig DR, Quinn JD (1990) The low-temperature destruction of organic waste by electrochemical oxidation, Inst Chem Eng Symp Ser 1990, 116 (Effluent Treat Waste Disposal)237; Chem Abstr 114: (1991) 149416v
305. Craig DR, Quinn JD, Richardson D, Page P, Steele DF, Efficient electrochemical destruction or organic wastes, 42nd Meeting, ISE, Montreux, p 5–02
306. Hickman RG, Farmer JC, Chiba Z, Mediated electrochemical oxidation techniques as an alternative to incineration in Ref [45a]
306a. Farmer JC, Hickman RG, Wang, FT, Lewis PR, Summers LJ (1991) Initial study on the complete mediated electrochemical oxidation of ethylene glycol, Report 1991, UCRL-LR-106479; Order No. DE91015523; Chem Abstr 117: (1992) 117670u
306b. Farmer JC, Wang FT, Hawley-Fedder R, Lewis PR, Summers LJ, Foiles L (1991) Initial study of halide-tolerant mediators for the electrochemical treatment of mixed and hazardous wastes, Report 1991, UCRL-LR-107781; Order No. DE91017528; Chem Abstr 117: (1992) 96683h
307. Farmer JC, Wang FT, Hawley-Fedder RA, Lewis PR, Summers LJ, Foiles L (1992) J Electrochem Soc 139: 654
307a. Farmer JC, Wang FT, Lewis PR, Summers LJ (1992) Electrochemical treatment of mixed and hazardous wastes: oxidation of ethylene glycol by cobalt (III) and iron (III), Inst Chem Eng Symp Ser, 1992, 127 (Electrochem Eng Environ 92) 203; Chem Abstr 117: (1992) 197713y
308. Zawodzinski Ch, Smith W, Martinez K, Studies of the electrochemical generation of Ag(II) ion and of the catalytic oxidation of selected organics, in Ref [45a]
309. Almon AC and Buchanan BR, in Ref [23], p 301
310. ref (3*–7*) in ref (309)
311. Farmer JC, Wang FT, Lewis PR, Summers LJ, LLNL, UCRL-JC-109633(1992), J Electrochem Soc, in press; cited in Ref [307]
311a. Tzedakis T, Savall A, Clifton M (1989) J Appl Electrochem 19: 911
311b. Clifton MJ, Savall A (1986) J Appl Electrochem 16: 812
312. Boreskie M, Vanderhoeven R, Downie Ch, Wilcock A (1989) Electrochemical treatment of disperse dyes: reduction of toxicity and reuse of effluent, Book Pap Int Conf Exhib, AATCC 59; Chem Abstr 115: (1991) 262531n
313. Halliday PJ, Beszedits S (1986) Canad Textile J, 103(4): 78
314. Park J, Shore J (1984) J Soc Dyers Colour, 100: 383
315. Berg H (1954) Chem Techn, 6: 585
316. Bersier PM, Bersier J (1986) Trends Anal Chem, 5: 97
317. Popova VI, Gorocha GE, Konovalchek OK, Khimiya I Tekhnol Kras (1977) Sintes Kras I polim mat, 59

318. Kiener LV, Uhrich KD (1987) Electrochemical removal of color from dye laden wastewater, Water and Pollution Control Assoc of South Carolina Fall Conf, Nov 24; cited in Ref [312]
319. Tincher W, Weinberg M, Stephens S (1988) Electrochemical removal of dyes and chemicals from textile wastewater, AATCC Annual Technical Conf, Knoxville, Tenn, through Ref [312]
320. Jones DL (1972) Electrolytic treatment of waste water, American Dyestuff Rep, 61(8) 28; Chem Abstr 78: (1972) 7551z
321. Endyus'kin PN, Selezenkin SV, Dyumaev KM (1983) Electrochemical treatment of wastewater from organic dye manufacturing plants, Zh Prikl Khim, 56(5) 1167; Chem Abstr 99: (1983) 76176b
322. Zhu, Hongli; Wang Shuhui (1986) Application of three-dimensional electrodes in wastewater treatment, Huanjing Kexue 6(6) 36; Chem Abstr 104: (1986) 192374k
323. Shen, Hao; Ye, Lei; Wu, Jianhua (1987) A method for dyeing wastewater treatment by a flowing type carbon-felt electrode, Huaning Baohu (Beijing) (1)15; Chem Abstr 106: (1987) 219043g
324. Abdo MSE (1979) Bull Fac Eng Alexandria Univ, Vol XVI 205; through Ref [325]
325. Abdo MSE, Rasheed S Al-Ameeri (1987) Anodic oxidation of a direct dye in an electrochemical reactor, J Environ Sci Health Part A 22(1) 27; Chem Abstr 106: (1987) 219052j
326. Danigel H, Gross H, Zumbrunn W (1989) Techn Mess 56: 285

Author Index Volumes 151-170

The volume numbers are printed in italics

Wan, P., see Krogh, E.: *156*, 93-116 (1990).
Warwel, S., Sojka, M., and Rüsch, M.: Synthesis of Dicarboxylic Acids by Transition-Metal Catalyzed Oxidative Cleavage of Terminal-Unsaturated Fatty Acids. *164*, 79-98 (1993).
Willner, I., and Willner, B.: Artificial Photosynthetic Model Systems Using Light-Induced Electron Transfer Reactions in Catalytic and Biocatalytic Assemblies. *159*, 153-218 (1991).

Yoshida, J.: Electrochemical Reactions of Organosilicon Compounds. *170*, 39-82 (1994).
Yoshihara, K.: Chemical Nuclear Probes Using Photon Intensity Ratios. *157*, 1-34 (1990).

Zamaraev, K.I., see Lymar, S.V.: *159*, 1-66 (1991).
Zamaraev, K.I., Kairutdinov, R.F.: Photoinduced Electron Tunneling Reactions in Chemistry and Biology. *163*, 1-94 (1992).
Zander, M.: Molecular Topology and Chemical Reactivity of Polynuclear Benzenoid Hydrocarbons. *153*, 101-122 (1990).
Zhang, F.J., Guo, X.F., and Chen, R.S.: The Existence of Kekulé Structures in a Benzenoid System. *153*, 181-194 (1990).
Zimmermann, S.C.: Rigid Molecular Tweezers as Hosts for the Complexation of Neutral Guests. *165*, 71-102 (1993).
Zybill, Ch.: The Coordination Chemistry of Low Valent Silicon. *160*, 1-46 (1991).

Springer-Verlag
and the Environment

We at Springer-Verlag firmly believe that an international science publisher has a special obligation to the environment, and our corporate policies consistently reflect this conviction.

We also expect our business partners – paper mills, printers, packaging manufacturers, etc. – to commit themselves to using environmentally friendly materials and production processes.

The paper in this book is made from low- or no-chlorine pulp and is acid free, in conformance with international standards for paper permanency.